Hot Rod Pioneers
The Creators of the Fastest Sport on Wheels

Related reading:

**The Golden Age of the
American Racing Car**
(Second Edition)
by Griffith Borgeson

(Order No. R-196)

For more information or to order this book, contact SAE at
400 Commonwealth Drive, Warrendale, PA 15096-0001;
phone (724) 776-4970; fax (724) 776-0790; e-mail: publications@sae.org.

Hot Rod Pioneers
The Creators of the Fastest Sport on Wheels

ED ALMQUIST

Society of Automotive Engineers, Inc.
Warrendale, Pa.

> **Library of Congress Cataloging-in-Publication Data**
>
> Almquist, Ed.
>   Hot rod pioneers : the creators of the fastest
>   sport on wheels / Ed Almquist.
>   p.   cm.
>   Includes index.
>   ISBN 0-7680-0232-X
>   1. Automobile mechanics—United States—Biography. 2.
> Hobbyists—United States—Biography. 3. Hot rods—United
> States—History. I. Title.
>
> TL139 .A43 2000
> 629.28'786'092273—dc21
> [B]   00-036527

Copyright © 2000 Edgar Almquist

    Society of Automotive Engineers, Inc.
    400 Commonwealth Drive
    Warrendale, PA 15096-0001 U.S.A.
    Phone: (724) 776-4841
    Fax: (724) 776-5760
    http://www.sae.org

ISBN 0-7680-0232-X
All rights reserved. Printed in the United States of America.

Permission to photocopy for internal or personal use, or the internal or personal use of specific clients, is granted by SAE for libraries and other users registered with the Copyright Clearance Center (CCC), provided that the base fee of $.50 per page is paid directly to CCC, 222 Rosewood Dr., Danvers, MA 01923. Special requests should be addressed to the SAE Publications Group. 0-7680-0232-X/00-$.50.

SAE Order No. R-228

*This book is dedicated to the "hot rod rebel" in all
of us and to the pioneer auto enthusiasts who created a
new motorsport, industry, and cultural phenomenon.
Their legacy enriches us all.*

A special thanks to my grammarian, Susan Fazzino, for her careful attention to my spelling and punctuation throughout my manuscript.

CONTENTS

FOREWORD xi
INTRODUCTION xiii
MY ENCOUNTER WITH HENRY FORD 1

IN THE BEGINNING 2
THE FOUNDING FORTIES 22
THE NIFTY FIFTIES 142
THE SEXY SIXTIES 264
THE SAD SEVENTIES 324
THE SWINGING EIGHTIES 352
THE HIGH-TECH NINETIES 360

INDEX 367
ABOUT THE AUTHOR 379

THE PIONEERS

Don Alderson 306
The Alexander Brothers 312
Ed Almquist 48
Doug Anderson 320
Art Arfons 242
Walt Arfons 242
Zora Arkus-Duntov 154
Joaquin Arnett 180

Joe Bailon 185
Lou Baney. 351
The Bannister Brothers. 170
George Barris 57
Dean Batchelor 86
Cliff Bedwell 181
Keith Black 346
Don Blair 94
Charlie Boucher 309
Adolph F. Braun 127
Craig Breedlove 280
Ray Brock 176
Ray Brown 76
Bill Burke. 74

Bill Campbell 214
"Honest Charley" Card. 182
Fred Carrillo 250
Carl Casper 281
Ed Cholakian 322
Art Chrisman 219
The Chrisman Clan 152
Don Clark. 168
Jack Clifford 300
Gordon Collett. 300
Cliff Collins. 130
Emery Cook 181, 321
Gary Cooper 107
Jackie Cooper 97
Ed Cortopassi 227

Jeg Coughlin. 295
Ron Covell 336
Harvey Crane, Jr. 290
Bruce Crower 140
Bill Cushenbery 310
Ted Cyr 255

Bob Daniels 233
Bobby Darrin 238
Robert De Bisschop 273
Jim Deist 225
Frank Dominianni 120
Howard Douglass 82
Albert Drake 169

Chris Economaki 279
Vic Edelbrock 32
Eddie Edmunds 131
Dimitri "Dema" Elgin. 320
Bob Estes 100
Mike Estlack 350
Earl Evans 56

Tony Feil 345
Ken "Posie" Fenical 354
Aaron J. Fenton. 190
John Fitch 254
Jack Flynn 297
Joe Fontana 350
A.J. Foyt. 327
Don Francisco. 89
Nye Frank 293
Kent Fuller 218

"Big Daddy" Don Garlits . . . 192, 321, 357
Willie Garner 184
Angelo Giampetroni. 326
Bob Glidden 351
The Legendary Granatellis 60

Gerry Grant 320
Carl "Pop" Green 8
Tom Grove 309
William Guentzler 355
Dan Gurney 274

Ted Halibrand 106
Jim Hall 300
Kenny Harman 130
Leonard Harris 285
C.J. "Pappy" Hart 162
Ernie Hashim 129
Chet Herbert 151
Fran Hernandez 73
Stuart Hilborn. 68
Eddie Hill 359
Tommy Hinnershitz. 42
John Holman 271
Roscoe "Pappy" Hough 16
Joseph Hrudka 330
Joe Hunt 128
George Hurst. 210

Ermie Immerso 228
Bill Ireland 309
Ed "Isky" Iskenderian 108
"TV" Tommy Ivo. 200

Johnny Jackson. 305
Kong Jackson 44
Dean Jeffries. 195
Bill "Grumpy" Jenkins 302
Howard Johansen 150
Jocko Johnson 257
The Justice Brothers 67

Conrad "Connie" Kalitta. 294
Chris Karamesines. 229
Jack Kelly 101

Contents

Bill Kenz 70
Frank Kurtis 36

Dick Landy 298
Bob Larivee 258
Bruce Larson 248
Bill Lawton 301
Roy Leslie 72
"Jungle Jim" Liberman 335
Els Lohn 286

Ed McCulloch 335
Tom McEwen 337
Pete McNicholl 284
Art Malone 198
Dave Marquez 146
Buddy Martin 301
Dick Martin 126
Tom Medley 121
Bobby Meeks 35
Jack Mendenhall 226
Eddie Meyer 19
Louie Meyer 19
Pete Millar 292
Ak Miller 144
Harry Miller 14
Bill Milliken 34
Don Montgomery 287
"Ohio George" Montgomery 206
Ralph Moody 271
Dean Moon 266
Gene Mooneyham 240
Dick Moroso 350
Bob Morton 132
Shirley Muldowney 321, 335
Paula Murphy 328

Barney Navarro 51
Tony Nancy 278
"Jazzy" Jim Nelson 171
Jim Nelson 138
"Dyno Don" Nickelson 307

Fred C. Offenhauser 96
Danny Ongais 300
Karl Orr 83

Wally Parks 66
Jerry Pennington 350
Robert "Pete" Petersen 90
Bob Pierson 104
Dick Pierson 104
Edward Pink 288
Hubert Platt 309
Hayden Proffitt 320
Don Prudhomme 296

Ramchargers 236
Pinky Randall 359
Calvin Rice 182
Roy Richter 24
Eric "Rick" Rickman 173
L.R. "Pete" Robinson 305
Gaspar "Gas" Ronda 281
Robert M. Roof 10
Ed "Beatnik" Roth 174
Bob Rufi 28
Otto Ryssman 159

Ed Schartman, Jr. 321
Carl Schiefer 92
Paul Schiefer 92
Joe Schubeck 232
Lloyd Scott 190
Lou Senter 53
Carroll Shelby 319
Joe Sigretto 359
Bill Simpson 282
Rodney Singer 221
Clay Smith 88
"Speedy Bill" Smith 189
Ronnie Sox 301
The Spalding Brothers 40
Bob Spar 256
Don Spar 256

Tom Sparks 160
Jere Stahl 323
Darryl Starbird 260
Ray Stillwell 246
Bill Stroppe 122
Connie Swingle 259

Bob Tattersfield 30
Al Teague 332
Clem TeBow 168
Tommy Thickstun 29
Doug Thompson 340
Mickey Thompson 164
Bill Toia 326
Jim Travis 220

Pete Van Iderstine 315
Linda Vaughn 217
Les Viland 102

Don Waite 177
Jack "Doc" Watson 216
Harry Weber 112
Joan Weiand 46
Phil Weiand 46
Hank Weidenhammer 318
Roger Whipp 133
Joe Wilhelm 276
Scott Wilson 309
Bud Winfield 6
Ed Winfield 4
Gene Winfield 105
Bill Winterbottom 249
Joe Wolf 166
John Wolf 331

Alex Xydias 84

Smokey Yunick 118

Hot Rod Pioneers

THE CARS AND THE SPORT

"Tin Lizzie" Speedsters 7
The Sanctioning Wars 12
Police Souped-Up Cars, Too 17
1910 Buick "Bug" Racer 17
The Four-Bangers 18
Roaring Roadsters 26
Flat Out on the Dry Lakes. 38
Front Facelifts 64
Fat-Fendered Oldies 78
Chevy "Stove Bolt" Six. 87
The Birth of an Industry. 98
Christmas Trees and Timing Devices. . 111
Holy Grounds of Speed 114
    Daytona Beach 114
    Indianapolis Speedway. 116
    Bonneville Salt Flats 117
An "Exhausting" Business. 124
Speed Equipment. 134
Flatheads Forever. 148

The Stock Car Racing Boom 156
A Typical Drag Strip Layout. 172
Auto Beauty-Shop Glitz. 178
Hemi Hopping. 183
Drag Racing Dynamics 186
Road Racing Finally Recognized. 191
East Versus West 196
Years-Ahead Styling with a New
    Fiberglass Body. 202
Rubber Battles of Racing. 208
Dragster Chassis Design 213
Pro and Con 215
The Big "Little Go's" 222
Supercharging Magic 230
Chevy's Hot One 234
Hot Rods, Not Shot Rods. 239
Road Warriors 245
Cool Coupes 247
Showtime 252

Sports Rod Specials 262
Land Speed Record Cars 268
Exhibition Cars 272
The Multiple Engine Trend 299
The SEMA Saga. 304
Drag Racing Goes International 308
Detroit "Muscles In". 314
"New Look" Accessories. 316
Ragtop Rage and Its Renaissance . . . 333
Street Rodding Makes a Comeback. . . 334
T-Buckets for Show and Go. 338
Durable Deuces 342
Gasoline Alley 347
Trick Dream Trucks 348
The Museum of Drag Racing 357
NHRA Motorsports Museum 358
International Drag Racing
    Hall of Fame. 362
Old Timers' Reunion 364

FOREWORD

Talk about bias! Some journalists would have you believe that all hot rodding began in Southern California! On the contrary, there have always been far more rodding-type enthusiasts scattered throughout America than those concentrated in Southern California.

Hot rodding, as we know it, was essentially *a nationwide phenomenon that occurred simultaneously throughout America right after World War II.* Even the esteemed auto journalist and historian Dean Batchelor said, "The modification of production engines started about the same time in the Northeast, Midwest, and West" (page 8 of *The American Hot Rod*).

Tom Medley said, "While it is generally agreed that hot rodding has roots dating back to the introduction of the automobile in America, and that much of the earliest hot rodding activities were in the Midwest and Eastern parts of the country, there is no doubt that the sport gained a major boost in Southern California (*Hot Rod History II,* page 145).

Some folks still say that drag racing began on the West Coast, but I don't agree. History actually informs us that the first legitimate drag race occurred in Daytona Beach, Florida, in 1905, with a Stanley Steamer driven by Louis Ross eliminating a Napier and two Mercedes cars.

Unfortunately, bias came from self-serving journalists who worked for West Coast magazines such as *Hot Rod* (first published in 1949). Let's not forget that other auto enthusiast publications existed before that, such as *Throttle* (1930s) and *Speed Age* (1947) magazines. We also must give credit where credit is due: Ed Almquist wrote the world's first widely circulated hot rod how-to manual (copyright 1946), followed by Bill Fisher with a speed manual (1948) and Roger Huntington with a hop-up manual (1951).

In hot rodding's early speed trial days, when it was truly a new daredevil sport, people were not always concerned about safety, especially because those typically homemade machines were raced on almost any back road. Mastery over those early machines empowered many men to break high-speed records while making lasting technological contributions to the new sport. But who would believe I would be called the "King of Drag Racing"? I am proud and lucky to be part of groundbreaking history and to now run a drag racing museum that keeps history alive.

In 1959, I was elected president of the American Hot Rod Association, which was founded by civic-minded hot rodder Walt Mentzer from Pennsylvania. We were focused on safety in all related activities, on acquiring public acceptance of supervised hot rod activities, on providing suitable tracks on which hot rodders could race their stock and modified automobiles, and on sponsoring an annual Championship Drag Race.

At the forefront of racing was experimentation. In 1957, in Cordova, Illinois, at the AATA "World Series" of drag racing, my brother Ed and I learned to burn 100% nitro-methane, and we "beat the pants off" the famous Cook and Bedwell dragster (the first to exceed 160 mph). Wow! The taste of that victory—I will never forget it! How could any racer forget a glorious moment like that? Who could ever forget the tires smoking, the crowd going wild, the thrill of competition, the passion for racing with all its sights, sounds, and deeply intoxicating smells? For me, that is heaven on earth!

Individuality and creative expression create a marriage that goes hand in hand with top-notch design and construction of any dragster. Like any drag racer, I was hellbent on breaking more records and stumbled onto a few "speed secrets" that made my reworked dragster national news. The smug Californians thought it was impossible for anyone to outdo them, but I proved them wrong.

It is amazing that there is no limit to what hot rodders and speed merchants have done (and still can do) with their pioneering innovation and bold foresight. If these guys did not initially come along with their gutsy, high-performance technology, the everyday automobile might still be clanking along in the dark age of mediocre performance. Instead, hot rodders have unleashed a new sport, a new frontier, and a new industry that, even to this day, remains miles ahead of the car manufacturers (except in the area of computer controls).

All of this, and much more, Ed Almquist has brilliantly and respectfully mapped in this unique, historical book that spotlights the motorsports world and the innovations that improved the safety, durability, style, and performance of the cars that you and I drive today.

Don Garlits
The "Swamp Rat"

INTRODUCTION

For most readers, the golden age of hot rodding began just after World War II, when the term "hot rod" was coined. Then, as with later generations, the challenge was not merely for more speed; it was for romance, creativity, individuality, camaraderie, and fun with all kinds of cars, trucks, and their hybrids.

I am honored to be one of the "hot rod pioneers" and am privileged to share in the fabulous growth period of the 1940s, 1950s, and 1960s. I am grateful for my younger friends, especially Don Garlits and "Speedway Bill" Smith, who urged me to tell of my own personal insider's view of the growing sport and industry, as well as the intimate profiles of fascinating personalities I have known, with whom I have worked, and even against whom I competed.

What started as a strictly historical book has become so much more than that. While doing research for this book, I discovered an underlying Horatio Alger theme. Did you know that many of the rich and famous in motorsports actually began without special skills and with little or no money? Nourished by enthusiasm and drive, these people often turned simple ideas and hobbies into avant-garde careers and businesses. In many cases, the "have-nots" became wealthy propagators of the "Great American Dream."

Consequently, I have tried to chronicle never-been-told facts about many colorful "motorheads," including their humble beginnings, dreams, grand plans, failures, secrets of success, and contributions in building a dynamic sport and industry. Many of these people likewise have significantly influenced the performance, safety, and appearance of today's production automobiles.

Many of us old-timers (now in our mortality years of our 70s and 80s), fused by a common ethos, were part of rodding's exciting mid-century roots and legacy. We and the later baby boomer generation gladly share in a windfall of personal photographs, tattered scrapbooks, old news clippings, and oral history—as a fitting apotheosis to hot rodding.

Finally, I apologize to the persons whose stories I was not able to tell here and for any omissions, possible ineptitudes, or inexactness of dates or happenings as supplied by key persons and sundry printed sources. I gratefully acknowledge the bench racing and formal interviews with notables such as Wally Parks, Don Garlits, Bill Burke, Lou Senter, George Barris, and my former racing parts partner, the late George Hurst, among others.

I hope this book will turn back your "memory clock" to a glorious time when automobiles were truly exciting and challenging. Perhaps it will even tweak or rekindle your motorsports hobby and spark some latent creativity, or perhaps it may help you to follow your "car dreams."

Two years, thousands of dollars, 1,200 letters, 380 phone calls, more than 100 personal visits, and 28,000 miles traveled were spent obtaining information and photos for this book. Nonetheless, I am happy to contribute my net profits from this book to the National Hot Rod Association (NHRA) and the Society of Automotive Engineers (SAE) scholarship funds and other charities.

MY ENCOUNTER WITH HENRY FORD

What a blast for a young sailor and car guy such as me to chat with the genius industrialist Henry Ford about his famous Ford flathead V8, the engine that would later become every hot rodder's "love child!"

It was a sunny midsummer afternoon in 1944 when Henry Ford, Sr., ambled up the gangplank onto our huge motorship that was named *Henry Ford II*. The vessel was shipshape for the transit of two Ford granddaughters en route to the family's summer home near Marquette, about halfway to Duluth—from whence the ship would double back loaded with iron ore for Ford's plant in Dearborn, Michigan. The feisty Mr. Ford, who was about eighty years old then, regally personified authority as he stepped into the monstrous engine room of our ship for one of his occasional surprise "inspections" of the mighty Sun-Doxford opposed-piston diesel engine, probably simply to "feel" its awesome rumbling power.

That particular day, I boldly seized the opportunity to speak to the great automotive magnate about souping-up my Ford V8-powered "Tin Lizzy." Despite my clumsy approach, Mr. Ford responded to my naive questioning. Coupled with his adamant advice to "avoid over-carbureting," he added his final words: "Call it modifying, not souping."

Despite the brevity of Mr. Ford's shipboard visit, he helped validate my "motorhead mentality." Thanks to one of America's first true rodders, I became a lifelong hot rodder and performance equipment pioneer.

Henry Ford (standing) with driver Barney Oldfield. Ford's 999 racer did a record 91.37 mph over a frozen lake in 1903, which helped promote the newly established Ford Motor Company. The monstrous engine pumped out 70 hp. Steering was done via tiller.

> By the time Henry Ford died in 1947 at the age of 83, his famous "bent eight" flathead—as the unplanned by-product of one man's vision—was already the basic evolutionary rodder's engine for the emerging new hot rod motorsport and speed equipment industry.

IN THE BEGINNING

America's love affair with the automobile and racing began more than a century ago—probably after the second car was built, for it was then that the first driver had someone against whom to compete.

Until the late 1920s, many of the high-performance automobiles that were not made in Europe were custom built in America for either wealthy American motorists or for professional racing teams. Even before the storms of the Great Depression, many car enthusiasts managed to cobble up their own hybrid speedsters and "Bugs"—and even roaring speedway racers—usually from stripped Model T or Model A Fords. Those fire-belching home-built cars later would be powered by Ford V8 engines souped up for full-throttle fun on road or track. Although unnamed, those freewheeling young drivers were our first true hot rodders.

Later, even the legendary Henry Ford was labeled a "hot rodder." His venerable flathead V8 engine became the rodders' darling and the hallowed precursor to the birthing of a vibrant high-performance industry.

Throughout the 1930s, many types of exciting grassroots competition occurred throughout the United States, ranging from wild amateur jalopy races on cow pastures to legitimate oval track roadster meets and later timed dry lakes speed runs in Southern California. However, the maverick evolution that eventually became the great hot rodding movement was halted as imminent war clouds loomed.
During the next decade, a generation of American youth faced the biggest challenge of their lives.

ED WINFIELD
The Reclusive Genius

Ed Winfield, "Rodding's Grandpa," was born in 1901, with genes that played magic with engines. At the age of 13, he showed signs of genius when he stripped and souped up a Model T Ford.

"That sucker ran at 80 mph, almost doubling its stock top speed," boasted Ed.

As a mechanic and an innovator, Ed had few equals. As a preppy driver, he was a consistent winner in his No. 1 dirt track racer at the famed Ascot Speedway in the 1930s.

While a novice in the Harry Miller Racing Facility—where the famous Miller race cars and engines were made—Ed first learned to calibrate Miller carburetors, progressing swiftly to complete assembly of racing engines. One of Ed's secrets in getting a remarkable 175 brake horsepower from a flathead Model T Ford was to make a special crankshaft with the throws arranged to provide even 180-degree intake strokes, thus producing a ram effect, even with a single updraft carburetor. At Muroc Dry Lake, Ed later modified a Model B Ford that made history by hitting 119.60 mph, a record in 1933 that held for more than a dozen years.

In 1921, Ed Winfield started his own business called the "Winfield Carburetor Company" where, for 18 years, he and his brother "Bud" manufactured the renowned Model SR downdraft racing carburetor.

A 1925 photograph of Ed Winfield shows him with cap and necktie, attire in which he frequently raced.

Ed Winfield is winning in the #9 Ford at the Ascot Speedway in 1924. Note the temperature gauge on the Model T Ford radiator.

In the Beginning

As a perfectionist, Ed insisted on machining his own cams. Therefore, delivery from the plant in Glendale, California, often was slow—sometimes taking months. Those who did business with Ed Winfield generally found it exasperating, if not impossible.

Although reclusive in his later years, Ed Winfield was a vast storehouse of mechanical knowledge for old-timers such as Eddie Meyer, Kong Jackson, and Ed Iskenderian. The legacy lives on.

Ed Winfield was a frequent winner at the old Ascot Speedway in the mid-1920s.

A pair of Winfield Model S carburetors on a Model A Ford racing engine.

"I sold the patent rights and was supposed to get a royalty. But it was the Depression then, and nothing sold," recalled Ed.

Later, Ed's updated invention was tested by Buick which, although impressed, would not convert to a better carburetor design because it would supposedly double Buick's existing unit manufacturing cost of $1.50.

During the Depression years, Winfield built a cam grinder, which only a few friends were privileged to see. He was among the first to design radical, long duration, high-lift camshafts that were both functional and durable.

Ed Winfield's reputation as a mechanical wizard attracted men such as George Riley, who had permission to copy Winfield's special cam design as his own. Apparently, Ed's experiments helped George Riley develop the Riley 4-Port head for the Model B engine, which at the time was one of the hottest conversions available.

Ed Winfield developed numerous Ford improvements and manufactured a high-compression cylinder head for Model A Ford engines, along with futuristic racing camshafts for the Model A and V8 Fords, which were later copied by other grinders.

BUD WINFIELD
Co-Designer of the Legendary Novi

Back in the 1940s, W.C. "Bud" Winfield was probably racing's "biggest figure," being 6-1/2 feet tall and weighing 250 pounds. However, he was short-shrifted by the historians who thought that his brother Ed was more of the mechanical genius. Bud's fame quickly reeled from the development of the famous Novi racing cars that garnered speed records at Muroc Dry Lake and Indianapolis.

The Novi story began in 1945 when Lou Welch, a Novi, Michigan, maker of truck governors, came to the Winfield brothers' shop with an idea for a new racing car for Indy. Welch hired Bud, hoping he would receive Ed's help for free.

When Ed was asked what he thought of the new proposed racing car, the pragmatist said, "It's okay, but there are three things wrong with it: the front wheel drive, the supercharger, and the V8 engine, but you don't need them to win. It'll take years to work the bugs out!"

Bud and Ed Winfield did the original Novi engine design in 1939–1940, with refinement done later by Leo Goosen, who incorporated some Offy features. The powerful V8 dual overhead cam engine delivered 450 hp in 1941 (830 hp in 1965) and cruised effortlessly at 6500 rpm, peaking at 8500 rpm.

The awesome supercharged engine was first used at Indy in 1941 in a Miller front-drive chassis that finished fourth. By 1946, a new lap record of 134 mph was set at Indy with the Novi V8 in a brand new front-drive chassis. In succeeding years at the Indy races, mechanical difficulties plagued the fabled Novi race cars. Finally in 1960, after spending a fortune, Welch gave up the nemesis and sold his cars, engines, and patterns to Andy Granatelli for a mere $10,000! After the Granatelli brothers updated the twenty-year-old Novi engine, they too failed in their early 1960s Indy efforts.

In the same way as their designers, the mighty supercharged Novi V8s eventually became historical lore of some of the glory days of motorsports.

A rebuilt Novi ran at Muroc Dry Lake in 1949 after windshield streamlining was done by Frank Kurtis. The tall man is Bud Winfield. (Photo supplied by Frank Kurtis)

The Novi is shown running at Muroc Dry Lake in March 1949, prior to racing in Indianapolis. It ran 225 mph (unofficial). Due to timing equipment failure, the car was credited with an AAA 207 mph.

"Tin Lizzie" Speedsters

Few cars have become the butt of more jokes than the unforgettable Model T Ford, which, ironically, our fathers and grandfathers drove. That is probably why the young guys of the 1920s enjoyed souping up their "T's" and outrunning the expensive makes. Imagine the humiliation of having your classy Cad trounced by a cheap little "Tin Lizzie" speedster?

Of the dozens of Model T speed equipment makers throughout America in the 1920s, the Fronty Fords were among the most popular. Then, the Chevrolet brothers of Indianapolis, who produced Frontenac equipment, claimed that a Fronty-equipped Model T Ford racer could be a consistent winner because it could negotiate a banked half-mile dirt track in less than 30 seconds and a one-mile track in 45 to 50 seconds. For straightaway speed trials of a mile or more, speeds of 100 mph and higher were promised (with a gear ratio of 3:1).

In addition to a rocker-arm cylinder head conversion and dual carburetion, other modifications included a special oversized crankshaft, lightweight pistons, pressure-fed oiling system, special water pump, magneto ignition, racing exhaust manifold, and chopped flywheel. A full-race "T" could deliver more than 100 hp at 3600 rpm.

In its day, a Model T race car, made from a cut-down chassis with a special racer body and weighing only 1,000 pounds, could hold its own in county fairs and professional racing circuits. Many early champions started their careers by driving hot Model T's.

The 1924 Model T engine with Frontenac overhead valve conversion head and dual carburetors had a high-pressure oil pump, water pump, magneto, and camshaft, all driven by a crankshaft extension unit.

This 1926 ad also promotes a single-seat runabout for $290, the lowest price ever offered for a Ford. The 15 million Model Ts that were produced from 1908 to 1927 helped put America on wheels.

Ed Winfield guns his little flathead Model T to a win at Ascot in 1925. Ed's secret was boosting high rpm breathing through a jumbo carburetor and "gulping" cam timing.

CARL "POP" GREEN
Gasoline Alley's Green Giant

"With all the improvements you rodders make, auto makers have a free testing laboratory," said Carl R. Green, a self-taught racing engineer who was a rather sophisticated hot rodder in his heyday.

Carl's lifelong speed career began in the late 1920s in Dayton, Ohio, where he raced and built both Model T and Model A racing cars in one of the first speed shops in America. When he reached "Gasoline Alley" of Paterson, New Jersey, he constructed several fire-spewing speedsters that were outlawed by local magistrates because they scared horses. During the 1930s and 1940s, Carl "squeezed the max out of Ford engines," thus making his dirt track big cars and roadsters the terror of Eastern racetracks.

"I was nicknamed 'Pop' because of my premature white hair," said Pop. "I loved to give advice to those 'car nuts' willing to listen, such as Almquist whom I trusted with my homemade camshaft grinder that few were privileged to see." In a small side-street garage, Pop could regrind any Ford V8 cam to "killer specs," but he took his time in doing it.

Pop was one of the few who built both car and engine. Proud of his surname, Pop painted his cars an uncommon green—some say to upset superstitious competitors who saw green as a "bad-luck" color. As a born innovator, he made some unusual modifications to the Ford V8 engine. One was the removal of counterweights and the installation of a special 180-degree crankshaft similar in design to that of a four-cylinder engine with the V8 then treated as a pair of fours that could be fired 1-3-4-2 or 1-2-4-3 in each

A dejected Carl Green walks toward his #8 Model A-powered car that Bob Sall drove at Woodbridge, New Jersey, in 1933. Nevertheless, Bob's high point standing in 35 events won the Eastern Dirt Track Championship that year.

```
CAMSHAFTS REGROUND  Consult The   "POP GREEN"
                    Old Reliable
        (Pioneer Racing Equipment Engineer)
  Because of Improved Grinding Methods, We Are Making A
  New Low Price For Grinding Camshafts of Ford V-8 60, 85 and Mercury
    V8-60 Cams Ground $24.00 - 85 and Mercury $30.00
       254, 260, 268, or 274 Degrees Valve Duration
                World's Fastest Cams

   RACING MOTORS, Axles, Steering Gears, Gear Cutting, etc.
         GREEN ENGINEERING COMPANY
    SHOP — 25 River Road, E. Paterson, N.J.    Fair Lawn 6-1173
             Res. 165 Trenton Avenue, Paterson, N. J.
```

Pop ran similar ads in the 1930s. He also advertised blueprints for building a midget racer, speeding up a Ford, and fuel formulas for $0.50 to $1.00. (Circa 1948)

Dirt track scenes such as this were common throughout the country in the 1930s. Among the most popular racing engines then were the reliable Model A and Model B four-bangers, often equipped with speed equipment from Pop Green. (Senter Collection)

bank. Either firing order required a special camshaft that Pop also would grind. Likewise, two cylinders could be fired simultaneously.

During retirement, Pop was the technical editor of *National Speed-Sport News*, a racing tabloid with a long-time publisher, Chris Economaki.

Hot Rods on the Highways
Written in 1949 by Carl "Pop" Green

The Hot Rod, a new name applied to an old-type car, dates back to the early part of the century. Younger people drove roadsters or cut down touring cars to race each other on dusty roads even when it was impossible to see their radiator caps. Then, speed limits through small towns and larger ones were as low as 8 mph, while ropes sometimes stretched across the road to stop fast cars whose approach was alerted by a cloud of dust. About 1915, when automobiles became more plentiful, the professional speed cop made his appearance and commercialized on the speed laws—often doing a profitable business catching unwary speedsters and rushing them before magistrates with whom they were double-dealing. By 1920, these speed traps discouraged the Hot Rod car for a while.

As time passed, roads improved, speed laws softened, and speed limits increased, all of which encouraged the return of the Hot Rod. Not being content with racing on regular racetracks, those "super hot" cars took to the highways and city streets. By the 1930s, accidents became so numerous that the American Automobile Association (not the racing division) asked for cooperation between law enforcement officials and courts to use their combined efforts to curb speeding and reckless driving. With traffic conditions as they are today [1949], a driver is much safer driving on a race track in competition with experts than driving on the highway with amateurs.

ROBERT M. ROOF
A Granddaddy of Speed Equipment

Some historians believe Robert Roof, a Midwestern racer turned engineer, was one of the first (if not the first) speed merchants in America. In 1917, Roof began making speed equipment for Model T Fords. Because of this, a new industry was in the making.

After a chance meeting with Henry Ford at a Michigan speedway, visionary Robert Roof hit upon the idea of a sixteen-overhead-valve Ford. Soon afterward, Roof abandoned his youthful career as a race car driver and turned his energy to developing a sixteen-valve cylinder head with an overhead camshaft. Because the patented conversion performed well in Model T's, Roof began making it for the four-cylinder Ford Ferguson truck-and-tractor engine—a tough little mill at the time.

In the 1930s, the modified Model B Ford four-cylinder block became a hot power plant. A Roof-equipped "RCO Flash" racing motor with a special manifold and two Winfield carburetors sold in 1946 for $650. Although that was a lot of money at the time, it was worth it for that extra power.

During midget racing's heyday, a Roof-equipped V8 Ford 60 was the "low buck" one to beat. Made for the Ford 60 and the larger Ford and Mercury V8s were Roof cylinder heads and a dual intake manifold with two loading chambers and a balance tube. Both models featured two spark plugs per cylinder and were made in two pieces with a gasket sandwiched between them. This eliminated a foundry core and was cheaper to make. Tolerable ratios were 10:1 or higher if alcohol was burned.

In the 1950s, the Roof family also marketed racing camshafts, oversized valves, pistons, and lightweight flywheels for other makes, including Willys. Often referred to as a pioneering speed shop, the Roof enterprise, called R & R Manufacturing Company, was located to the last in Anderson, Indiana.

Robert Roof, an amicable guy, looks mean at the wheel of his dirt racer in Indiana. Seen in the photo are two Winfield carburetors projecting from a Model B Ford engine. Tires are Allstate balloons. Roof, knowing that he was onto a good thing with his speed equipment, posed with his Model A-powered, dirt track racer.

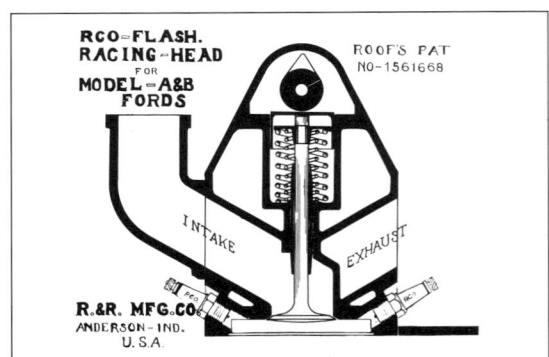

Patent drawing of Roof's overhead-valve racing head.

In the Beginning

In 1908, Robert M. Roof, Sr., drove this hot speedster to Detroit.

Roof-equipped V8 Ford 60 with a Delco distributor from a sixteen-cylinder Cadillac fired two spark plugs per cylinder. A set of polished heads cost $75. A complete ready-to-race 1946 Mercury V8 engine was only $675.

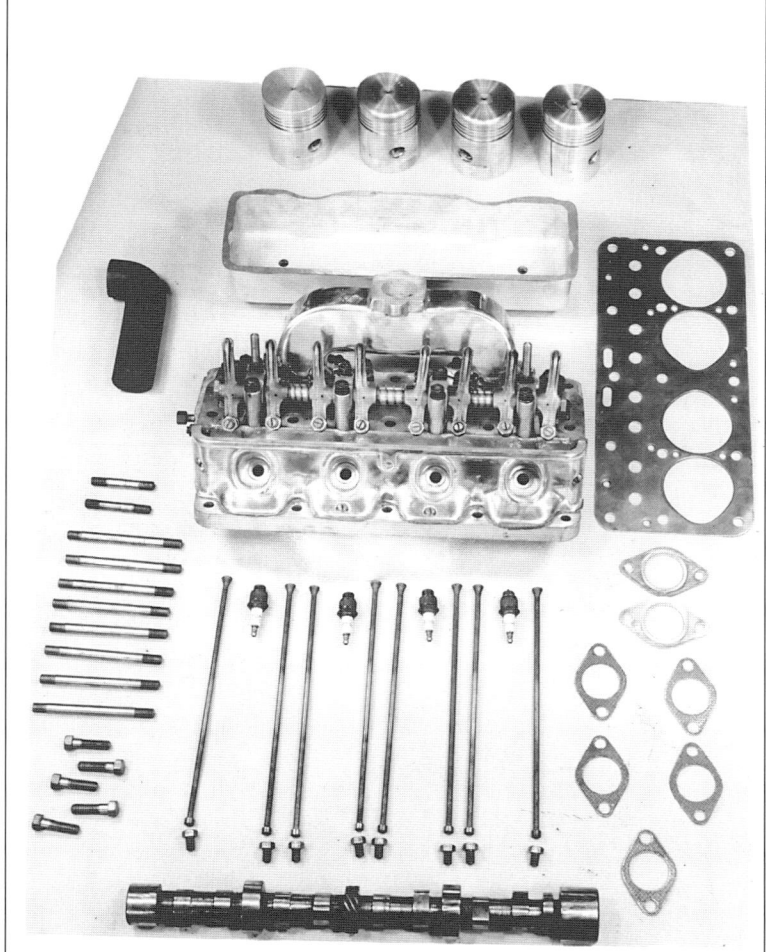

Roof made racing equipment for Willys as did Alexander, who sold this conversion kit that includes a reworked OHV head and full race cam for only $110 in 1938.

Roof's Model B Ford racing engine produced 140 hp at 5500 rpm. The valve cover exposes the single overhead camshaft that activates two valves per cylinder.

The Sanctioning Wars Rulers of Racing Won Over Rivals

After the world's first organized auto race from Paris to Rouen was held in 1894, rule-making organizations were spawned. These included the Federation Internationale de l'Automobile (FIA), a major European governing body of motor racing.

A year later, discipline was still lacking when America's first automobile race occurred with pokey, marathon-like speeds. (The 54-mile Chicago run took more than 82 hours.) Because international rules could hardly apply to the more diverse racing in 1911 in the United States, the American Automobile Association (AAA) Contest Board stepped in to sanction the first 500-mile Indianapolis race. Ray Harroun won this race in a Marmon Wasp, averaging 74.59 mph. The wondrous Golden Age of automobile racing in America was later interrupted by the Depression of 1929.

When thoroughbred competition formulas were replaced by "Free Formula," the AAA hoped to stimulate the sport by lowering racing costs that had been escalating and by making it easy for production car engines to compete. However, high-performance machines were severely handicapped with restrictions such as minimum weight, reduced engine size, and no superchargers. As the economy recovered, so did the rule-makers' sixth sense. By the mid-1930s, rigid and arbitrary limits were lifted, thus allowing for improvement of the breed and the return of pure, race-bred machinery.

After World War II, the Contest Board of the AAA remained the strong-armed control body. However, as outlaw roadster, midget, big car (sprint), and stock car racing continued unbridled, the tiny Southern California Timing Association (SCTA) formed in 1937 and began running time trials at the dry lakes. In 1944, the Sports Car Club of America (SCCA) began organizing road races—a rarity in America. Four years later, the SCCA ran the first U.S. Grand Prix for Formula I cars.

Moonshine-running good ol' boys, who later raced stockers at Daytona Beach, established the National Association of Stock Car Automobile Racing (NASCAR) in 1947, which soon involved major car makers. Of the other oval track groups that formed, the United States Auto Club (USAC) became a heavyweight.

Official AAA observers confer with 1930 AAA Indy race champion Billy Arnold at Daytona Beach, Florida, after he had established six stock car speed records in a Chrysler Eight Roadster. The Atlantic Ocean is in the background.

In the Beginning

This modified production roadster, an ancient European cousin of hot rodding, is being gassed-up at the Dieppe Circuit in France—probably for the Grand Prix, circa 1910. Road racing in America did not begin until 1944 at New York's picturesque Watkins Glen, which soon became the center of American road racing.

Famous World War I flying ace Eddie Rickenbacker is riding in an official Cadillac 16-cylinder pace car beside a Maserati two-seater race car, prior to a long-ago AAA-sanctioned Indianapolis Speedway event.

The great postwar hot rodding movement erupted and eventually crested into America's most innovative and fastest motorsport. Hoping to unify and regulate the nation's many clubs and neo-drag racing activities, organizations such as the National Hot Rod Association (NHRA) were formed in 1951. This occurred at approximately the same time that the first drag strip "officially" started. Many regional associations formed but eventually folded. Even the once mighty American Hot Rod Association (AHRA), founded in 1955, faded away in 1984. When the Dixie-based International Hot Rod Association (IHRA) was established in 1971, it was called an "outlaw," as were other lesser-known sanctioning groups popping up in the United States. Rivalry, accusations of cheating, confusing rules, and lack of standardization within classes—especially among drag strips—threatened the new sport.

After the Le Mans disaster in 1955, the AAA disassociated itself from racing, declaring the sport too dangerous. In its wake, the USAC took the reins, but other sanctioning battles continued. In an attempt to oversee all racing, the organization known as the American Automobile Competition Committee (ACCUS) was formed.

Another power fight loomed when NASCAR jumped into drag racing in the mid-1950s. Critics said that the NHRA's quick tie-in with FIA (to certify a challenge of a long-standing 5/8-mile European record) was a deliberate attempt to slow NASCAR's drag racing ambitions. By 1960, a compromised NASCAR had joined the NHRA in presenting the first major Winternationals.

Today, NASCAR is the mighty kingpin that holds together the wheels of big-time stock car racing. The USAC continues to rule most Indy-type competitions, and the NHRA continues growing as motorsports' largest sanctioning body. They all have more classes than a university!

Finally in 1993, a historic event occurred when drag racing was officially recognized by the FIA World Council with the FIA Drag Racing Commission firmly established. However, as the world turns, so will auto racing technology and the need for more regulations, unified rules, formulas, and classes. Stringent stewardship and battles will probably follow.

HARRY MILLER
In Pursuit of Mechanical Perfection

With his brilliant "years-ahead" thinking, Wisconsin native Harry Miller, born in 1875, became one of the greatest designers in American motor-racing history. His gift for pioneering avant-garde innovation belied his shy personality and humble background—all of which carried him through peaks of wealth and back to near-poverty.

From the 1920s and onward, the ultimate in racing was seen in Harry Miller's advanced engine configurations, which made his screaming four-banger the forerunner of the famed Offy engine. Despite painstaking hand techniques but through volume and standardization of parts, Miller's thoroughbred racing cars sold for as little as $10,000. That was a small fortune in the 1920s, but was far less than today's "one-off" racers that cost up to $500,000 to put on the track.

In 1935, Preston Tucker (later, the creator of the Tucker car) brokered a deal with Ford Motor Company, in which Harry Miller was hired to hurriedly design and build 10 racing cars around the then-new Ford V8 engine for the Indy 500. Although the flashy Miller-Fords stole the glory of the day, only one finished the race. (The exhaust manifold heat

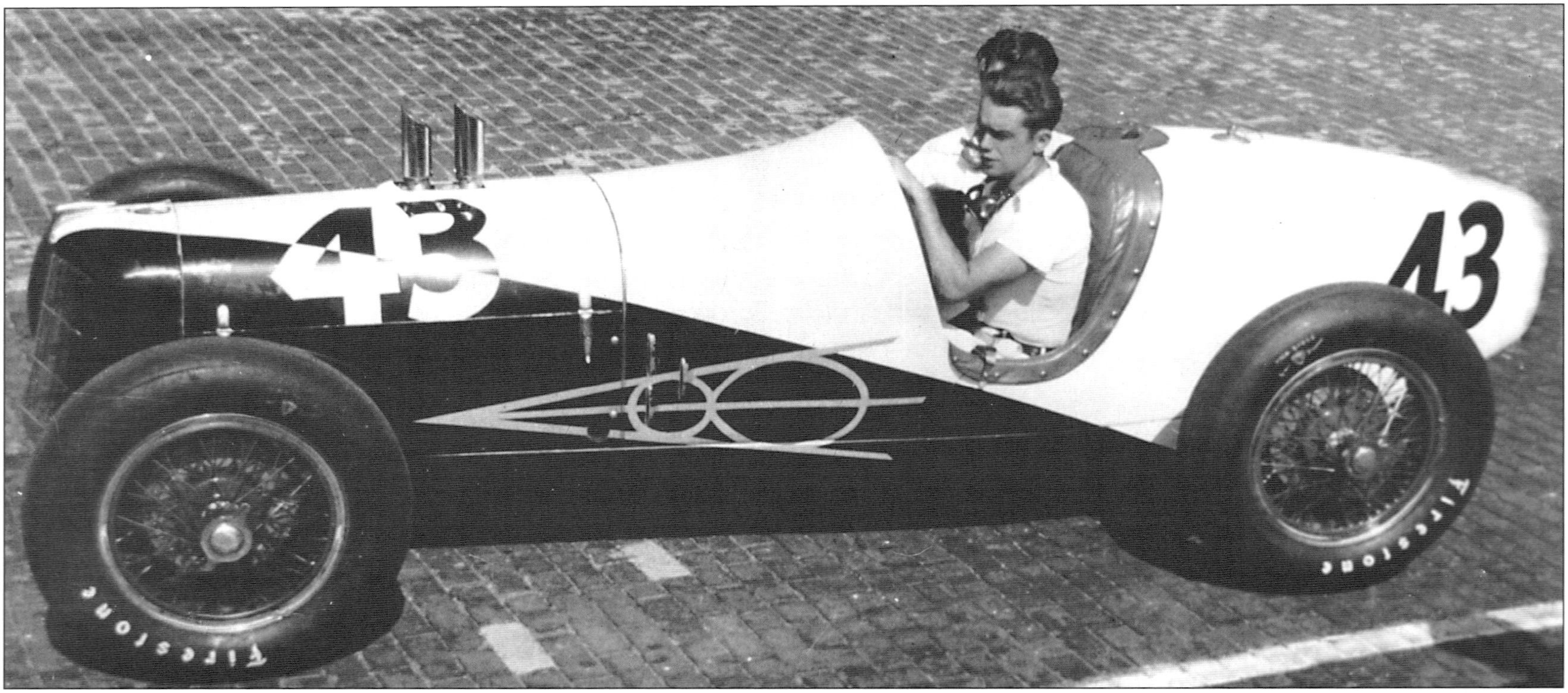

Dirt track star Ted Horn appears dejected as he sits in one of the ten jinxed front-drive Miller-Fords raced at Indy in 1935. Ted finished sixteenth but won the AAA National Championship the following year. Like many early big-time racers, Ted got his start driving hot rodded Model T and Model A Fords, often on dusty outlaw tracks.

In the Beginning

had caused the steering gear to seize.) Nevertheless, credit should be given to Harry Miller for doubling the horsepower of the flathead, while most of the engine remained stock. The only changes were a reground camshaft, 92:1 aluminum heads, a larger oil pan with cooler, magneto ignition, exhaust headers, and a multiple carburetor manifold—all standard hop-up modifications at that time and today.

Despite its mildly souped-up engine, the front-wheel drive, independent suspension, low height, and body streamlining, the Ford Special (similar to many other Miller cars) represented radical departure from racing technology of that era.

Ironically, the rush project had been authorized by Edsel Ford without his father's knowledge. Therefore, old Henry Ford decreed its immediate termination. The year 1935 pre-dated the first serious effort to hop-up the famous flathead V8 on a grand scale and demonstrated Harry Miller's hot rodding spirit, mistakes and all.

Some of the reversed engines in the 1935 Miller-Fords had dual carburetors. Others had log-type manifolds (shown here) with four Miller single throat carburetors.

Miller's "swan song" was this unusual rear engine four-wheel drive design, shown here with the body removed. The boxed-in frame side members contained four fuel tanks. Although the output of the big six-cylinder centrifugally super-charged Miller was 246 bhp on 81 octane gas, it failed at Indy. However, at Bonneville in 1940, George Barringer piloted the Miller to 12 International Class D records with a top speed of 158 mph.

ROSCOE "PAPPY" HOUGH
Fair Deal

The legendary Roscoe "Pappy" Hough did not quit racing until he was past 60. His career winnings topped $2-1/2 million.

"When I started racing in 1917, we called our stripped-down cars 'jalopies' or 'bugs.' Later we called them hot rods," said the colorful Roscoe Hough, who from 1919 to 1962 won over $2-1/2 million in more than 1,000 races. This was a remarkable achievement, considering there were no huge purses or big-time sponsors back then.

Born in 1903, Roscoe began racing Model T jalopies at the age of 16 in Midwestern cornfields. "Winners then got $5 to $10—which looked really good to me because I was making only 35¢ an hour at my job," Roscoe reminisced.

In 1934, when midget car racing was gaining popularity, Roscoe built two midgets with V8 Ford 60 engines. He became adept at stretching the rules—such as using transferable letters for a quick switch to the fastest car. Roscoe earned the nickname "Pappy" after he cut a fair deal for fellow drivers by outfoxing a cheating promoter.

"In those days, we cobbled up our own racing parts. I even made my own cams by grinding the base smaller to get a higher lift. Later I manufactured over 700 quick-change gear boxes, and them suckers were the toughest made," Roscoe remembers.

"Pappy" Hough, with a trailerload of midgets, was an exponent of the "team concept." (Circa 1946)

In 1949, Roscoe began racing stock cars, and, at the age of 47, he began competing in Late Model NASCAR. He was runner-up for the Grand National Sportsman title in 1950 and was National Short Track Champion in 1951.

After racing roadsters and midgets, Roscoe switched to stock cars. His first big wins were in a 1937 Ford and the 1950 Ford pictured. "In the Sportsman's class, the outside of the engine had to be stock, but boy, did I jazz up the inside...boosting compression to about 10:1, stroking to 4-3/8", and cutting the flywheel to practically nothing," Roscoe recalled.

In the Beginning

Police Souped-Up Cars, Too

The speed and reliability of flathead Ford V8s made them ideal police chase cars from 1932 onward, until the OHV interceptors were first introduced. This 1935 cruiser came with an oversized radiator and higher compression. Later-model 1953s could reach 100 mph. (Photo by Krauss)

1910 Buick "Bug" Racer

This 1940 photo shows two Buick engineers looking over the ancient Buick Bug that barnstormed America in 1910, winning half the road races and speed dashes. Its huge four-cylinder 325 c.i. engine developed 144 hp. With famed Louis Chevrolet as its driver, it averaged 71.8 mph in a 200-mile race in Atlanta.

A former chief engineer at Buick, A. DeWaters (left) built the car with drivers Bob Burman and Louis Chevrolet. Charles A. Chayne, another chief engineer at Buick, is on the right.

The Four-Bangers: The Poor Man's Offy

Of the five million Model A Fords made from 1928 to 1932, several thousand are still run by die-hard collectors, and many are shown at antique car shows. In its time, the tough Model A and B Ford four-cylinder engines were the "cheap choice" for dirt track and roadster racing; overhead conversions were the "top dogs" of that brawny era. In the early days of dry lakes racing (before flathead V8s), the Model A was used in its entirety. However, the rugged Model B block was preferred because of larger main bearings. Its counterbalanced crankshaft often was drilled for pressure-fed oiling, and a higher pressure oil pump (sometimes with a dry sump) was added. Dual carburetion, a racing grind camshaft, and larger valves with double springs and port enlargement were common modifications.

Harry Miller, Hall, Ambler, Roof, R & R, Riley, and Morton & Brett were among the many OHV head and dual carburetion manifold makers. Today, they are long gone, but fond memories of them remain.

This 1929 Model A Ford with a boat-tailed speedster body has a Burns dual intake manifold with two Stromberg V8 carburetors, which improved breathing for racing but probably made the vehicle over-carbureted for road use.

This illustration appeared in a 1930 trade magazine, announcing Harry Miller's OHV head conversion for Model A Fords. The ad disclaimed any connection with any other Ford head. Another ad indicated Leo Goosen was the designer. Later, Cragar, who purchased Miller's facility, cut the price of the OHV from $137 to $112.

Before he became famous, actor Robert Stack liked weekend runs at Muroc. Here he adjusts the Winfield carburetors on his Model A roadster. (Circa 1939)

EDDIE AND LOUIE MEYER
They Did the Gamut

From the Roaring Twenties and well into the Fifties, the Meyer brothers were hard to beat as drivers, mechanics, and speed merchants. In Southern California, Eddie Meyer raced midgets, roadsters, and boats. Louie Meyer, the younger of the two, polished his mechanical and driving abilities for three big Indy wins.

Eddie was born in Metz, France, in 1893, and he won his first race in his own homemade Model T Ford race car in a 100-mile road race in Ontario, California, in 1919.

Louie was born in New York City in 1904 and began wrenching for Indianapolis racer Frank Elliott, who ran a newly introduced front-drive Miller. There, Louie honed his driving skills. In 1928, he took his first Indy 500 win and clenched the AAA national driving championship as one of the youngest men ever to win the coveted crown.

When inboard boat racing began to flourish in Southern California in the 1930s, Eddie Meyer helped develop the Ford V8 for the

While testing flathead racing parts, Eddie Meyer drove his Deuce like "Hammers of Hell." Few drivers could afford good crash helmets during the 1930s.

Eddie Meyer dual intake manifold for flathead Ford and Mercury has an exhaust-heated, high-rise section that can be removed for racing. High-compression heads were also made for 21- and 24-stud blocks.

225 Class. Eddie got hooked on racing on water, and he and his son Bud won dozens of trophies over the years. Eddie, who earned the title "Pappy" Meyer for his long-running racing career, became the first driver to officially exceed 100 mph in a 135-Class inboard.

While building some of the more competitive flathead Fords in Southern California, Eddie Meyer began manufacturing flathead speed equipment in 1938. With $200 borrowed from his mother, Eddie had patterns made for an attractively finned dual intake manifold. With Meyer equipment, Eddie's son Bud set the SCTA roadster record in 1939 at Muroc Dry Lake in a 1932 Ford roadster.

Long after Louie's driving career ended, he went into partnership briefly (in 1946) with Dale Drake, making the famous Offenhauser

Hot Rod Pioneers

racing engine. After a tantalizing offer by Henry Ford, Louie sold out and joined the Ford racing team.

In 1963, Louie and his son, Louis "Sonny" Meyer, Jr., helped develop another milestone—the high-tech Indy Ford V8 racing engine for the Ford Motor Company. With four valves per cylinder and four cams, it was the first purebred American engine designed for championship racing in many years.

The Eddie Meyer organization, which had made high-compression cylinder heads, dual intake manifolds, and permanent mold racing pistons for both large flathead Ford V8s and 60s, dropped out of manufacturing when overhead valve V8s outran the flatheads. Bud Meyer said, "The speed equipment business became a rat race, so we went into the specialized world of exotic foreign car repair such as Mercedes, BMW, and Rolls-Royce." Soon the *Los Angeles Times* voted the Meyer shop as Hollywood's mechanic because of its affiliation with countless movie stars.

At age 65, Louie retired to a farm in Kerklin, Indiana, where he won his last trophy at a Phoenix electric car race, when he was 88 years old. He died in 1995, and Eddie Meyer died in 1983, at the age of 89. Bud Meyer continues to operate the shop.

Louie Meyer, Jr., (Sonny) continues the Meyer tradition as a racing mechanic, wrenching for top teams such as Menard.

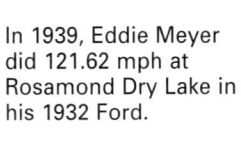

In 1939, Eddie Meyer did 121.62 mph at Rosamond Dry Lake in his 1932 Ford.

Eddie Meyer in front of his shop with "T" roadster identified as an AHRA record holder. (Circa 1960)

Louie Meyer with his riding mechanic and brother Eddie Meyer. Louie won the Indy 500 in 1928, 1933, and 1936. His Ring-Free Miller averaged 109 mph in the last classic.

In the Beginning

Johnny Parsons, a roadster-racer, drove a Meyer midget in 1951. He rolled this car 10 times and walked away unhurt. Johnny ran at Indy later that year.

THE FOUNDING FORTIES

After struggling through the horrors of World War II, Americans ventured into an era of exciting new beginnings. Millions of returning servicemen who had put their dreams on hold now saw life anew from a hopeful perspective that included a sporty car, a good job, and a family of their own.

Rationing had ended, and gasoline again was plentiful. The peacetime economy was booming in America. Happy cruising days were here again, and a lot of catching up had to be done. The World War II generation was going places—to big band concerts and to drive-in eateries where car-hops waited on customers curbside. The tab for a hamburger and french fries was less than 50 cents.

Around 1946, the new auto hobby's car was given the snappy name "hot rod"—by whom, no one knows. In every part of America, youthful car buffs formed the nucleus of an excitingly fresh and innovative car culture.

Because race-winning speed secrets often were hoarded, the time was ripe for the first how-to hot rod book written by Ed Almquist in 1946. Afterward, a proliferation of car magazines began with *Speed Age* and *Road and Track* in 1947 and *Hot Rod* in 1948—at a price of only "two bits" (25¢) per copy.

A bright new hot rod industry had come to life, and daring pioneers were turning their backyard hobbies into businesses to form the foundation for today's gigantic high-performance industry. Hot rodders set the trends for auto makers.

With jobs being plentiful, workers began earning enough money to buy a car, even with the 75-cent hourly minimum wage. A Ford Model A, or Deuce, was ideal for reworking into a hot rod and could be bought "for peanuts." A new car sold for less than $1,000 and a new house for $6,000. War-surplus jeeps, which were sold cheaply to farmers as practical and everyday land rovers, set the stage for the recreational vehicles (RVs) of today.

Customizers trashed running boards and began frenching headlights and lumpy fenders into bodies for a smartly restyled, custom appearance. The totally new shoebox 1949 Ford and Mercury soon would overtake Chevrolet as the top sellers, but old models remained a rodding favorite.

Throughout the country, abandoned airfields became "drag strips" for standing-start acceleration runs, sometimes riling the judicial system. Speed trials moved from Daytona Beach and the dry lakes to the Bonneville Salt Flats, the site of the first major hot rod speed competition in 1949. There on "holy ground" for hot rodders, the Xydias-Batchelor Streamliner set a top speed record of 193 mph—faster than any hot rod had ever gone.

In 1948, enthusiasts flocked to see the first hot rod and customs show in Los Angeles. However, Easterners had to wait a few more years to view their first hot rod show in Linden, New Jersey, where one dollar paid the admission fee or bought five gallons of gasoline.

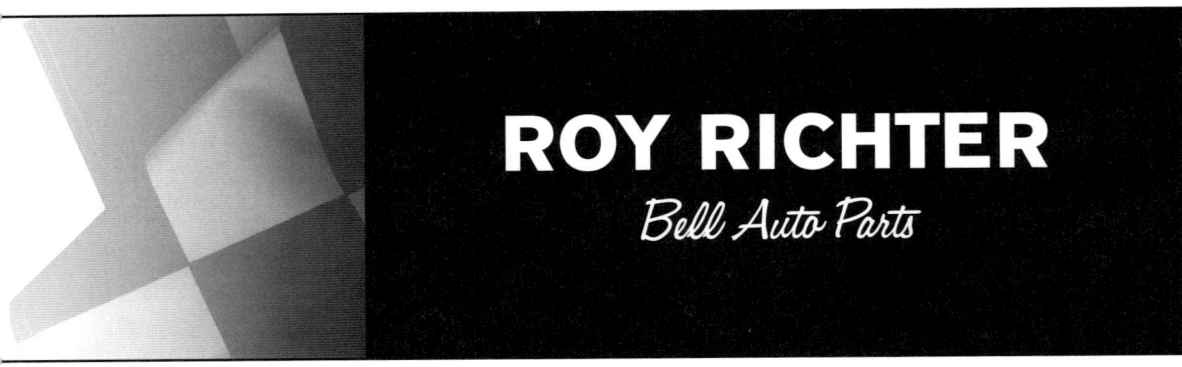

ROY RICHTER
Bell Auto Parts

Early race drivers, who wore little or no protective headgear, would put their lives on the line every time they raced. Consequently, Roy Richter recognized the need to pioneer a full-coverage helmet to replace imported hardhat types that offered no more protection than the box in which they came. In 1954, the first production Bell helmet was made in a one-car garage at the rear of Bell Auto Parts. However, the hand-laminated fiberglass helmet with a foam liner was rejected after failing crude tests made on human cadavers by the prestigious Snell Foundation. Later, to save his reputation, Roy bought out Toptex to form Bell-Toptex, Inc., thus acquiring the patent rights for a non-resilient, energy-absorbing liner. Since the early 1960s, many Indy winners have worn "Bells." By the 1970s, the full-face design had become the standard in the helmet industry.

Years ago, when every racing locality in the nation had a particular hangout for the racing fraternity, George Wight's Bell Auto Parts was one of the first speed shops and was the hot spot in Southern California. There Roy Richter and "Roscoe" Turner began building a midget race car from salvage parts and a 1918 vintage Saxon four-cylinder flathead engine.

After a few local races in California, Roy, who was born in Dupo, Illinois, returned to the Midwest to resume racing with his midget in the Detroit area. In early 1936, Roy began a race car fabrication shop with a few hammers and some dollies for forming body panels. After Johnny Wohlfield succeeded in running an outboard-engined, Richter-built midget, competitors sought Roy's expertise. Later, Roy moved back to California, where top drivers such as Sam Hanks and Johnny Parsons utilized his talent.

After World War II, Roy purchased Bell Auto Parts with $1,000 and his customized 1939 Ford convertible. At that time, the struggling business had little inventory—a few used racing car parts, a case of oil, some old machine shop equipment, and a louver press for punching louvers on hoods or trunks at 10 cents per louver.

However, Roy had bigger aspirations, and he contacted area speed equipment manufacturers for extra discounts in exchange for advertising space in a proposed mail-order catalog. When the first Bell Auto Parts catalog was distributed nationally, it brought quick cash orders from retail customers and small speed shops. Soon, the catalog resembled a "general store" of racing and hot rod parts.

Roy Richter (right) with his helper, Pete Smiley (center), did some repairs on the Offy midget that Henry Banks (left) drove in 1946.

The Founding Forties

During the postwar prosperity, racing swelled with the pent-up demand for speed, and Bell Auto Parts rode that wave. By the 1950s, "hot rod" street machines became the new craze sweeping the country. Roy developed another business called Cragar Industries. To capitalize on the car enthusiast market that was exploding at that time, Roy developed custom steering wheels and "Mag" wheels that had broad sales appeal. With Tom Shedden, Ray Lavely, Art Bagnall, and Frank Heacox as company team players, sales expanded with 60,000 dealers nationwide.

By 1967, speed equipment had become big business, and Roy helped found the Specialty Equipment Market Association (SEMA). By the end of the next decade, Roy had three prosperous businesses. However, because of his failing health, he was forced to sell Cragar Industries and Bell Helmets, which continues to make the famous Bell helmets. In 1978, Roy sold Bell Auto Parts.

Over the years, Roy Richter's companies did hundreds of millions of dollars in business and amassed 250,000 square feet of business space with 300 employees. Such an accomplishment was not bad for a man who started with only a tiny building and a part-time employee.

TV star Dick Smothers puts on a Bell helmet in preparation for a racing event.

Each year after the Bonneville Speed Trials began in 1949, Bell Auto Parts pitched its tent called the "Palm Tree." It offered racers shade, cool water, and, of course, parts.

Roaring Roadsters The Poor Man's Racer

It's almost laughable that an early article in *Speed Age* magazine referred to the "Roaring Roadsters" as a new development in auto racing. Actually, a form of roadster racing began with the dawn of motor car competition, when the open roadster was practically the only body style available.

"Old-timers didn't know it, but they were building so-called 'hot rods' shortly after the automobile was first invented," said Don Radbruch, whose book, *Roaring Roadsters*, tells about the men and machines of an almost forgotten era (1929–1956).

Post-World War II track roadsters rarely were more than stripped-down stock cars, often with Model T or deuce "high boy" bodies. Later, they more closely resembled state-of-the-art racing cars.

At first, the AAA and other elite racing organizations, as well as many veteran big car and midget car drivers, looked disdainfully on the hot rodded roadsters and hoped that those "outlaws" would go away. However, as roadsters began banding into competitive groups, promoters seized on their public appeal, and roadster racing acquired respect and recognition.

Larger organizations tried to preserve the true "roadster" name, but as racing authority Russ Catlin once said, "Greedy promoters, capable only of spelling words of three letters, insisted on using the euphonious term, hot

From 1924, Southern California was a hotbed of roadster racing. This faded photograph shows Ed Winfield's bob-tailed Model T Ford bouncing off the dirt at the old Ascot Speedway. (Circa 1924)

Here, Lou Senter in his #66 Ford is trying to pass a Chevy jalopy at the Ascot dirt track. (Circa 1940)

In 1948, Jim Murphy drove this fast track roadster, which was built by Lou Senter. Its flathead Ford V8 engine sported a McCullough blower that gave a power boost only at higher rpm's.

The Founding Forties

Similar to his brothers Tim and Fonty, Bob Flock, pictured here in 1949 at Williamsgrove, Pennsylvania, raced roadsters before becoming a top stock car driver. This Ford V8-powered T-bodied roadster was owned by Red Vogt, who was NASCAR's first superstar mechanic. (Frank Smith photograph)

Troy Ruttman, one of the top drivers in the California Roadster Association, eventually won the 1952 Indianapolis crown. Here, Troy holds a trophy he won driving Del Baxter's Sharp-equipped Ford V8 flathead.

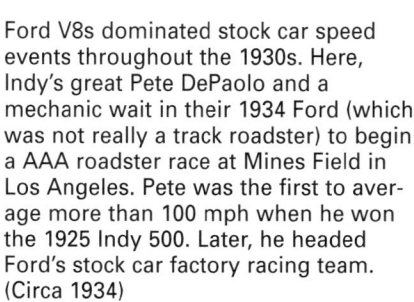

Ford V8s dominated stock car speed events throughout the 1930s. Here, Indy's great Pete DePaolo and a mechanic wait in their 1934 Ford (which was not really a track roadster) to begin a AAA roadster race at Mines Field in Los Angeles. Pete was the first to average more than 100 mph when he won the 1925 Indy 500. Later, he headed Ford's stock car factory racing team. (Circa 1934)

rod... so, now after repetitious use, the public has caught on and must live with the vulgar term."

In California, the popularity of roadsters was firmly established, both with amateur time trials supervised by the SCTA and in professional circle-track roadster circuits, chiefly Gardena. East of the Rockies, grass-roots roadster racing was poorly organized. However, by 1947, strong associations were formed, mainly in Virginia, Florida, Ohio, Indiana, Michigan, and Illinois, soon followed by other states.

Rules and specifications were as varied as their names, with some groups requiring roll-bars or hoops, and others accepting anything that ran. Despite the confusion, attendance was good, and roadsters began outdrawing midgets. Fans loved this new renegade type of racing, which was done primarily on half-mile tracks. It was both a thrill show and a competitive action, aptly packed into the same program. Many early roadster drivers seemed fearless, but those who lacked experience paid for their mistakes by careening into fences and banging up their cars. Some of the great graduates of roadster racing were Floyd Roberts, Bill Vukovich, and Mauri Rose, all of whom later captured Indy 500 championships.

By the next decade, the great roadster racing tracks would give way to the excitement of fender-bending stock car racing. From then onward, roadsters would indeed be called hot rods.

BOB RUFI
A Legend in a Streamliner

By the time Robert Rufi, a carpenter, was 20 years old in the late 1930s, he and his friend Charlie Spurgin said they were using the word "hot rod" freely as they ran the salt with a four-cylinder Chevy-powered modified on the dry lakes of Southern California. At that time, their hot car was one of the fastest in its class.

"At the third race at Muroc of 1937," said Rufi, "I got through the timing course at 114-plus mph and broke the record for four-cylinder Chevys of 111.11 mph held by Lee Chappel."

When the SCTA stated that a new streamliner division would be added the next year, Rufi and Spurgin added a canvas tail over the rear engine and set a streamliner record with a two-way average of 115 mph.

After brief research at a local library, Bob consulted with record-holder Ernie McAfee on chassis construction for a new streamliner, which resulted in a teardrop body shape and an electrical conduit chassis. After adding twin Winfield carburetors that doubled the size of the jets for an experimental fuel mixture of 10% gas, 10% benzol, 80% alcohol, and a jigger of acetone, the first test run took place on Lincoln Boulevard near what was later known as the Los Angeles Airport. When a motorcycle cop pulled over Rufi and looked over the car, he shook his head and said, "I won't give you a ticket, because I can't figure out what name to give the damn contraption you're driving."

When the Army Air Corps shut down Muroc for racers, Bob took his innovative streamliner to Harper Dry Lake in May 1940. There he made a 143.54-mph pass and a streamliner record with a two-way average of 140 mph, which held for 10 years until Stu Hilborn finally broke the record. At the last meet of the 1940 season, Rufi, who was the SCTA champion, lost control of his car and was hospitalized with a fractured skull and a broken right arm. One year later, he sold the car, and soon a legend faded from the rodding scene—except for the exhaust system, which Rufi kept over the years.

Bob Rufi waits in his Chevy four streamliner before making a 140-mph record at Harper Dry Lake in 1940. The removed cockpit canopy rests beside the car. In 1946, Bob Rufi was vice president and Wally Parks was president of SCTA.

A motorcycle cop stopped Bob's little homemade streamliner after he made this sneak tune-up run on a lightly traveled highway. The narrowed tubular frame had no springs or radius rods in the rear. Axle fairings and wheel discs (from an old World War II airplane) improved the aerodynamics, as did the sleek canvas-covered body. The tailpipe exiting the body of the car gave it a jet-like appearance. (Circa 1940)

TOMMY THICKSTUN
Early Dealer Option to Rare Collectible

An unsung hero among early hot rodding peers was Tommy Thickstun. He was a bright, young engineer, whose legacy as a high-performance pioneer served as a benchmark reference decades later for speed equipment and auto makers. Tommy Thickstun's developmental work harkened back to the 1930s when he and his mechanic friend Frank Baron set out to improve carburetion efficiency. Their efforts produced the first dual-carburetor intake manifold for Ford and Mercury V8 engines with 180-degree porting. This was an advancement over single plenum designs and resulted in better fuel mixture distribution that yielded increased horsepower and speed.

From the Thickstun Manufacturing Company in Inglewood, California, marketing was done through new car dealers and service stations because no speed shops or related publications existed in those days. Thickstun equipment, known as a "dealer option," was installed in the vehicles of the Los Angeles Police Department. Vic Edelbrock, Sr., became one of Thickstun's first dealers, although their working relationship was short-lived.

Tommy frequently collaborated with talented Frank Baron who, like other team players, received little recognition for his contribution to the Thickstun odyssey. Among their early experiments was the pop-up piston that extended as much as 7/10 inch above the engine block's surface into the cylinder head. This not only raised the compression ratio but improved breathing through larger valves and special heads. This necessitated additional carburetion as supplied by the new four-carburetor intake manifold conceived by Frank Baron. Lou Baney and Louis Senter would later refine the pop-up piston concept even further.

As war clouds loomed, Tommy Thickstun was pressed to duty as an aviation engineer. In 1946, Tommy Thickstun, who was to be married a week later, died from a heart attack at Lake Elsinore, a favorite vacation spot. He was only 34 years old.

The Thickstun Manufacturing Company's patterns for the dual manifold line, which by then included Chevrolet, Plymouth, and Dodge, were acquired by Bob Tattersfield, who produced them later under the Tattersfield name. Thickstun products are now rare collector's items that occasionally surface at major swap meets and car shows as a reminder of one man's vision.

A Thickstun finned air cleaner feeds air to two carburetors astride this rare Thickstun high-rise manifold.

Legendary engineer Tommy Thickstun is shown testing dual carburetion. He is credited with producing some of the first dual intake manifolds and high-compression cylinder heads for Ford and Mercury flatheads in California.

BOB TATTERSFIELD
Continuing the Thickstun/Baron Legend

Bob Tattersfield makes adjustments as actress Colleen Townsend leans over the tanker.

Destiny struck when Robert Tattersfield, an enterprising young hot rodder, acquired the defunct Thickstun Manufacturing Company's assets in 1947. Tattersfield continued where Tommy Thickstun and Frank Baron had left off. The clever Tattersfield started working on projects that the three had partially developed.

Born in 1915 in Seattle, Washington, and raised in Los Angeles, California, Bob Tattersfield's car interest blossomed in the mid-1940s when the hot rod craze was beginning. When Bob was in his 1946 Ford coupe with its souped-up engine and handsomely chromed equipment, a man in a huge Auburn stopped next to him at a traffic light and coaxed him to a race. Because it was midnight and no other cars were in sight, Bob took him on and beat him. It was then that Bob realized that his challenger was Clark Gable, the famous movie star and an avid racing fan.

When Bob met Bill Burke, they both collaborated on Bill's idea for a rear-engine belly tank racer. Using a large-sized surplus P-38 airplane fuel tank for a body and powered by a 1946 Mercury V8 engine, it was hopped up from 95 hp to 230 hp.

As Bob Tattersfield's business grew, so did his aspirations. For promotional purposes, he entered the Tattersfield Special, an Alfa-Romeo race car in the 1947 Indianapolis 500. It was driven by Cy Marshall, who finished in eighth place.

Although the business prospered, production ceased in 1952, when the boom in high-performance equipment was about to begin. In 1968, Bob was badly hurt in a Jeep accident. One year later, he died at age 53. To this day, the Thickstun/Tattersfield/Baron names live on in the fond memories of hot rod history buffs.

Bob Tattersfield sits in the Tattersfield-Baron Special.

This belly tank streamliner, which was built by Bob Tattersfield, Frank Baron, and Bill Burke, made a Class C record of 140.40 mph at Bonneville in 1949 with a top (unofficial) run of 161 mph. Drivers were Bill Burke, Ed Robucks, George Davis, and Bob Morton. The tail fin helped stabilize the car.

The Founding Forties

Talk about clever marketing for the 1940s! This trailer displayed Tattersfield equipment at speed events in the Los Angeles area.

The rear-mounted 1948 Mercury engine in this belly tank streamliner was equipped with a Winfield cam, Tattersfield heads, and Baron-designed manifold with four alcohol-converted carburetors.

Engine builder Frank Baron collaborated with Tommy Thickstun on early projects such as pop-up pistons to increase compression. The photograph showing Frank working on his 1931 Model A was for a Waverly Tune-up Oil ad.

The original (and first) four-carburetor manifold for a Ford flathead racing engine designed by Frank Baron was later produced with Tattersfield. (Circa 1948)

Almost a half century later, Baron Racing Equipment for flathead V8s for the nostalgic market is again being produced by engine builder Frank Baron's son Tony. The venturi effect of the velocity stacks copied from European motorcycles is said to add slightly more horsepower.

VIC EDELBROCK
Excellence in Overdrive

Vic Edelbrock, Sr., posed in front of his Deuce for this SCTA racing program that sold for 10 cents at the Muroc Lake Time Trials in 1940.

Can you imagine how miffed the legendary Vic Edelbrock, Sr., was when a magazine wrote: "Almquist is to the East what Edelbrock is to the West?" That was in 1950, when Vic was the biggest honcho in the speed business. The Edelbrock name that once swung above a tiny Los Angeles repair shop is now listed as a hot stock on the NASDAQ.

After the family's Wichita grocery store burned, Vic quit school and became a grease monkey in a local garage. To escape the Depression, Vic headed to California and set up shop in Beverly Hills with his brother-in-law.

Vic and his wife Katie, who worked as a maid, saved their pennies and later opened another repair shop in Los Angeles. In 1936, their only child, Vic, Jr., was born. The business thrived, and Vic's pent-up yearning for racing surfaced.

After the war, Vic began building midget racers, which were then the rage. Vic's mighty midgets, with Bobby Meeks as head wrench, toured the Los Angeles area tracks almost nightly, with driving greats such as Walt Faulkner, Billy Vukovich, and Roger Ward. They often beat the Offys with their first and only V8 Ford 60 midget to win at Gilmore Stadium. Their secret was simple—20% nitro in methanol camouflaged by the scent of orange oil.

Soon afterward, Tommy Thickstun was having success with the first dual intake manifold for flatheads, and Vic began making his own twin-carbureted manifold—dubbed the "Slingshot." Although fewer than 100 were cast in aluminum, rodders were eager to try the new speed trick.

As business flourished, Vic left the bullpens of midget racing and began producing high-compression heads, along with an improved 180-degree dual intake manifold for the bigger Ford and Mercury flathead V8 engines. After the introduction of the small block Chevrolet engine in 1955, Vic's breakthrough came when he obtained one horsepower per cubic

Removal of the top, headlights, and fenders would turn Vic's 1932 "daily driver" into a dry lakes racer on weekends in 1939. He once reached a top speed of 121 mph.

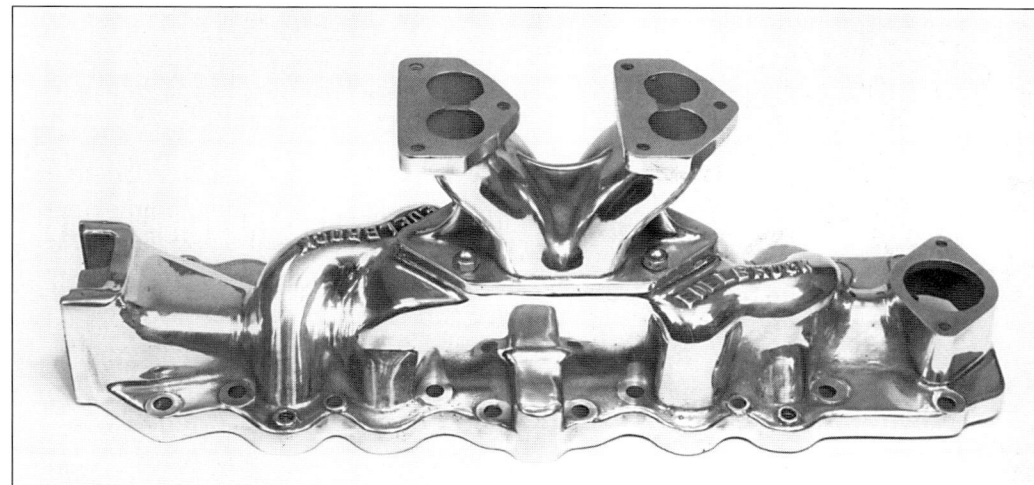
The "Slingshot" manifold for a Ford flathead was Edelbrock's first product.

Vic Sr. and Vic Jr., with aluminum castings ready for machining.

inch from a 283 Chevy engine, which featured his new "Cross Ram" manifold. Back then, that feat seemed like magic.

After graduating from college with a business degree, Vic Jr. excelled in marketing. In 1959, Vic Jr. married Nancy Crook, a college classmate who once vowed that she would never marry a hot rodder. To this day, Nancy has never regretted her lifetime of super rodding with Vic Edelbrock, Jr.

After Vic Sr. died, Nancy Edelbrock said, "Vic Sr. was a very special man, a man of few words. He was a very intuitive person and a deep thinker."

In 1962, the Edelbrock Company's future fell on the shoulders of Vic Jr. who, fortunately, had the vision to continue his father's dreams but on a much grander scale. At a Chicago trade show in the early 1960s, I first met Vic and his wife Nancy. Over breakfast, I tried to calm their pre-show jitters. Later, Nancy thanked this fellow hot rodder for encouraging the Edelbrocks to live up to their good name.

Edelbrock's machining facilities were still small in the 1950s.

Bobby Meeks (left) watches as Fran Hernandez checks a spark plug on this three-carbureted Ford flathead. The small dyno was Edelbrock's first in 1949.

Hot Rod Pioneers

Vic Edelbrock, Sr., timing badge, 1941.

At his early speed shop, Vic Edelbrock, Sr., poses with his mighty midgets.

A pivotal point came in the golden years of the 1960s and 1970s when the corporation introduced Tunnel Ram and Streetmaster manifolds and other hot products, many of which Don Waite and Jim McFarland, the Vice President of Research and Development, helped spearhead. In 1987, Edelbrock moved to a massive plant in Torrance, California, and added state-of-the-art computerized manufacturing equipment.

In 1990, an automated sand cast aluminum foundry was acquired, while another facility produced mufflers and suspension components. Also added was a new line of shock absorbers considered the most technologically advanced in the world.

The Edelbrock Corporation now offers more than 1,000 products with a distribution network consisting of 13,000 wholesalers. Vic Jr., the CEO, and his daughters Christi and Camee join Nancy in the Edelbrock mission—to change the racing and automotive world for the better.

Edelbrock continues to be the industry standard in performance by which others are judged. Today, the Edelbrock enterprise is the largest high-performance entity in the world.

Bill Milliken

Most racers are remembered for winning. However, Bill Milliken was once immortalized for losing. "Milliken's Corner" was humorously named for the spot in which Bill flipped his Bugatti at America's first postwar road race in Watkins Glen. This wintry photo, taken in Buffalo, New York, shows Bill seated in the same type of 1935 Bugatti prior to the 1946 Pike's Peak hill climb in which he finished sixth. Today, hot rodding aeronautical engineer Bill Milliken is co-author of a practical 900-page book entitled *Race Car Vehicle Dynamics*, which covers vehicle control, stability, and handling.

BOBBY MEEKS
Another Unsung Hot Rod Hero

Robert Meeks, who was born in time for the beginning of hot rodding, was Vic Edelbrock's right-hand man for more than a half century. All totaled, Bobby put together more killer flatheads and winning engines than he can count.

In 1934, at only 13 years of age, Bobby began working for Vic Edelbrock in a newly opened Los Angeles repair shop. Then, the sport of midget racing was hot, and so was Vic's little V8 Ford 60. When Bobby became head wrench, the new car was an instant winner, feared even by Offy drivers.

In early Russetta Timing Association meets, the engines that Bobby built held nine records in eleven classes. The record-setting So-Cal belly tank and streamliner in 1949 and 1950 were powered by Bobby's hot flatheads.

No job or customer was too big for the small shop. Once they built a 3/4 race engine for Clark Gable's 1949 Ford convertible, hoping to attract other movie-star business.

Among many other firsts in Bobby's career were his pioneering of the use of nitro fuels, building the world's fastest flatheads, and utilizing the West Coast's first dynamometer for developing Edelbrock's famous speed parts. Although retired, Bobby still keeps his toolbox handy at Edelbrock's new plant for special projects that require his much sought-after expertise.

Probably no man in the history of hot rodding has played so many major roles in the success of the early racing teams as has Bobby Meeks.

For Bobby, a 1996 inductee into the Dry Lakes Hall of Fame, love of racing's sights and sounds continues to be his cornerstone.

Legendary car builder Bobby Meeks' career spanned six decades—from the early dry lakes to the Indy 500. Here he "dynos" perhaps his last flathead. It delivered 168 hp—almost double its stock rating

Here, Bobby Meeks is rebuilding the V8 Ford 60 engine with which Vic Edelbrock Sr. (leaning) consistently beat Offy midgets.

FRANK KURTIS
A Mover and a Shaker

The biggest hit at the Indy Speedway was not a roaring race car but a gorgeous convertible sports car custom built by Frank Kurtis. Called the "Uncrowned Prince of Hot Rod Heaven," Frank Kurtis was by 1948 the gifted creator of some of America's swiftest hot rods and prettiest customs.

As a blacksmith's son from Colorado, Frank quickly inherited the desire to shape and form metal. After the family moved to Los Angeles, Frank lied about his age and landed a servicing job at a Cadillac dealership. He was only 14 years old then.

In his spare time, Frank built a racer with a Model T engine and scraps from junkers. By the time he was 18 years old, he had won rave reviews for his futuristic streamliner, which was a real beauty. Five years later, he had built a novel race car that cleaned up at the local dirt tracks.

In 1937, after Frank opened his own race car repair shop behind his home, he spruced up the great Rex Mays' midget race car, making it a top winner. Since World War II, Frank had built 600 midgets and 128 Indianapolis racers in his two plants in Glendale, California. Johnny Parsons drove the Kurtis Kraft entry to its first Indy 500 victory in 1950.

By 1953, Kurtis Kraft 500s were legendary, with 22 of the 33 starting cars being Kurtis ones at the speedway. Kurtis Kraft cars took home many first place spots in 1950, 1951, 1953, 1954, and 1955. The wins represented 88% of the Indy records for those years.

Frank Kurtis built many one-of-a-kind custom cars including a production sports car, which was later to become the Muntz Road Jet in 1950. The first Kurtis roadster-type

The 500-S competition sports car was practically the same in appearance as the racing model, but with fenders and lights added. In do-it-yourself kit form, the starting price for the frame kit was $596; a complete kit car cost approximately $3,000.

The popular Kurtis Indy race cars had wide, squared-off bodies, which gave the cars a roadster appearance.

The Founding Forties

Production line of Kurtis midget racer bodies at the Kurtis factory in 1949.

Indy car was built for Bill Vukovich, who won the Indianapolis 500 in 1953 and 1954.

Frank retired in 1968, and his son Arlen continued the family business and began reproducing his father's cars of the 1940s and the 1950s.

A national TV show in 1994 featured Arlen and paid tribute to Frank Kurtis. Keeping the Kurtis name alive in the world of speed certainly would make Frank Kurtis proud.

Years ahead in styling, this Kurtis Custom inspired the Muntz Road Jet of the 1950s.

The Kurtis Kraft midget race car was king from 1945 to 1950, when the mighty midgets reached their all-time high in popularity and then took a downward slide to near-oblivion.

Flat Out on the Dry Lakes

Nothing changes much in the vast dry lake beds that dot the barren deserts of the great Southwest. The same clear sky and hot relentless sun, which bakes the mud to a rock-like hardness, makes the flat and powdery surface perfect for speed record attempts.

Even in the early 1930s, the dry lakes were the same. Had you been there then, when the first organized amateur speed trial was held at Muroc Dry Lake, you would hear the roar of Model T, Fronty Ford, and four-banger engines powering open-wheel cars with their modified or stripped stock bodies. Later, flathead Ford V8s became the dominant racing engine and weekend playthings of old-time notables such as Winfield, Kong, Chapel, McGurk, and others.

By the 1946 season's last contest run by the SCTA, more than 400 hot rods had run the dusty mile course in four classes: roadsters, modifieds, streamliners, and unlimited. Some record times were 128.66, 133, 140, and 137.24 mph.

"I like to think of those early dry lake cars as the grandfathers of the thousands of street rods on the road today," remembers Al Drake, a hot rod historian. "They not only created an interest in modified cars that has grown over the decades, but they also created the style that later cars would emulate, including cars that never got within a couple thousand miles of a dry lake."

This early dry lakes time trials event shows the "run what you brung" variety of cars that included roadsters, modifieds, tankers, and stocks—many stripped of windshields, fenders, lights, etc., to lighten them. (Circa 1948)

The Founding Forties

A tanker waits for a run at a 1946 dry lakes meet. The timing equipment, similar to the starting line stand, is primitive.

Drake continues, "The early rodders hopped up that faithful four-banger, often adding chrome goodies to the engine. They stripped off fenders and bumpers to make a square car more streamlined, and they did what they could to make a tall car lower. These are tricks that rodders still use when building a car today."

"When I went to the SCTA dry lakes meet held in 1977 at El Mirage," says Drake, "I saw what had been going on for roughly fifty years: Dawn suddenly appearing as the sun rose over the mountains, the cars lined up, a hot engine with open exhausts breaking the stillness, a roadster kicking up a rooster-tail of dust as it made the long run toward the clocks, the shift, the noise of the engine rising in the distance, then the loudspeaker voice giving a speed. The sun climbed upward and the day became much hotter, but the enthusiasm for speed was not diminished."

"It was a great time to be alive," remarks Drake. "I was still young enough to drive hundreds of miles a day, sleep on the ground, stay up all night talking about rodding's past. My quest led me to all kinds of interesting people, like Karl and Veda Orr, Ak Miller, Wally Parks, and others with vaguely familiar names that need to be better known. My timing was fortuitous because these guys, who remembered rodding thirty or forty years earlier, were getting older."

Seventeen years later, the result was the publication of two books by Al Drake: *Hot Rodder! From Lakes to Street*, a collection of oral histories, and *Flat Out: California Dry Lake Time Trials 1930–1950*, a definitive history of dry lake activities.

This faded photograph taken at Muroc Dry Lake in 1936 shows Kong Jackson leading in this match race. Impromptu racing ended with SCTA sanctioning in 1938.

Which way? The driver evidently received good instructions because he garnered top SCTA honors for the day in this flathead Ford V8 streamlined modified roadster. (Photo by Dean Moon)

39

THE SPALDING BROTHERS
Famous for Ignitions and Cams

In 1936, after Tom and Bill Spalding first raced their newly built Ford V8-powered 1929 roadster, they could have cried with disappointment. Even with a Harmon cam, a homemade two-carb manifold, and 21-stud heads milled in the machine shop of their high school, the engine cut out at 4500 rpm.

After determining the problem was ignition failure, the brothers experimented and hit upon the idea of converting stock Ford distributors to utilize the reliable Lincoln-Zephyr dual-coil setup. After learning their reworked distributor could turn an additional 1500 or more rpm without misfiring, the youngsters realized they were on to something big.

Starting in the corner of their parents' garage, the brothers began doing a tidy business rebuilding ignitions for the local hot rodders.

Eventually, various refinements in the Spalding ignition allowed it to operate at speeds exceeding 8000 rpm, which was unheard of in those days.

After serving in World War II, Tom continued in the ignition business throughout his working life. In 1972, he was granted a patent on electronic ignition that was used on many Indy winners. Its almost unlimited sparking ability helped the Offy again be competitive, even against Ford V8 turbos.

Mechanical technology was Bill's forte. By 1948, he was making racing camshafts. Instead of the usual guessing game when designing master cams, Bill developed a machine/process that simplified combining elements such as radii, involutes, and other mathematical profiles.

After the brothers went their separate ways, Bill worked for 23 years as a technical representative for Ford Motor Company.

"During the 1973 fuel shortage, I suggested to Ford that they use an 'overdrive' behind the automatic transmission to improve fuel economy, which they did," said Bill. "But I have yet to receive recognition or even a thank you from the top brass of Ford."

Bill Spalding in his Riley V8 overhead was a regular at the dry lakes. Previously, Bill and Tom hit a record 128 mph with a flathead V8 at Harper Dry Lake. (Circa 1940)

The Founding Forties

The Spalding "Flame-Thrower" dual-coil/dual-point ignition was guaranteed never to cut out. Each coil and each four-lobe breaker cam sparked four cylinders. A previous model was a converted Ford and Lincoln-Zephyr distributor with two coils.

A masterpiece in motion, this neat Chevrolet six T-roadster was built in 1948 by Tom and Bill in only three weeks. Its Wayne-converted Chevy 248 c.i. engine boosted about 300 hp with a 14:1 compression ratio 12-port cylinder head, Spalding camshaft, and a Spalding twin-coil, 12-volt ignition. The car ran 150 mph and won every trophy dash in which it was entered. (Circa 1948)

One of the pioneer teams in the streamliner class was the Spalding brothers. Their pre-war steel-bodied streamliner called "Carpet Sweeper" weighed 1400 pounds but turned 130 mph with almost a stock Ford V8 engine. (Circa 1939)

TOMMY HINNERSHITZ
The Flying Farmer

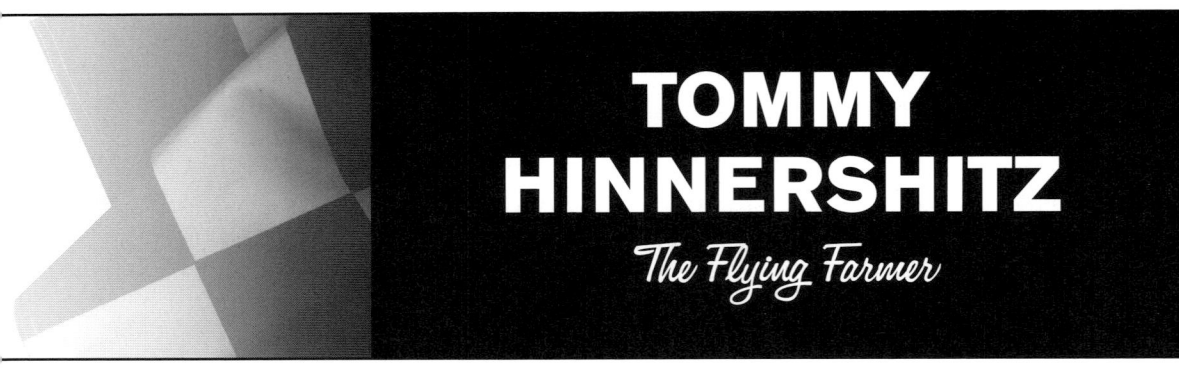

From the glory days of dirt track roadster racing, self-taught hot rod mechanic and driver Thomas Paul Hinnershitz became the victor with seven Eastern AAA sprint car championships.

After he jalopy-raced in the early 1930s, Tommy barnstormed with sprint cars, often at outlaw board tracks and dusty fairgrounds. Then, the stocky Pennsylvania Dutchman used an alias to avoid being fined.

In his prime, Tommy was "hell on wheels," always driving high on the outside lane to stay in front. He did it with such finesse.

"I tried to see a track as having only two corners—not four," said Tommy with the blue eyes.

Old-timer Tommy, a hearty, jovial individual and a crowd-pleaser like other early racers, found that placing in the top ten barely covered travel expenses and that sleep was hampered while traveling the long distances between the major events. At that time, open wheel racing was not only a financial fiasco—it was definitely dangerous business.

"I had some really close calls. The scariest was when a tire blew out and threw the car upside down, landing crosswise over a drainage ditch. That saved me from being squashed," said Tommy.

After beating drivers half his age, Tommy quit racing in 1960. At the age of 48, he opened a speed shop in Reading, Pennsylvania, which featured Almquist racing equipment.

Tommy Hinnershitz receives a well-deserved trophy for winning the main feature at a Midwestern race. In addition to seven AAA spring titles, Tommy captured hundreds of midget, roadster, and big-car wins.

Mechanical troubles and a crash spoiled Tommy's three attempts at winning the Indy 500. Here he sits in the Marks Offy in 1941.

The Founding Forties

Despite a close call with the fence, Tommy Hinnershitz won the sprint car feature event at the fair in Allentown, Pennsylvania, in 1959.

KONG JACKSON
He Sparked Hot Rodding

A walk through legendary Charles "Kong" Jackson's shop becomes a flashback in time to half a century ago when Kong dual-coil ignitions first sparked the fastest Ford V8 flatheads in the world. There time stands still with the same machinery that produced the Kong ignition systems of 1946.

When Kong, the brilliant innovator and hot rodder, saw the stock Ford ignition system as the weak link in the power chain in 1937, he was a regular dry lakes participant. After considerable experimentation, Kong developed a dual coil ignition setup, much to the surprise of his fellow competitors. During World War II, Kong put his plans on hold and served in Africa and Italy. Upon his discharge, Kong resumed his ignition project with the aid of Ed Winfield, a lifelong friend.

"For some reason, Winfield took a liking to me and taught me a great deal about ignition systems and camshafts," stated Kong. "In fact, he taught me how to make a camshaft master using the machine that he kept hidden behind his shop. As a result, I still have some of the original Winfield camshaft masters."

In 1951, Kong's hot rodding was again put on hold when he was called by NATO to teach maintenance to the Danish Air Force. After a year, and for the next ten years, Kong's sharp mind was put to work on classified U.S. defense projects, including the Polaris Guidance System.

Pride and precision go hand in hand with Kong speed equipment. Kong's dual-coil ignition systems feature precision-machined components, including sealed ball bearings with either automatic or manual advance. In the early days, the retail price for the system was $67. The same units now sell at swap meets for $800 to $1,200. Beginning in 1955, Kong's "Rotofaze" dual-coil ignition was the hot setup for later model cars.

Kong also manufactured intake manifolds for flatheads that were adaptable for two, three,

Three Winfield carburetors are mounted on a Kong open plenum manifold. The spark plugs on the Kong aluminum heads are located high to improve flathead efficiency.

Kong Jackson with his first production ignition system in 1947. His left hand holds a cable for manual spark adjustment from the dash.

The Founding Forties

or four carburetors or fuel injectors or a supercharger.

"Despite the manifold having an open plenum directly feeding all eight ports, it ran as smooth as glass from idle all the way up," said Kong.

Kong also manufactured high-compression cylinder heads for Ford flatheads, which were of a rugged one- and two-piece design.

"Since Ford flatheads refused to die, I began making an improved line of intake manifolds, high-compression heads, and ignition systems in 1980," said Kong.

Today, even as he approaches 80 years of age, Kong is experimenting with new concepts for more horsepower and better mileage for Henry's flathead and other engines.

Muroc's dry lake bed with its hard-packed surface helped Kong's 1932 roadster reach 110 mph in 1941. The 21-stud flathead Ford V8 engine had a high Winfield dual manifold, Winfield cam, and filled heads. At the time, Kong was waiting for the next run.

The #175 Ford flathead had four Stromberg carburetors mounted on a homemade, welded tubular manifold. Kong (seated) is at Harpers Dry Lake in 1940.

In Italy in 1943, G.I. Kong Jackson was working on a P-38-J Air Force plane. During World War II, the military made good use of hot rodders' mechanical abilities.

PHIL AND JOAN WEIAND
The Ultimate Team

Although the Weiand name elicits visions of the birth of hot rodding as a sport and industry, the Weiand saga is an inspirational story about a dedicated husband-and-wife team, united in love, tragedy, and triumph.

On weekends in the late 1930s, Phil Weiand joined his fellow rodders for the Los Angeles dry lakes speed trials. On one fateful day, a tire blew out and overturned Phil's Rajo-Model T Ford. Phil was pinned beneath the metal that almost crushed him to death. Although Phil was left in a wheelchair for the remainder of his life, he remained true to his passion for hot rodding.

One day, while experimenting with new hop-up ideas, Phil discovered a desirable ram effect that could be achieved by lengthening the runners (ports) in the intake manifold, thus improving carburetion and low-end torque. With $250 borrowed from his mother, Phil had wooden patterns made to produce his first product—the "high riser" dual intake manifold for Ford and Mercury V8s, which became the forerunner of the modern hi-ram.

Hot rodding blossomed following World War II, and so did Phil's proficient ability to develop various products for Ford and Mercury. These included high-compression cylinder heads and other related speed equipment.

Beside being a brilliant innovator, Phil, who once bartered his mandolin for a Model T Ford, was an early proponent of automated manufacturing. Calling in specialists, he employed a "design team" that reduced development time for new products.

In 1955, Phil acquired a small aluminum foundry and made castings for himself and for several competitors. Because money was

Phil Weiand sits next to the beloved roadster that replaced the Rajo-Model T that almost killed him.

Joan Weiand was recently hailed as "the performance industry's most influential female executive." Weiand now also manufactures superchargers and belt drives as shown.

extremely tight then, production runs were usually numbered in the dozens rather than the 1,000-piece runs of today.

When GMC G-71 superchargers appeared in the late 1950s, Phil was quick to develop manifolds and drive systems for Roots-type blowers.

When Phil had become a founding member of SEMA, Joan had organized the original paperwork. Phil was inducted into the SEMA Hall of Fame in 1975. Twenty years later, Joan was inducted.

"Phil had a heart of gold, and he was so generous," said Harvey Goldberg, a Weiand sales manager in the 1970s. "It is a wonder that he made any money at all."

Following Phil's death in 1978, Joan took over the reins of the company.

"Frankly, I was so scared," said Joan, who witnessed a slump in business due to stringent emission laws. "I was tempted to sell the business. But, in my heart, I wanted to continue what Phil had worked so hard to build."

Despite this, Joan and her staff overcame the rigors of the competitive, male-dominated industry and saw the industry increase tenfold.

It is no wonder that the world-recognized name of Weiand is known as "the ultimate driving force."

The early DeSoto OHV engine was a sleeper until clever hop-up artists such as Ray Brown awoke it. Phil Weiand was one of the first to produce an intake manifold that accommodated four Stromberg 97 carburetors. With a good roller cam, domed compression-boosting pistons, and other tweaks, the stock 160 bhp output could almost be doubled.

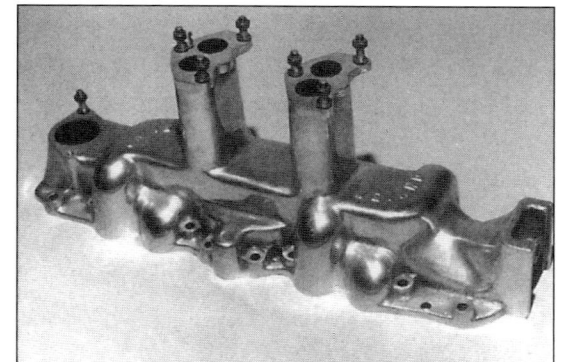

The original "High Weiand" dual intake manifold began 50 years of trend-setting innovations from Weiand. In 1999, Weiand surprisingly became part of the Holley performance family, furthering an industry-wide merger trend.

ED ALMQUIST
The "Imagineer"

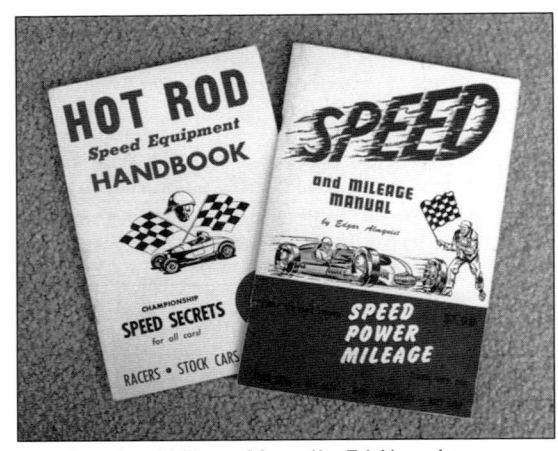

The *Speed and Mileage Manual* by Ed Almquist was America's first widely circulated how-to hot rod book. Written in 1946, more than two years before *Hot Rod Magazine* appeared, its copyright was granted to Ed while he was in the Maritime Service. The "little" $1.98 book helped promote interest in the new motorsport. A million copies of Ed's later book, *Double the Performance of Your Car*, showed even unskilled motorists how to maximize power and fuel economy and improve vehicle appearance.

"Car nuts—especially those over 50 years of age—will remember Ed Almquist," explains automotive historian Albert Drake. "Ed's life is the fascinating story of a man who grew up with the automobile, who combined his mechanical abilities with an entrepreneurial spirit and good old American gumption and hard work, and it all paid off for him."

"In the early 1950s, hot rodding magazines carried ads for Ed's speed equipment, including Almquist heads, manifolds, cams, headers, Hollywood mufflers, and other accessories for the Ford flathead V8 and other engines. Almquist was king of the mail-order speed business in those days when it was difficult to find racing equipment outside big cities. California speed merchants held the West Coast markets, but Almquist equipment was the brand of choice on Midwest and East Coast circle tracks and drag strips."

According to Drake, Don Garlits even ran Almquist equipment on his early dragster, and Almquist-equipped cars held numerous records nationally.

Ed Almquist was born in a log cabin in Zim, Minnesota, a tiny hamlet with a dozen hard-scrabble farms. Because few toys were around in those Depression days, youngsters indulged in their own imaginations. For 10-year-old Ed, that meant kicking up the dust on dirt roads and cow pastures by driving a homemade Maytag-engined "soapbox racer," or playing "racer" with the farm "Bug" tractor whenever his parents were not there. By age 14, Ed was honing his innovative skills on his hopped-up Model T Ford salvaged from a junk yard.

Ed Almquist never grabbed a big racing win, but his little $1.98 book idea won big in speed equipment sales. (Circa 1950)

Ed did not have the money to attend college. Therefore, in addition to working as a celery farmer and part-time railroad brakeman and racing roaring jalopies on weekends, Ed took extension courses in diesel engineering from the University of Minnesota. As World War II intensified, Ed joined the U.S. Maritime Service and eventually became a licensed marine engineering officer. In his free time aboard ship, Ed wrote down some engine souping and customizing tricks that he learned from racing hot shots back in Minnesota.

Ed's first venture into marketing came by a happy accident. On a hunch, Ed borrowed a mimeograph machine and turned his notes into two small books. While still in the service, Ed advertised his first book, *Speed and Mileage Manual*, with a $12 ad in *Popular Mechanics*. The book was an immediate

The Founding Forties

success. Because the results of that little ad and other classifieds brought a bonanza of orders plus requests for speed equipment, Ed knew that the new motorsport called "hot rodding" was to be his life's calling.

When he was discharged from the service, Ed moved to Milford, Pennsylvania. Similar to many other speed pioneers who had little or no formal engineering training, Ed began designing what would become a complete line of more than 100 speed and custom equipment products, many secretly run on Almquist-sponsored race cars that Ed called his "rolling test labs." One of Ed's first innovations was a comical-looking but effective automatic water/alcohol vapor injector that he assembled in his basement using a Mason jar for a reservoir. It cost 50¢ to make but sold

Beginning in 1946, millions of Almquist's books and catalogs distributed throughout the United States helped propagate the hot rodding sport. Customers included the rich and famous such as Gary Moore, Clark Gable, Herb Shriner, and many U.S. racing greats such as the Pettys.

by the thousands for $6. Another hot product was a unique "log manifold" that adapted two carburetors to virtually any car. With progressive linkage (at which competitors laughed), it worked like magic and gave even snappier acceleration than a four-barrel carburetor—all for only $12.95! More sophisticated products followed, including winning stock car camshafts, more intake manifolds, a "telescopic" ram tuning kit that was years ahead of its time, and dual exhaust headers with mufflers so loud that Ed had to design a tail pipe silencer to keep them legal.

In 1954, with Jonas Anchel as a partner, Ed formed another wholesale division called Sparkomatic Corporation. Around 1959, they entered into a partnership with George Hurst and Bill Campbell, who were making engine conversion mounts. The timing was right for an

Designed by Ed a half century ago, the Sparkomatic spark plug, the forerunner of today's multiple electrode plugs, had six ground electrodes instead of one. Thus, it would fire without missing and would last six times longer than a conventional plug, even in high-compression engines.

Ed Almquist holds a prototype of his famous Ram-Jet, a gas-saving invention proven so effective that Ed easily won a court case initiated by the State of Colorado in 1979. The judge refused an injunction because laboratory tests proved the patented "gas-sipper" improved crankcase ventilation and engine efficiency. (Circa 1970)

aftermarket floor shifter; therefore, after each designed his own floor shifter, the short-lived but friendly partnership was dissolved. After coming up with a wild marketing idea for a clever "fits-all" universal mounting kit that helped dealers minimize inventory, Ed received two patents on floor shifters that won instant industry acclaim.

After Ed sold Almquist Engineering and his interest in Sparkomatic Corporation in 1966,

Hot Rod Pioneers

Here Ed Almquist checks a Ford V8 flathead ignition with Almquist racing equipment.

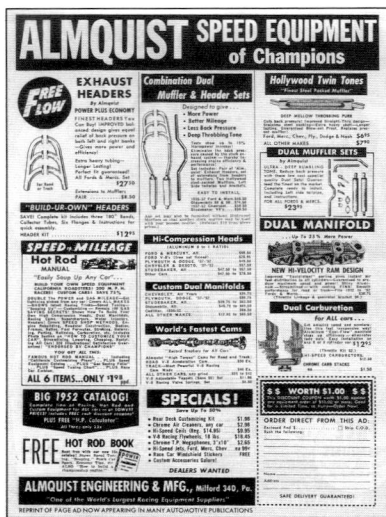

Almquist magazine ads in the 1940s, 1950s, and 1960s pioneered many high-performance products. Similar to a few other entrepreneurs in motorsports, Ed Almquist parlayed a $100 investment into millions of dollars in sales.

he delved full time into slippery lubricant chemistry. He improved an earlier breakthrough, heralded as the first nonsettling "moly" oil additive that permitted racers to safely use a thinner oil, thus freeing power formerly lost to fluid friction. In 1979, the U.S. Department of Energy proved that Ed's new formulation increased power and mileage dramatically while reducing wear by 50%. Today, top racers continue to add Power-X to their motor oil.

For 30 years, Ed has been manufacturing anti-friction engine treatments that he says are similar to "horsepower in a can." He is a frequent guest speaker in the Far East for his wearproofing lubricants, which are used by governmental agencies. Ed won't retire—he continues revving in high gear at full throttle and currently is perfecting another invention.

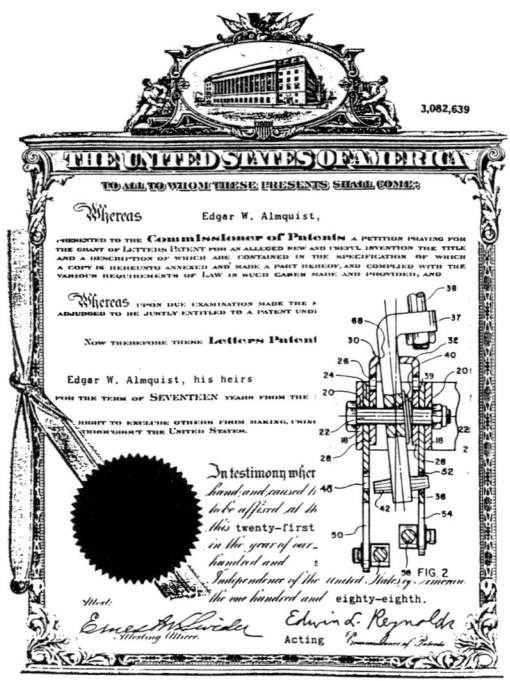

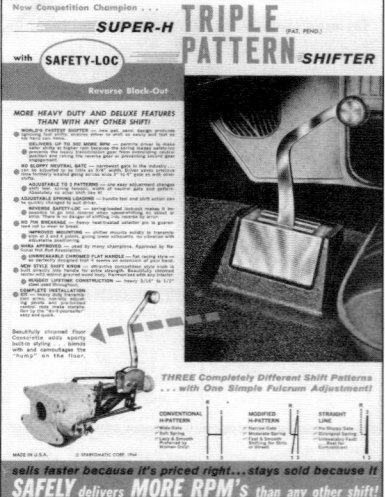

Over the years, millions of early hot rodders and daily motorists have utilized Ed's patented stick shift and fulcrum that can still be found in OEM floor shifters, including some Hurst shifters. Ed scooped the industry with the first universal shifter, and soon Sparkomatic became one of the biggest suppliers of floor shifters to mass merchandisers such as Montgomery Ward, Sears, Firestone, J.C. Whitney, and Western Auto.

BARNEY NAVARRO
"Oldfield" of the Hot Rod Industry

Similar to many of hot rodding's founding fathers, Barney Navarro was a quiet achiever who got his kicks from technical achievement rather than driving glory. In his heyday, fellow speed competitors called Barney Navarro the "Oldfield" of the hot rod industry. Whenever Barney masterminded a challenge, he always came out a winner, especially as an accomplished product designer. Whenever Barney eloquently spoke about one of his engine theories, he sounded like an astute physics professor rather than a former hot rod racer turned manufacturer.

Before high school graduation in Glendale, California, Barney completed his first hot rod. It was a homely 1929 Hudson with a six-cylinder F-head engine that Barney had already converted to dual carburetion by welding on two new intake ports.

By 1941, he ran 107 mph at Muroc Dry Lake in a borrowed T-roadster. Its 1939 Ford V8 engine,

Barney's modified T-roadster Ford V8 engine was destroked 3/4 inch, giving it a 176-c.i. displacement. The twin-carburetor setup was boosted by homemade oxygen injectors. A one-way run at El Mirage circa 1952 was 137 mph. With a "Jimmy" blower, top speed was 147 mph.

which was owned by Barney, was fitted with milled re-domed and flycut iron heads and a Weiand high rise dual intake manifold, which Barney acquired as payment for machining ten of Phil Weiand's first castings.

After serving in the U.S. Army Air Corps, Barney resumed racing and installed a GMC blower in a flathead hot rod with a homebuilt manifold and V-belt drive. This was a first.

Being one who never followed convention, Barney developed an improvement over existing Ford dual intake manifolds by using the same sequence as the Ford factory and

The six-cylinder, stock-block Rambler engine that powered this Indy race car developed 640 hp at 7000 rpm. The turbocharger is located outside the body, as was customary in the 1960s.

Hot Rod Pioneers

by refining the plenum and passages, which improved fuel mixture distribution. The pattern also made his three-carburetor manifold perform better than others.

Barney's aluminum high-compression cylinder heads were different from other makers because their improved combustion chamber design permitted increased intake charge cylinder filling for extra power. Although Barney's products were among the best, the demand for flathead equipment declined, and Barney departed from the heads and manifold business to make a heart-lung machine and an electronically powered concrete cutting saw.

In 1966, Barney caught the racing bug again and began building a turbocharged Rambler engine for an "Indy" racer as a surprise entry at Indianapolis. Although the engine produced more than 600 hp, Barney had to build his own fuel injection system and a totally new turbocharger control valve to increase the useful boost range of the turbos.

The AMC-backed "Indy" Navarro racer made its debut at Indianapolis in 1968 but failed to qualify. It marked the first time that an American Motors' entry had been in a national championship race of such caliber.

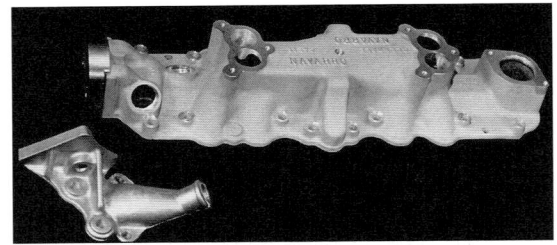
Shown here is a Navarro dual intake manifold for Ford and Mercury cars from 1949 to 1953. A three-carburetor manifold also was available for Ford and Mercury cars from 1932 to 1953.

A recent rebirth of old car interest motivated Barney to go back into short runs of manifolds and heads for flathead Ford and Mercury V8s. This was a throwback to the peak days of flathead hot rodding in Southern California when four out of five dry lakes' records were held with Navarro equipment.

Today, Navarro users say Navarro equipment is the preferred nostalgia choice for customers of the 1990s, although the cost today is ten times more than the original.

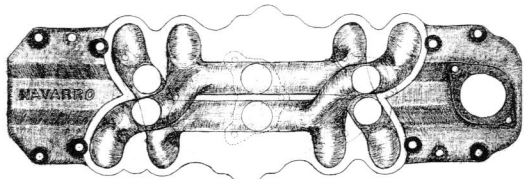
This drawing shows a porting pattern that differed from other makes of dual and triple carburetor manifolds.

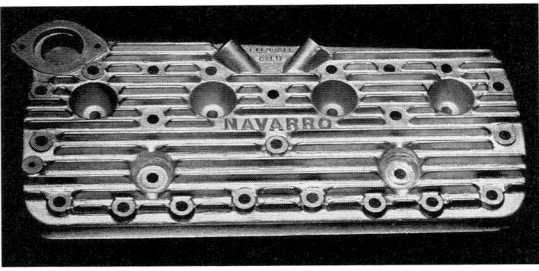
This Navarro cylinder head has a 9-1/2 to 1 compression ratio and fits 1949 to 1953 Ford and Mercury cars.

Barney's modified car with a 21-stud flathead reached 101 mph in 1941.

LOU SENTER
Racers' Head Guru

Likable folk hero Louis Senter was one of the most versatile pioneers of hot rodding, but he never received the star-quality hype that he deserved from the media. On dirt tracks and country roads in the 1930s, years before he made record runs at Bonneville and Indy, Lou raced four-bangers and flatheads. When Lou was not behind a racing wheel, he was creating "go-fast" products by the dozens, often scooping the competition.

After serving in the Navy, Lou worked for Eddie Meyer's engine works, eventually opening a small Los Angeles machine shop with his brother Sol. Its used machinery came from the sacrificial sale of Lou's personal car. In 1948, after moving to larger quarters, Lou took on local engine builder Jack Andrews as another partner. Their Ansen business, specializing in Cadillac and Olds V8 engine swaps, was one of the first true speed shops in the nation.

Although skeptics said that it could not be done, Lou created a lighter crankshaft (minus its center counterweights) for Ford flatheads that produced an additional 500 rpm. With the help of Lou Barney, Ansen built some of the country's fastest flatheads using pop-up pistons that protruded into the cylinder head for added high-compression efficiency. Unknown to Lou, some of the engines ended up in bootlegger's getaway cars in the Southeastern states, and the chase was on for police departments to order even more potent engines for catching the renegade rum runners.

When the partnership was dissolved in 1950, Lou became sole proprietor of Ansen Automotive, and he raced dragsters, sprint cars, and midgets. As part owner, he was

Lou Senter's racing career began at age 12, when he won a soapbox derby in this homemade push car in 1930.

A youthful Lou Senter stands beside his handsome 1932 roadster. The Ford engine had rare 21-stud Winfield heads. Later, the car would serve as Lou's "test lab." (Circa 1949)

Hot Rod Pioneers

influential in the opening of the Saugus Drag Strip, which held Sunday meets where winners took home $25 war bonds. One day, in the midst of selling alcohol fuel from his pickup truck, Lou was challenged to a race by concessioner Bob Corbett. Because Lou was determined to show that he was the head honcho, he quickly punched the accelerator on his rattle-trap truck and sent alcohol spewing across the racetrack.

In 1966, Lou built a noisy, rear-engine Offy and took it to Indy; however, it failed to qualify. Later, Lou worked with Indy driver Jim Hurtubise on a roadster which, at that time, was the fastest front engine car to run in Indy. Today, the engine reposes nostalgically in a bright yellow track roadster, garaged behind Lou's mansion.

As the infant hot rod industry grew, Lou became one of the first to develop floor shifter conversion kits that moved the column shift to the floor for more positive shifting. Other products that Ansen fathered were forged pistons, aluminum and forged steel connecting rods, safety bell housings, custom wheels, and a bevy of racing products.

Lou holds one of the trophies his 1953 Studebaker collected at Bonneville when driven by Gary Cagle to a record 247 mph. (Circa 1954)

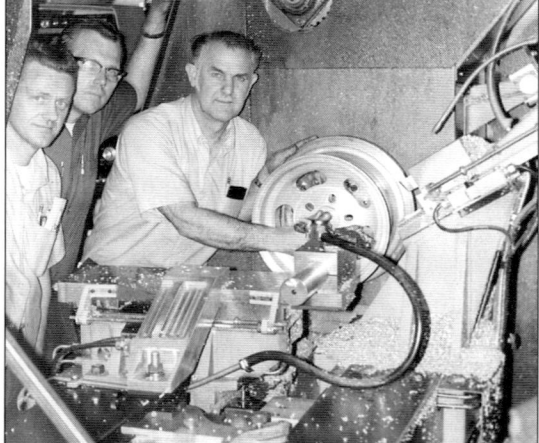

In 1963, R&D engineer Bob Johnson (left) assisted Lou in developing this automatic wheel turning tracer machine that did the work of today's CNC lathes. The aluminum wheel shown was produced on a centrifugal casting machine where the mold was rotated at high speeds while hot metal was poured into the mold.

Always trying something different, Lou Senter built his first dragster using a chain drive with a Packard engine. It was one of the first gassers to break 150 mph.

The Founding Forties

The firm was sold in 1969 to acquisition-hungry Whittaker Corporation, and Lou Senter became its non-gratis consultant. On one occasion, Lou, with a smirk on his face, bragged that he could double the company's growth in two years if enough working capital was provided. The stiff-necked neomanagers scoffed at Lou, called his bluff, and offered him a $1 million bonus if he could meet the challenge within three years. Lou floored the experts by doubling the sales volume in less than the agreed time, and he proudly pocketed the loot.

During a business slump, Whittaker's board of directors rashly sold Ansen piecemeal at a bargain price. Els Lohn, a former competitor whose company was named Eelco, wisely purchased the prestigious Ansen brand name and many of its products.

Carrying on the family's tradition, Lou's daughter Marsha won many trophies with the 1/4 midget racer that Lou built. His grandson, Shane Scully, drove a TQ before graduating to full-size midget racing. The 1978 SEMA Hall of Fame recipient is now in his seventies, but Lou Senter's brain continues to cruise at a speed few can match—even with his idea for a small electric off-road vehicle.

This modern Ansen facility in Gardena, California, was the design, manufacturing, and distribution center for hundreds of products in 1964. Lou Baney, Ed Pink, and Dick Horasian would also get their start with Ansen.

In the Ansen Special A/FD, Kansas driver Rod Stuckey turned 188.66 mph to win the 1962 AHRA Winter Championship. His low e.t. of 8.51 was a new record.

EARL EVANS
Terror on the Tracks

Known as one of the "Grand Old Men" of the speed industry, Earl Evans started dirt track racing in 1923 in an Outlaw Circuit that had broken from the AAA sanctioning body. Although he was older than most of the other rodders when he competed at dry lakes events, Evans was a prime mover in early California rodding.

After World War II, Evans built a tiny shed-like foundry in the rear of his speed shop in Whittier, California. There, he cast high-compression heads and multiple-carburetor manifolds for Fords and Mercurys. Earl continued to be a "weekend warrior" and in his spare time set numerous records using his own products on his experimental cars. These cars were usually open-wheeled, wing-tank lakesters. Although the meticulously machined Evans Racing Equipment was extremely popular in Southern California, it failed to become a national player, possibly due to marketing that was not aggressive. As flatheads faded from the scene, so did Evans Equipment.

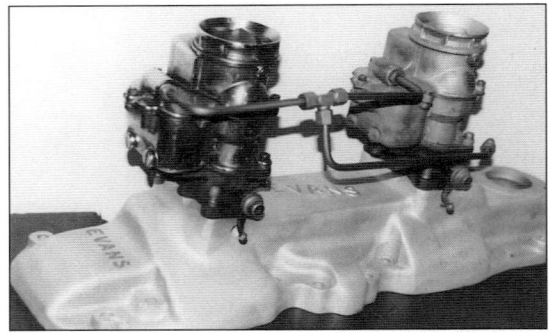

In addition to 8:1 to 10:1 ratio cylinder heads, Evans made a hot two-carburetor intake manifold for flathead Ford V8s.

One of the fastest unblown open-wheeled cars at the time, Evans' blue-and-white belly tank lakester held several records, including the C Class at 185 mph at Bonneville in 1951. Its 1946 Mercury engine was over-bored 3/8, stroke increased 3/8, ported, and relieved with a compression ratio of 9-1/2 to 1. It had a Smith & Jones cam, Potvin ignition, and Evans heads and manifold.

GEORGE BARRIS
King of the 'Kustomizers'

The Hirohata Mercury is probably George Barris' most famous custom.

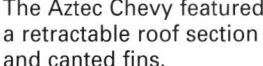

The Aztec Chevy featured a retractable roof section and canted fins.

Many decades ago, a popular car magazine declared that George Barris changed the face of Detroit cars. Another article brazenly accused Motown stylists of swiping hot rod customizers' ideas and applying them to production cars.

There is little question that talented customizers such as George Barris have profoundly influenced the automotive world. At one time, the Barris shop, which attracted worldwide attention, was touted as the leading edge of custom car culture. In a busy little California shop, the Barris brothers (George and Sam) first began performing their unique brand of auto alchemy. Not only did they customize cars to customers' specs, but they frequently developed an entirely new concept vehicle. For more than half a century, the trend-setting Barris' creations graced the pages of car magazines.

George was born near Chicago in 1925. As a toddler, he moved to Sacramento, California, with his parents, who opened a small Greek restaurant. After making balsa wood car models and hanging around a local shop, George, who was not old enough to drive, tackled his first custom—his mother's cast-off 1925 Buick.

While in high school, George and his older brother Sam restyled their knockabout 1936 Ford convertible with such dramatic improvements that raves from their teenage pals encouraged them to turn their hobby into a business. In 1945, George opened a shop in Lynwood, California. His brother Sam joined him after a stint in the military.

In Lynwood, the "Barris Kustom" began evolving as a radically lowered vehicle with frenched or fadeaway fenders, minimum

Hot Rod Pioneers

chrome trim, and a chopped top. Many of these restyling features were later incorporated in production vehicles.

Depending on your point of view, George was either ahead of his time or slightly balmy. He was a modern Leonardo da Vinci with boundless creativity and enjoyed the challenge of transforming boxy stockers into beautiful custom cruisers. Chopping became a trademark, along with styling originality. "Channel and drop to keep 'em close to the ground" was the Barris formula.

The Barris brothers worked effectively as a team, with George coming up with the ideas and Sam executing them. Unfortunately, Sam died of cancer at the young age of 42—a great loss to the young business and the custom industry.

Car shows and the media helped spread the Barris influence throughout the country. So did model companies that made toy replicas of Barris show cars and movie specials. Long-time affiliations also began when Ford, Chrysler, GM, and American Motors contracted to have experimental and show cars built. Barris also created dozens of exotic cars for television and motion pictures.

Over the past half century, thousands of vehicles have had the "Barris Coachworks Treatment," which continues today in a modern facility in North Hollywood where creative ideas are both born and executed.

George came a long way from the toy car models of his boyhood. He will be fondly remembered as the "King of the Kustomizers."

The "Kopper Kart" was a favorite Barris pickup.

Sam Barris' Mercury featured frenched headlights, a custom grill, and a chopped top. Twin spotlights were a "must" accessory in the 1940s and 1950s.

Elvis Presley stands before a customized Cadillac Limo. Other customers that received the "Barris Treatment" included John Travolta, Ringo Starr, Donny Osmond, Zsa Zsa Gabor, and Senator Barry Goldwater.

The Founding Forties

Lou Senter helped customizer George Barris build the Munster and Coffin cars featured on television and in the movies. Here Loe Senter (left) poses with actor Fred Gwynn (right), who portrayed TV character Herman Munster during the 1960s.

George Barris stands with the trick creations he made for the "Batman" TV series. He also designed the "Munsters Coach," the "Beverly Hillbillies," and dozens of other exotic TV and movie cars.

THE LEGENDARY GRANATELLIS
Been There! Done That!

Always fast, Joe is "smooching" his sister-in-law for the camera while both are sitting in the Grancor Special. (Circa 1946)

Often called the "Katzenjammer Kids," the famed Granatelli brothers—Joe, Andy, and Vince—were the "Speed Kings" of the Midwest after World War II. The pecking order ruled with Joe as patriarchal boss, Vince as top wrench, and Andy as relentless promoter. All three had the "Granatelli touch" of talking to engines.

In the Depression days in Chicago, all three brothers boosted the family income. (Joe even repaired cars in the streets.) After serving in the Army, Joe cupped his lucky winnings from a local dice game for payment on a small gas station. Andy, Joe's partner and hotshot salesman, once sold a tire to a man who did not even need one. In return, the man offered him a job. Soon afterward, Joe and his brothers began hopping up Ford engines for local roadster races. The 1932 Ford roadster that Andy drove quickly became "the one to beat."

The clever speed merchants adopted the corporate name of Grancor. By 1947, the Grancor enterprise acquired larger quarters.

"After we talked a relative into a $500 loan, we began making wooden patterns for high-compression cylinder heads and a dual intake manifold for flathead Fords," Joe said. "From then on, the business took off 'like a bat out of hell.'"

As speed wizards, the brothers modified Ford V8 engines, calling them Grancor-Fords. A do-it-yourselfer could obtain a "crate" engine for $750, or a complete engine with installation could run $1,200. Putting that much money into a $2,000 car in those days was ludicrous, but customers kept coming. With the Granatelli mystique, 22 Grancor-Fords soon thundered through Chicago.

In the mid-1940s, Tom McCahill and the Granatellis made a deal, whereupon Grancor would soup up McCahill's 1949 Ford engine in trade for an article penned by Tom for *Mechanix Illustrated*. However, if any problem occurred with the $121 speed treatment, Tom threatened that everyone would read about it. Luckily, the result was a bonanza for Grancor.

Engine swapping soon became a Grancor specialty with the crossbreed "Fordallac," a

Now a Ph.D., Joe was proud of the first garage he opened in 1940. Those coveralls should be washed!

The Founding Forties

Joe, Andy, and Vince in their busy Chicago speed shop. Stacked behind them are Grancor manifolds and heads for Ford V8s. (Circa 1949)

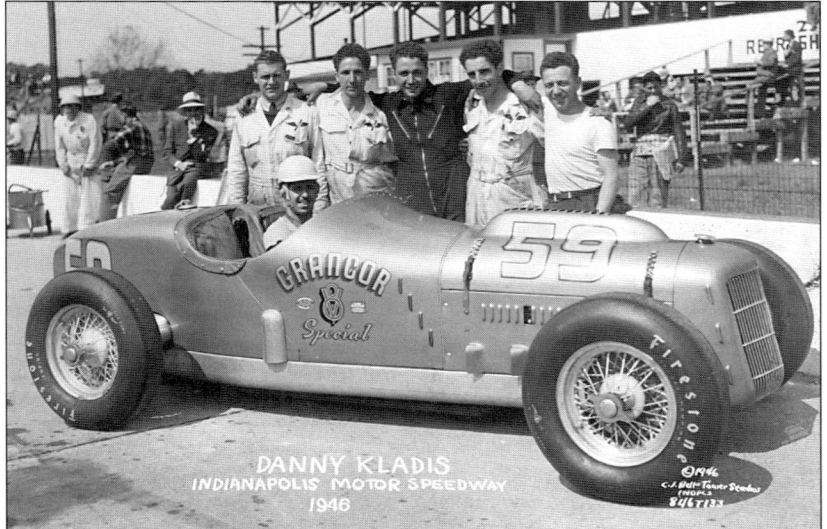

The Granatellis' first Indy car in 1946 was powered with a Ford flathead V8 engine built by Vince, Andy, and Joe, standing behind driver Danny Kladis. (Bell and Tower photo)

muscular Cadillac engine squeezed into a 1949 Ford that delivered 250 hp and more. The "Grancor boys" cranked out almost 100 of the hot bastardized Fords and sold them for $3,800 each—really big bucks in those days!

With the Granatelli brothers' sudden cash flow, they became wholesale distributors (WDs) for emerging speed equipment manufacturers.

"We set up dealerships throughout the nation and even helped honest Charley Card start his speed shop," Joe recollected.

Because there were hundreds of hot rods and no place to race in the Chicago area in 1947, the Granatellis began the Hurricane Hot Rod Association on a shoestring. Soon, roadster racing at Soldiers Field was the "in thing." The Granatellis held tightly to the purse money and brought in heavy-footed drivers such as Jack McGrath and Dick Rathmann.

"Racing is a show, and stock car racing became our money-winner," said Andy, who next promoted drag races. "People want action and thrills. They really don't want to see anyone hurt, but they love the crashes with the bumping and tearing of metal."

After Andy traveled to California for a set of timing clocks for $1,000, the Granatellis obtained an abandoned Navy airstrip in Half-Day, Illinois, solely for drag racing. The airstrip fanned outward in all directions similar to spokes on a wheel—perfect for running several drag races at once. On opening day, bedlam broke loose because the Granatellis were prepared for only 6,000 racing enthusiasts, not the 26,000 that came. The crowd

Hot Rod Pioneers

Long before he held hundreds of land speed records or became the godfather of STP, Andy Granatelli resembled the rookie driver he was in 1948.

knocked down the Granatellis' new fence and partially destroyed the new $12,000 asphalt surface. Soon all the refreshments were gone, the portable johns were flooded, and everybody was out of control. Nonetheless, drag racing had arrived in the Midwest.

Back in 1946, the Granatellis decided to enter their first Indy 500 race with an obsolete two-seat Miller racer. After installing a Grancor-equipped flathead Ford V8 engine, the Granatelli jalopy became the first car after World War II to race with a stock engine at Indy. At the old Speedway's Gasoline Alley, the Granatellis earned their nickname of the "Katzenjammer Kids" because of their cartoon-like swiftness in replacing an engine.

Their first Indy driver was Danny Kladis, a midget race driver who lapped at slightly over 123 mph, thus winning the accolades of driver Mauri Rose in beating his Maserati. Even with worn tires, the car ended with a qualifying average of over 118 mph, which was then a four-lap record for semi-stock and flatheads. Of the twelve race cars that finished the race, the Granatellis' hot rod lasted until the fifty-second lap when it stalled.

In ensuing years, the Granatellis entered the Indy 500—usually with high hopes that ended in despair. Their old Miller Ford took twelfth place in 1947. A year later, Andy crashed, sustaining a broken arm and 11 missing teeth. In 1950, the Kurtis Offy finished tenth. In 1952, Jim Rathmann drove a Grancor Special to a second-place win, and Freddie Agabashian took fourth place the following year.

By 1956, the Granatelli brothers were grossing $14 million per year—an astounding figure in those times. Although they were one of the top speed merchants in the country, they sold the business and moved to California. After purchasing Paxton, they began promoting the "sluggish" McCullock superchargers at Daytona.

In 1961, after Andy and Joe drove a Chrysler 300-F with two Paxton superchargers to a two-way NASCAR record of 172.166 mph, Andy's showmanship won him Chrysler's advertising spot on television, wherein he trumpeted the Flying Mile/Fastest Car for 1961. Then, a struggling Studebaker, which planned to launch the Avanti with an improved Paxton supercharger, contracted with the Granatellis to make record runs at Utah's salt flats. After 10 days, the 12 Granatelli team cars had racked up an astounding 357 speed and endurance records.

The Grancor-equipped flathead engine was mounted backwards to connect to the front-wheel-drive 1935 Miller chassis. With further modifications and four carburetors for 1947, the crossbreed took twelfth place at Indy.

The Founding Forties

Shown here is Parnelli Jones in the first turbine car to run at Indy. Despite breaking down a few laps from victory, it was the "sentimental winner" in 1967.

This Avanti is setting one of more than 350 speed and endurance records at Bonneville. (Lester Nehamkin photo)

In the early 1960s, Joe became president of Paxton and Andy became the corporate head of Studebaker's Chemical Division, whose principal product was STP, a viscosity improver that "beefed up" motor oil.

"To advertise, we freely dispersed millions of STP decals and stickers until the family's STP emblem was on lunch pails, kid's bikes, and vehicles all across America," Andy boasted.

In 1961, the Granatellis purchased and improved the remaining Novi race cars, whose 550-hp engines had been designed in the early 1940s by the Winfields. The beautiful, high-winding Novis did everything—except win races at Indy.

For the 1967 Indy race, the Granatellis entered their turbine-powered "Space Age" racer developed at Paxton under Vince's directives. With Parnelli Jones as driver, the car literally blew off the field at 196 laps, only to suffer transfer case failure a mere 10 miles from Victory Lane. After the race in 1968, the revolutionary machine became controversial when the USAC officials imposed a restriction on the air inlet of turbine engines that cut available power proportionally. The Granatellis fought the rule change but lost. Subsequently, the Indy races for the Granatellis ended.

According to Joseph R. Granatelli, Jr. (also known as J.R.), son of Joe and now head of Paxton, "People often consider me the next Andy Granatelli. Andy was always the marketer, Vince was the designer, and my father Joe was the make-it-happen engine builder who made the power that turned concepts into reality."

Nobody summarized it better for the Granatelli brothers than Andy did in his book, *They Call Me Mister 500*, "There are no positive barriers in racing, just hurdles you have to clear...The thing is, you must get in there and race. That's what it's all about. That's what makes it happen. Forever racing. Forever the Impossible Dream."

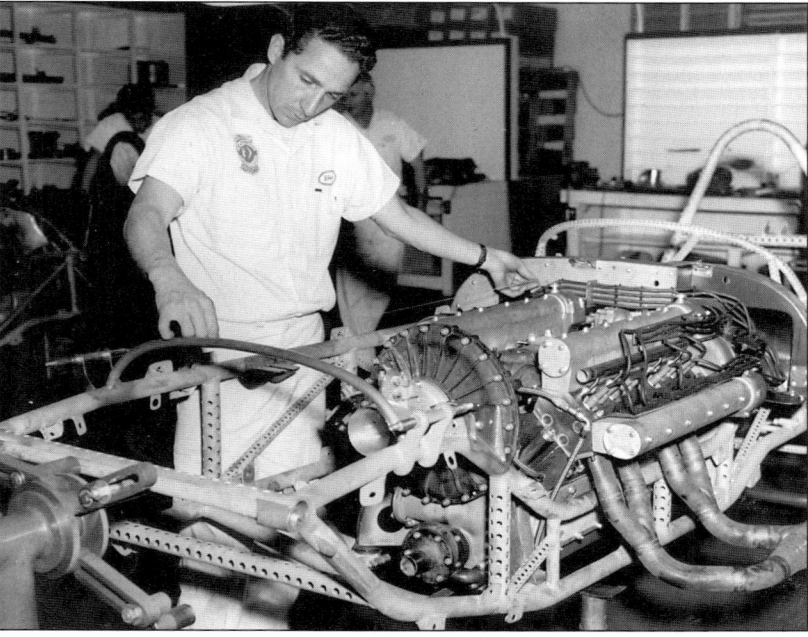
Vince works on the 550-hp Novi engine in preparation for another ill-fated try at Indy in 1961. The high-winding Novis failed in all brickyard attempts.

Hot Rod Pioneers

Front Facelifts

Similar to a person's face, the grille on a car is its key to instant identification. It is also the reason customizers make grille restyling a favored expression of individuality. Bolting on a new custom grille or simply one from a later model or another make of car quickly can create a new and different appearance.

Front restyling also included "nosing," or the removal of hood ornaments, chrome strips, and medallions and curving the corners of the hood bonnet. Headlight treatments varied from frenching on new bezels to adding hooded shades or canted quad sealed beams. Bumpers were either removed completely or integrated with the grille in later modeling. Louvers on long hoods and rear decks also added tasteful distinction.

Over the years, hundreds of pleasing frontal combinations with grille patterns of bars, teeth, egg crates, and mesh have been used to create handsome show stoppers, as shown in these photos.

The Founding Forties

Giving Years-Ahead Styling

65

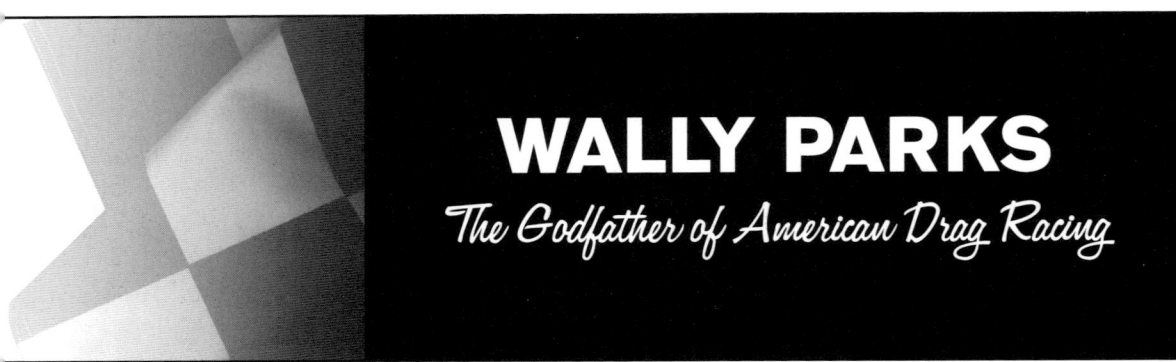

WALLY PARKS
The Godfather of American Drag Racing

Wally Parks has probably done more for the advancement of organized drag racing as a major motorsport than any other person. A true pioneer, Wally was born in the heart of Cherokee lands in Goltry, Oklahoma. He has been in the forefront as an SCTA founding member, a Bonneville Nationals' organizer, the first editor of *Hot Rod Magazine*, and NHRA's founder and first president in 1951. Now in his eighties, Wally Parks remains a passionate advocate for the sport.

As a teenager, Wally and his friends raced on the Mojave Desert's dry lake beds with their fenderless Model T and Model A roadsters. At that time, the sporting vehicles were not yet referred to as hot rods. After graduating from high school, Wally worked at the Southgate GM assembly plant that produced M-1 tanks during World War II. As a tank test driver, he racked up 15,000 miles. When he was in his early twenties, Wally and his family moved to Southern California, where he became fascinated with auto racing similarly to his close neighbor, Eddie Meyer, who was building racing engines.

Wally first heard the words "hot rod" used figuratively from a fellow G.I. when he had been shipped to the South Pacific as a U.S. Army Tank Battalion retriever commander. Upon his discharge from the service, Wally returned to his former job and picked up his hobby, with even more enthusiasm and dedication.

Likewise, other returning servicemen across America resumed life where they had left off, many with such renewed interest in cars that the phenomenon soon to be called hot rodding was born. As the new and exciting postwar sport of dry lake racing emerged, Wally was ready to be a part of it. Unfortunately, illegal street racing was harming the young sport.

Yielding to law enforcement pressures, major local clubs banded together and created safety rules for desert racing and daily driving. This was done through the reactivation of the SCTA, which Wally Parks helped found in 1937. Nine years later, Wally was elected its president.

Wally Parks (right) and Ray Brock (left) celebrate Ak Miller's win at Pike's Peak in 1958.

The Founding Forties

In April 1947, SCTA spokesmen Wally Parks, Ak Miller, and Rex Mays allied with police and politicians to promote safe driving and to upgrade the image of hot rodders because most were law-abiding drivers. These meetings and the favorable publicity bolstered the sport, and Wally's personal campaign resulted in SCTA's presentation of the first Bonneville National Speed Trials on Utah's salt flats.

In 1949, Wally, in a calculated career change, became the first editor of the newly launched *Hot Rod Magazine*. With little journalistic experience, Wally was a natural talent and paved the way for the commercial success of the magazine. By utilizing the magazine's editorial pages for uniting hot rod clubs across the nation into one self-disciplined sanctioning body, Wally and a few close associates formed the NHRA in 1951. Wally was its founding president.

In 1954, with a program titled "Safety Safari," Wally traveled more than 13,000 miles across the United States with a small team of rodders, teaching hot rod groups how to promote safer and better organized events.

"We functioned like missionaries going out among natives, spreading the good news of sportsmanship, fellowship, and safety," Wally said. "Also, we helped some clubs acquire approved drag strip sites."

In 1957, Wally was appointed editorial director of all Trend, Inc. (Petersen) automotive magazines, a position in which he served until 1963, when he resigned to assume full-time duties as president and chief administrator of the fast-growing NHRA. For four and a half decades, Wally helped to change a hot rod tinkerer's pastime into a professional drag racing sport. The NHRA, whose major activity is drag racing, now has more than 80,000 members and is the world's largest motorsport organization, with Wally as board chairman.

Wally Parks' sterling sense of mission led him from being a zealous crusader to a powerhouse negotiator and uncompromising leader. His dedication to the welfare of drag racing and the specialty industry is unsurpassed. An inductee into numerous halls of fame in motorsports, Wally Parks confirms what only one man can do for auto racing.

The Justice Brothers
Zeke Justice (right) shows off his highly modified motorcycle, while Ed Justice (left) sits in the first midget race car they built at their Kansas home in the late 1930s. After that, both crewed midgets and Indy cars for Frank Kurtis. Later, the imaginative Justice brothers pioneered what became well known as J-B Car Care Products—financed by the sale of their fast Kurtis-built midget.

STUART HILBORN
Racing's Fuel Injection Pioneer

Pictured in 1966 is Stuart Hilborn, the father of racing fuel injection. Also shown is young Stu Hilborn at El Mirage Dry Lake, half a century earlier.

Although we all dream, few people do so with any hope of their dreams coming true. Fortunately, Stuart Hilborn had the vision and courage to make his dreams a reality. With a brilliant but simple concept, Stu managed to build a workable, bolt-on, constant-flow fuel injection system. This technological breakthrough earned him the title of racing's fuel induction maestro.

Stuart Hilborn and driver Bill Vukovich at their first win at the Indy 500 in 1953.

After World War II halted his dry lakes racing, Stu joined the Army Air Force and served as gunnery instructor, always dreaming of a better way to feed fuel to an engine. In 1946, after months of experimentation, Stu found the answer.

"Skeptics kept on saying that my constant-flow system wouldn't work because the fuel injected was not individually timed to each intake valve's opening," Stu recalls.

Although lakes' contestants razzed Stu and his crew about their rag-tag instruments for measuring atmospheric pressures and temperatures, they drooled as Stu's invention pushed speeds upward. During one steamy summer day at El Mirage in 1948, Howie Wilson drove Stu's lakester to 150.50 mph, for a new Class B Streamliner record. *Ford Times* magazine promptly published: "Cracking the 150-mph mark was then as much of a feat as aviation's plunge through the sonic barrier."

After Stu's new partner Jim Travers, a wizard with Offy engines, proved that the Hilborn system provided a 10% gain in horsepower over carburetors in a 105 c.i. Offenhauser, Stu tooled-up for production of fuel injection systems for the small Offy midget racing engines. However, because of financial constraints, Stu could do only a short run of 10 F.I. systems for the bigger 270 Offy engine.

Stu's big break came at the 1949 Indy 500 race, with his injector that qualified five entries that had previously failed with conventional carburetors. Despite the fuel injector's resounding success, few racers had confidence in the conversion. Total acceptance came the following year, when two-time champion Mauri Rose used the injectors in his Keck Offy. By 1951, Stu had begun to make bolt-on fuel injection systems for Ford and Mercury V8 engines, and Hilborn F.I.

systems began fueling winners for nearly every kind of racing vehicle.

With rotary valves in place of butterflies to control air flow, Stu in 1978 designed a fuel injector for the English Cosworth engine. In later years, both Chevrolet and Ford copied the Hilborn system—adding only electronic controls.

Although Stuart Hilborn has long been acknowledged as a pioneer inventor and esteemed contributor to the racing and performance industry, it was not until 1996 that he was formally inducted into the SEMA Hall of Fame.

Stuart Hilborn's lakester is the first dry lakes' racer to break the 150-mph barrier.

The first Hilborn constant-flow fuel injection system was relatively simple. Each cylinder was fed by its own air-intake/throttle body. A nozzle behind each butterfly valve sprayed in the fuel at pressures of up to 35 psi. A fuel metering valve positioned between the fuel pump and nozzles was calibrated to open with the butterflies, increasing fuel flow with engine speed and throttle position.

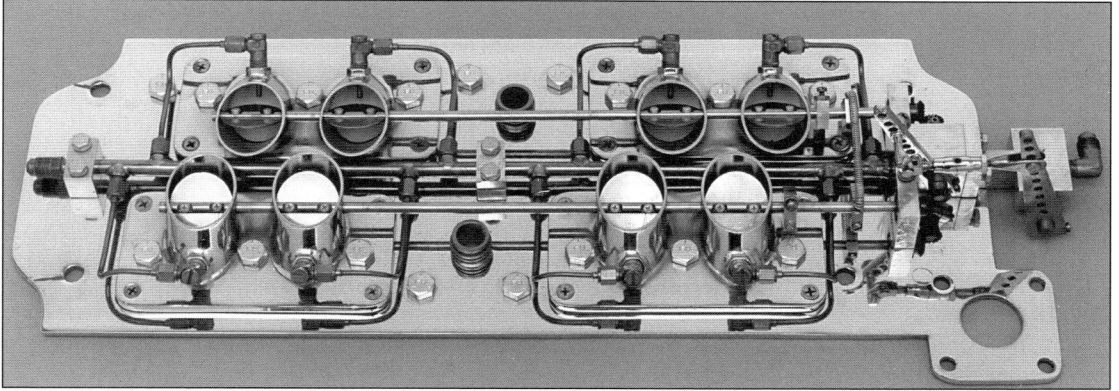

Stu Hilborn's wrecked car after it flipped several times, putting Stu in the hospital.

Stu Hilborn (left) in 1946 with Eddie Miller (right), old-time Indy racer, who made the four-carburetor setup on Stu's first Ford V8 lakester.

The blown Chrysler dragster, like many others, used Hilborn injectors in the 1960s. The turbocharger system worked well, despite its rather crude "Rube Goldberg plumbing."

BILL KENZ
Streamliner Builder

Born in 1905 in Denver, Colorado, William Kenz was another old-time hot rodder who was typecast even before the term had caught on in mainstream America. Tall and lanky, Bill quit school after the seventh grade and taught himself to be an expert mechanic and machinist. All his life, he was innovative, often scheming to squeeze more speed out of cars—beginning with a rattle-trap Model T and then a souped-up Model T roadster that he raced in nearby DuPont Speedway.

After Bill took sixth place in a Pike's Peak hill climb against specially built race cars, he was hooked on chasing checkered flags. Bill's early 1930s experience would prepare him well for an illustrious career of postwar hot rodding.

While working as a mechanic for a Ford agency, Bill soon became the local hop-up hero when he recommended milling and installing cylinder heads to boost compression to compensate for high-altitude power loss. In 1938, Bill left to form a partnership with Roy Leslie, which swiftly grew from a small speed shop into a large automotive parts distributorship.

Using zeal for racing as an excuse to promote their business, the partners campaigned midgets and roadsters to many area triumphs. During the first 1947 Bonneville speed finals, Bill entered an odd-looking 1932 Ford pickup truck with a semi-concealed extra flathead V8

The pickup truck was powered by two Ford V8 (59 A blocks) flatheads with 9:1 compression, four Stromberg carburetors, Spalding magnetos, and no flywheel in the rear engine with the front flywheel lightened to 13 pounds. There are no radiators. Engine cooling was from the water tank beside the cockpit. Instead of a transmission, a "dog" clutch is used on the rear engine; a regular friction clutch is used on the front engine.

Bill Kenz with his 1932 pickup truck, which was powered by two flathead engines that drove the rear wheels.

engine that astonished spectators by reaching a two-way average of 140.95 mph, beating many of the hot roadsters.

Because the So-Cal Special, a single-engine streamliner, was then the fastest car at the salt flats, Bill came up with the idea for a twin-engine streamliner to challenge it.

Neither partner could draw blueprints; thus, they developed the car from rough sketches drawn on the garage floor with chalk. It took eight months plus an estimated $10,000, including labor, to complete the Kenz–Leslie streamliner, also known as the Clymer Motorbook Special. It was named after their friend Floyd Clymer, who helped subsidize the effort in exchange for advertising.

On the opening day of the 1950 Bonneville Nationals, Kenz unveiled their handsome new streamliner, which futilely challenged the competition. In 1951, Kenz finally discovered the problem was lack of oxygen, which was corrected by larger intake air scoops that fed the two modified Ford V8 flathead engines. Willy Young then drove the streamliner to its first record, reaching over 230 mph and earning the title of "Fastest Car of the Year."

The sleek streamliner continued to be a consistent performer as it increased its speed and roared to a 270-mph Bonneville record in 1957. This was remarkable, considering that the engines were considerably modified and normally aspirated without benefit of supercharging, with only 20% nitro added to the gasoline.

Bill Kenz and Roy Leslie ended their car's eight-year campaign on the salt after reaching a speed nearing 300 mph with three V8s in the same chassis. Today, the famous car is the oldest existing American-built streamliner, and it remains a vibrant tribute to hot rod ingenuity.

The body shell was hand-formed from 0.064 half-hard aluminum into sections that were welded into smoothly contoured pieces.

Innovative Bill Kenz (left) poses with the handmade 18-foot squat streamliner that was "hunch-designed" without benefit of wind tunnel or other sophisticated testing. Skirts cover all wheels when racing.

ROY LESLIE
Rocky Mountain Racer

Possessed with a passion for racing, Roy Leslie was so competitive that he could not bear having anyone in front of him, even when traveling on the highway. That first "road-raging" spirit urged Roy into every type of racing. Then fame came to Roy as co-builder and driver of the famed Kenz–Leslie streamliner, once the world's fastest Ford-powered car.

A crucial turning point in Roy's life came when Roy teamed with Bill Kenz in 1938. Because their talents complemented each other, they became inseparable in business and in hot rodding. Despite the stress of coupling competition with business, they had great respect for one another. Roy once said, "Bill Kenz has micrometer fingers that can feel within thousands of an inch."

Both partners became well known for pioneering multiple engines in their Kenz–Leslie streamliner, which was one of the first to exceed 200 mph at Bonneville and hold many records. Subsequent experiments with two Ford flatheads in a fuel dragster were disappointing. However, Roy's son, Ron, later drove a supercharged 390 Ford dragster in record mile-high strip runs in Denver.

Today, Kenz and Leslie Ltd. is the largest specialty-parts warehouse distributor in the Rocky Mountain region—a legacy to the founders, Kenz and Leslie, who in their time ran the fastest hot rod in the world.

What was once the "world's fastest hot rod" reposes, somewhat restyled. In 1955, a third Ford flathead was installed in the cockpit's previous location, and a new enclosed cockpit was built into the rear fin. Running at Bonneville as the "Bob Jones Skyland Ford Special," the streamliner failed to top its previous 255-mph record. Later, Roy Leslie drove it almost 300 mph on the salt.

FRAN HERNANDEZ
Funny-Car Baptizer

In 1949 in Goletta, California, tires were smoking and nerves were jangling when tobacco heir Tom Cobbs challenged Edelbrock's speed shop manager Francisco "Fran" Hernandez to a contest of speed. It was not exactly a grudge race, but the Hernandez coupe roared to victory like a "lion on the streets." Onlookers went wild, and history was made.

A year earlier, in his 1932 three-window Ford coupe, Hernandez had topped the Russetta Timing Association lakes record for coupes by clocking 144 mph. Credit was given to the nitromethane fuel, which was used successfully for the first time in a four-cycle internal combustion engine.

After being a tool maker at Northrop Aircraft and a Navy machinist, Hernandez began punching a clock for Fred Offenhauser. Shortly thereafter, Hernandez opened a business with Fred Offenhauser, Jr., the nephew of the famous Indy Offy engine builder.

Because of Hernandez's fluidity in Spanish, he was heralded by Bill Stroppe for Ford's Mercury Division for the Mexican Road Race. In 1956, he became one of the head men of the 50,000-mile Ford run at Bonneville. After organizing Mercury's performance program, Hernandez entered five 1964 Comets in the East African Safari. His team was the first American team to enter this grueling automotive contest, covering more than 3,000 miles through several African provinces filled with muddy byways, washed-out bridges, stone-throwing natives, and animal traffic.

In the mid-1960s, Mercury took aim at serious drag racing in an effort to attract young buyers. With almost half the country's population under 25, the drag strip was the best place to show off the potential of our vehicle for future sales, said Fran.

When the new "funny car" racing became the rage, the first Comet funny car was designed in Hernandez's office, with the wheelbase drastically altered to improve weight transfer. Dyno Don Nicholson, Jack Chrisman, and "Fast Eddie" Schartman were Mercury's hot drivers in the exciting quarter-mile match race circuit. Then, Hernandez was credited with creating the slang term "funny car," a name that stuck.

Fran Hernandez, the rodder who spearheaded the Mercury racing programs in Dearborn, Michigan, will go down in history as a major player in the growth of the hot rod sport.

Fran Hernandez is ready to make a record run of 147 mph in the Hernandez and Meeks Class C Roadster at the 1950 Bonneville Nationals.

BILL BURKE
Tank Daddy

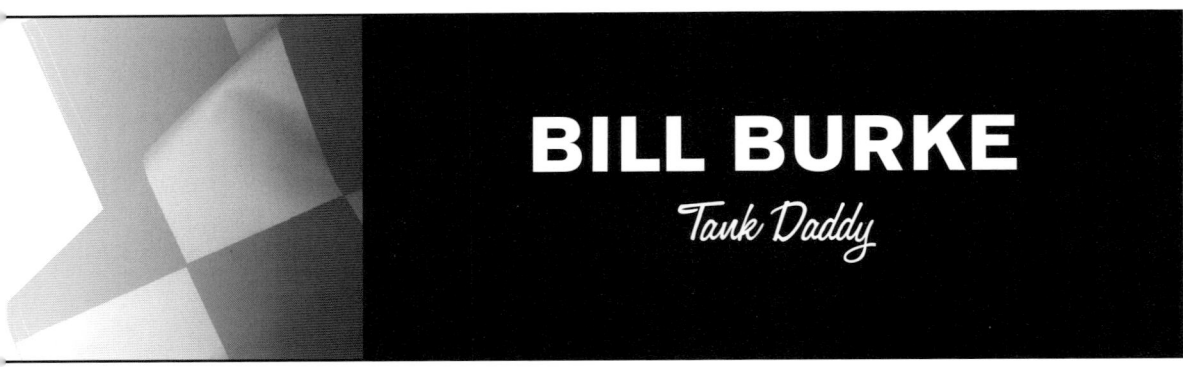

The first rear-engine tanker was built in 1947 by Don Francisco (left) and Bill Burke (right). Bill drove it to many records, including 158 mph at Bonneville in 1949. It was the first hot rod to hit 150 mph in 1947.

Few people personify hot rodding history as much as William Burke, who was always fired up by fast, unusual cars. Tall Bill, who pioneered many firsts, invented the wing-tank bodied lakester in 1946. He was the first to drive over 150 mph with a rear-engine belly tank and one of the first to build a fully enclosed streamliner.

A Los Angeles youth, Bill assembled his first lakes-only machine in 1938, which was a Model B powered 1927 T that hit a 110-mph record at Muroc. That year, Bill and his friends also formed a car club called "Road Rebels," which later expanded to become the Western Timing Association.

When World War II began, Bill enlisted in the Navy and was sent to the South Pacific. Some airplane wing tanks sparked him with an idea for a lightweight competition car. After returning home from the service, Bill went to work putting wheels on a "belly tank." The first effort was rather crude with a Model T frame narrowed to fit inside a surplus aircraft wing tank with a souped-up 1934 Ford V8 flathead. The unusual combination made runs over 100 mph, including a 1946 run at 137 mph. The following year, Bill located a larger (300-gallon) belly tank, which afforded room for a rear-mounted engine. Then Bill teamed with Don Francisco, and together they refined the lakester and set records for three years. In 1949, the tanker was the first hot rod to break 150 mph and was dubbed the "World's Fastest Hot Rod." Within a few years, Bill had built 13 tankers. The later models used aircraft wing struts for frame rails rather than Model T's.

Bill's gregarious and fun-loving nature made him a born salesman. He huckstered advertising that helped keep *Hot Rod Magazine* alive during its formative years.

Chopped Austin coupe with a streamlined nose had a Ford V8 flathead that set a 141-mph record in 1951.

Bill Burke built the world's first wing-tank lakester. Its Ford V8 engine had a Thickstun manifold and milled heads. It reached a record 137 mph in 1946.

74

The Founding Forties

Bill was elected president of the SCTA in 1950 and served as an official starter at the second Bonneville speed trials. He developed a partnership with Mickey Thompson and ran a fiberglass clone of the handsome Cisitalia Fiat Coupe for a record of 149 mph in 1953. By 1963, Bill was running a stable of cars at Bonneville, including an Avanti, a Camaro, a Corvette, and a blown Ford and Chrysler.

Hot rodding would not have been quite the same without Bill Burke. From the beginning, he acid-tested every rule in the book, with homemade machinery that sped from Muroc to Bonneville. Now in his golden years, Bill intently watches his sons and his grandson as they beat his records.

"It will be great to have three generations in the 200-MPH Club," Bill said.

Ray Brock hands Bill Burke a drink after his 1952 record run of 136.90 mph at Bonneville. The tiny fiberglass streamliner was powered with an 80-c.i. Harley cycle engine. The Class F record stood for 10 years.

Clarke Cable and Bill built this sleek aluminum-skinned streamliner with a supercharged Chrysler mill in 1966.

At the 1960 speed trials, Bill rocketed himself into the 200-MPH Club with a glass streamliner called the "Pumpkin Seed." The car had an injected Falcon 6 engine that made the Class D record at 205.64 mph. Later, in the same car, Bill ran 264 mph with a blown Tempest four-cylinder engine.

This 1968 was the world's fastest Corvette when Bill's son Steve set a record of 201.87 mph at Bonneville in 1972.

Going for a record with as few inches as possible, Burke built this little (587 pounds) "super-shaker" fiberglass streamliner that turned 151 mph with an 88-c.i. Harley-Davidson coned engine.

RAY BROWN
Buckle-Up Pioneer

Ray Brown, still the veteran hot rodder and entrepreneur, is now senior vice president for Superior Industries (a business started in a garage by Louis Borick), the world's largest manufacturer of aluminum wheels and a major wheel supplier to GM and Ford Motor Company.

Before he was a safety advocate, clean-cut Ray Brown was a dyed-in-the-wool hot rodder. In 1928, Ray was born in Sioux Falls, South Dakota. As a California transplant, he pursued his passion for things mechanical and began working for Eddie Meyer, master builder of championship car and boat engines. At the age of 21, Ray opened his own small shop for building competition engines. In 1953, his modified 302-c.i. Chrysler engine powered a car that set six national and six international Class C Streamliner records at Bonneville. In his heyday, Ray was considered the foremost authority on souping up the popular Chrysler Hemi OHV engines.

After a few close calls while riding at the California dry lakes, Ray in 1946 installed surplus military aircraft safety belts in a 1932 roadster. After noticing that many passenger car injuries resulted from victims striking dashboards, going through windshields, or being thrown from vehicles, Ray reasoned that a seat belt would protect passengers from catastrophic injuries during a crash. For his family's car, Ray made a seat belt, which probably was the first seat belt ever made for a passenger car.

After friends began asking Ray to make similar belts for their cars, he realized that seat belts had a huge market potential. Similar to other hot rod entrepreneurs, Ray needed start-up and operating capital; therefore, he mortgaged the family property.

By early 1950, Ray began producing his first short run of lap-type seat belts. The attractive spring-release buckles were made to exacting specifications, and the nylon-belt webbings were color-coordinated for existing vehicular interiors. In 1951, Ray placed his first small ad in *Hot Rod Magazine* for the "Impact Auto Saf-Tee Belt."

One press release stated: "A person's chance of being killed in an auto accident is five times greater when he is thrown from a vehicle." Another declared: "A seat belt will keep the crash victim in the car." Due to consumers' and automakers' indifference, initial response was slow.

This Deuce 1932 roadster, owned by Ray in 1945, raced a year later at the dry lakes to 123.87 mph. Its flathead V8 engine was bored and stroked with an Eddie Meyer dual intake manifold, Clay Smith super race cam, Jahns pistons, 9-1/2 to 1 compression ratio heads, and homemade headers.

The Founding Forties

Ray's belly tank lakester with a Ford flathead ran at El Mirage in 1949, topping 180 mph in 1950. At Bonneville in 1952, it averaged 197.88 mph for the Class C Lakester record. This was the beginning of the end for Ford flathead engines in serious salt competition.

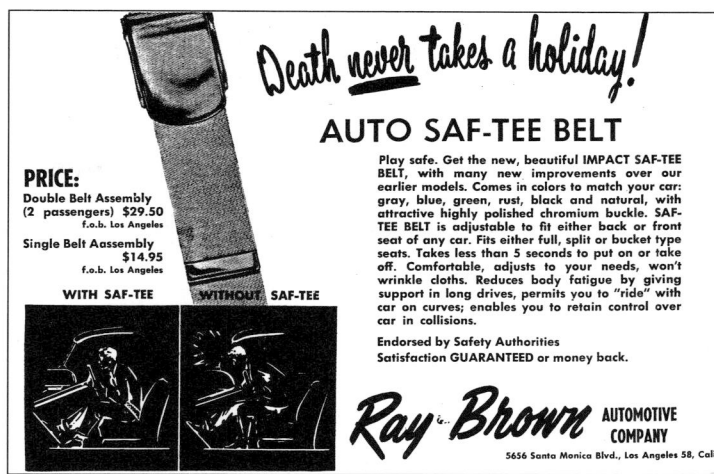

An early seat belt advertisement by Ray Brown's company.

This 22,400 square-foot factory in Pacoima, California, served as headquarters for both racing engine and seat belt manufacturing in the early days.

"Fancy gadgets and radios were more important to them [the general public] in those days," recalled Ray Brown.

Despite rebuffs from the automotive industry, Ray became one of the world's largest seat belt manufacturers within three years. (Two shifts operated daily with 150 employees.) By 1962, Ray was the first contractual supplier of seat belts and harnesses for the U.S. government.

By 1964, legislation finally forced U.S. car makers to equip new cars with front seat belts, thereby igniting a market of enormous potential for rear seat belts as well as front and rear seat belts for older vehicles. As new companies competed in the safety regime, inferior products were manufactured and quality control was almost nonexistent. With Ray's input, the industry established a policing and testing body within the American Seat Belt Council (ASBC). Because Ray was an erudite mediator and honest politician, he became its first vice president.

With the mandatory installation of both front and rear seat belts in the new models of American cars, Ray soon saw a diminishing aftermarket. In 1966, after selling his business, Ray joined Superior Industries and worked with his friend, Louis Borick, the founder and CEO of Superior Industries, the largest cast aluminum wheel manufacturer in the world.

Thanks to Ray Brown's pioneering vision, passenger-activated restraints have evolved into sophisticated lap-and-shoulder belt systems. In 1985, New York became the first state requiring front-seat occupants to utilize the restraining systems.

Ray Brown, a SEMA Hall of Fame recipient and past president of the ASBC, played a significant role in the development of automotive seat belts and in the formation of Federal Motor Vehicle Safety Standard #209. He is also another "little guy" who showed the big vehicle manufacturers the ropes!

From the upper echelon of the automotive aftermarket, Ray Brown, a marketing executive, now rides the highway of life with its safety signs that forewarn: "Buckle up! The life you save may be your own!"

Hot Rod Pioneers

Fat-Fendered Oldies

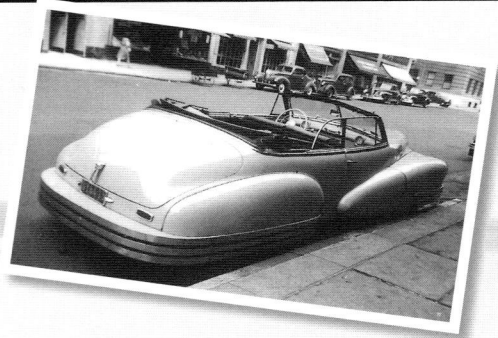

While famous coachbuilders such as Derham, Darrin, and Coachcraft were building lavish one-of-a-kind custom-tailored cars for kings and movie stars, the early post-World War II backyard builders were developing another kind of automobile customizing that was less formal and often even more creative.

Although some of the restyling and updating of 1930s and 1940s cars involved extensive body work such as sectioning and top chopping, much was accomplished through ingenious swapping and blending of body components from later or other model cars–sometimes simply installing a new grille as car factories often do today to freshen models from the previous year. The following examples show what passed for classy, custom cars many years ago.

Going topless meant something different in the 1940s than it does today. The doorless racing-style cockpit did not provide much protection against the elements.

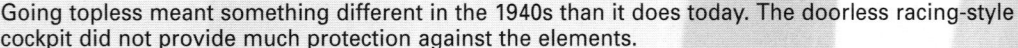

The Founding Forties

Designed by Norman Timbs, this car represented ultra-streamlining in the mid-1940s. It was 100% custom built with a tubular frame, independent suspension, and rear-mounted engine. The rear half of the aluminum body could be raised by an aircraft hydraulic system controlled by pushbutton.

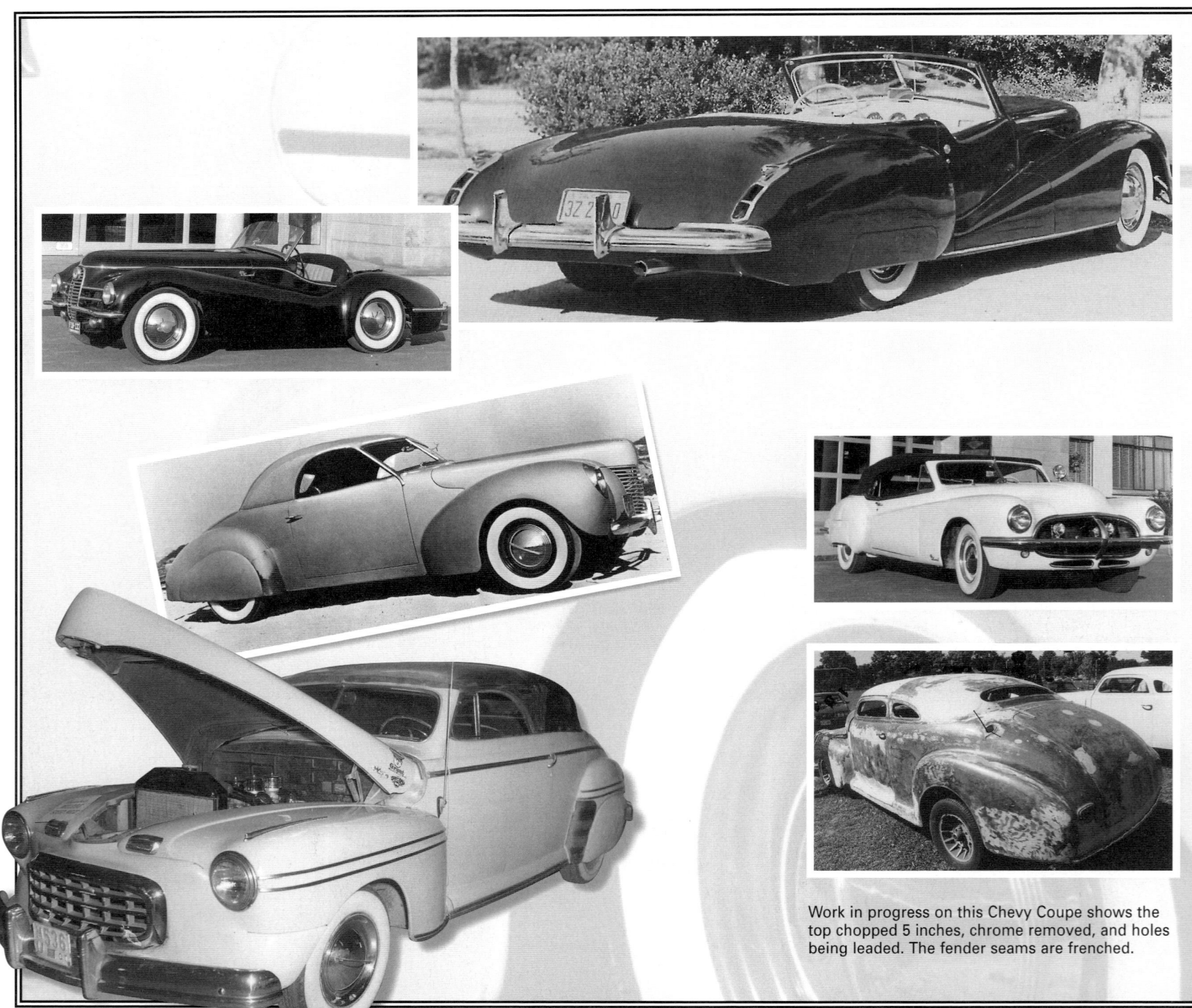

Work in progress on this Chevy Coupe shows the top chopped 5 inches, chrome removed, and holes being leaded. The fender seams are frenched.

The Founding Forties

Early Fords and Mercurys were restyler's favorites. This nicely customized 1941 Mercury, with lengthened hood and fade-away fenders, had a removable split top.

81

HOWARD DOUGLASS
Maker of Specialty Exhaust Systems

Howard Douglass proudly displays the steel pack muffler that started him on the road to big bucks. In the same way as other brands, the Douglass muffler was offered with both steel packing and glass packing for sound dampening, with the latter particularly effective in modulating high-frequency sound waves. (Circa 1949)

Douglass exhaust systems once were famous for their rugged construction and pleasant "musical" tone. However, in the 1930s, Howard Douglass cobbled noisy homemade mufflers, one at a time, while using a $38 welder. One day, as Howard was installing a dual exhaust system on an old Ford Coupe, newspapers in the rear rumble seat of the car caught fire from the welding sparks. Consequently, the City of San Gabriel soon shut down Howard's backyard shop. Howard was undaunted as he dusted himself off and began Douglass Muffler Manufacturing at a different site.

After publishing his own speed equipment catalog, Howard opened a warehouse and one of the first tubular header plants. By 1951, business was booming, but room was so scarce that Howard had to store header systems in his backyard and sleep under the stars with his inventory until the shipping truck arrived in the morning.

Howard Douglass' retail store (center) installed mufflers "While 'U' wait" in 1947. A few years later, two manufacturing plants were added. This montage shows Howard (right), his wife Ruth (left), and key employees who included a young Don Francisco, who later became technical editor for Petersen Publishing.

The Founding Forties

At one time, Howard Douglass was perhaps the largest specialty exhaust system maker in the country, with more than 200,000 mufflers shipped annually.

"In 1969, the market began suffering due to smog regulations," said Howard. "So in the early 1970s, I sold out. That's the time when big conglomerates were buying into the high-performance industry."

After his retirement, Howard started a financing business. Some people, such as Howard Douglass, never retire. They just keep on living out their destiny.

This photo is typical of a country gas station in 1925. Howard Douglass worked here with his father after school.

Howard Douglass (seated) receives advice from a friend before giving his roadster a push-start in 1936.

Karl Orr

Legendary Karl Orr, who raced Model T's in the Midwest, was the SCTA Dry Lakes Champion in 1942. He raced both modifieds and roadsters such as this fenderless 1932 Deuce, parked in front of his speed shop, one of the first in California. Karl's wife Veda was one of the first women competitors and produced *SCTA Racing News,* a mimeographed newsletter she mailed to 1,000 servicemen in World War II.

ALEX XYDIAS
So-Cal Speed Shop Founder

Even when Alex Xydias was a sergeant in the Air Force, the power didn't go to his head. Instead, he tried to put "power into hot rodding," which is where the young former lakes racer saw his future. On March 3, 1946, the day of his discharge, Alex opened his first So-Cal speed shop, specializing in Ford flathead equipment that was sold nationwide, primarily through mail order.

As a friendly, "hard-core L.A. hot rodder," Alex attracted friends such as industrial designer Dean Batchelor who later became a top auto journalist and collaborator for several So-Cal projects, of which one would become legion.

So-Cal's flagship was the benchmark for streamliners of its day. The aerodynamic aluminum body was designed by Dean and constructed by Neal Emory; the chassis was built by Alex and Keith Baldwin. The flathead engines were Bobby Meeks' masterpieces. Equipped with a small V8-60 engine, the car turned 152 mph. With a Mercury V8, it went 210 mph at the 1950 Bonneville speed trials.

As time passed, Alex and the So-Cal racing team became known for its stable of red and white cars, which captured more than 20 records at Bonneville and the dry lakes and the drag strips of Southern California.

Similar to his father, who was a prominent producer of silent films, Alex in 1954 began filming documentaries that featured clips of

The So-Cal streamliner was probably the most famous hot rod of its day. It set the best time of 193 mph at Bonneville in 1949, raising the old record by more than 30 mph. Dean Batchelor is the driver, Bobby Meeks is in the center, and Alex Xydias is on the right.

Sergeant Alex Xydias and his customized 1934 Ford weather the change of seasons—the hard way—in 1945.

The So-Cal Speed Shop, one of the first shops of its kind in the country, opened in 1946. Although it was only a small 20 ft. x 20 ft. Sears prefab building, its big sign forecasted the coming of many record-breaking dry lakes and drag cars. A larger facility closed in 1961.

Indy, Daytona, Sebring, Pike's Peak, Bonneville, and the first four NHRA Nationals. Of these, "The Hot Rod Story" is the most popular.

In 1963, Alex's latent literary talent was utilized as editor of *Car Craft* magazine for the Petersen Publishing Company. After more than a decade, he also became the associate publisher of *Hot Rod Industry News* and *Hot Rod Magazine*. In addition, Alex helped produce the first SEMA trade shows. In 1977, he began a decade-long partnership with Mickey Thompson and produced, with his wife Helen, the dazzling SCORE off-road shows. In 1982 Alex Xydias was inducted into the SEMA Hall of Fame—a much-deserved honor to a top-notch pioneer who helped hot rodding grow into a show-stopping, megabucks motorsport.

The handsome So-Cal 1934 Ford held the Class C Competition Coupe record of 172 mph at Bonneville in 1953 running the Fox & Cobb blown flathead V8. In 1954, Alex built a supercharged Ardun OHV for the coupe. It ran 178 mph at Bonneville and set the track record at Pomona with a speed of 132 mph.

The So-Cal lakester belly tank competed from 1951 through 1953 and ran the fastest times in Classes A, B, and C in Bonneville in 1952, which was the last year the flatheads reigned. Top time in Class C was 198 mph, with the Mercury engine fed with 40% nitro and 60% methanol. Co-owner Clyde Sturdy is left of Alex Xydias.

DEAN BATCHELOR
Rodding Writer

The sport of hot rodding might not have progressed swiftly in the early days without writers such as Dean Batchelor, whose car hobby became a life-long vocation. After racing, designing, and building, Dean became a highly respected automotive journalist and historian.

Born in Kansas in 1922, Dean began photographing cars on the streets of California, where he was raised. When World War II intervened, he served in the Air Corps and was shot down in Germany. There, he spent a year as a prisoner of war. When he returned to California, Dean earned a degree in industrial design and teamed with his friend Alex Xydias for designing and building the So-Cal streamliner. Dean drove it to a record win in 1949. Subsequent team efforts in 1951 and 1952 set records that remained until 1995.

Dean worked for Petersen's *Motor Life* before switching to *Road & Track* as editor-in-chief. Before Dean passed away in 1994, he had written the *American Hot Rod.*

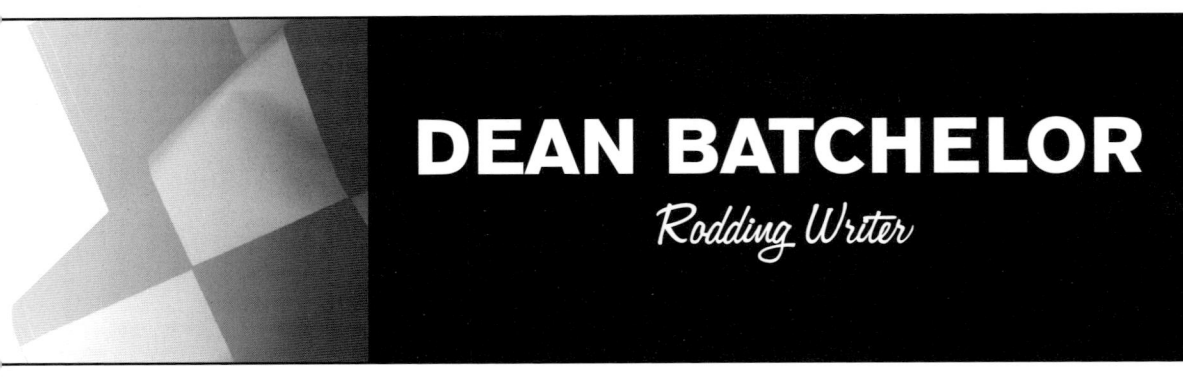

More than 350 years of racing heritage are represented in this photo. The So-Cal special team proudly shows off a trophy for the fastest time of the 1950 Bonneville meet at 210.87 mph. Left to right are: Keith Baldwin, Bobby Meeks, Ray Charbonneau, Dean Batchelor, Miss Trophy girl, Alex Xydias, and Bill Dailey (seated). The streamlined aluminum body was designed by Dean. Alex Xydias and the crew did the wrenching.

Chevy "Stove Bolt" Six

Nicknamed both "Cast Iron Wonder" and "Stove Bolt Six," the lowly Chevrolet six-cylinder engine, with its cast iron pistons and slotted head bolts, finally received some respect in 1948. At that time, a young mechanic named Wayne Horning first boosted the stock output of the Chevy engine to an astounding 220 hp, which pushed John Hartman's roadster to a sizzling 140.18 mph on the dry lakes. A year later, Marvin Lee's B streamliner set a record 153.545 mph at El Mirage—also with a Wayne Chevy engine.

Then, it was said that a good hop-up artist could rework the head (fill and mill to raise compression to 9.5 to 1), add oversized intake valves, a 3/4 race cam, dual carburetion, and headers, and produce 1 hp for every 3.5 pounds of engine weight in the 150-hp range for less money than any other hot road engine! A do-it-yourselfer could double the Chevy's horsepower for a few hundred dollars and probably even beat fast Ford flatheads.

Wayne, Nicson, and McGurk were among the many makers of speed parts for the 1937 through 1954 Chevy (and GMC) six engine. Eventually, they all faded from the racing scene, as did the workhorse sixes.

The Wayne Chevy Track engine featured individual (polished) porting with six intake ports on the left side and six exhaust ports on the right side, jumbo-sized valves, rocker arms of 1-1/2 to 1, dual coils, and compression ratios up to 14 to 1. In 1949, the engine cost $950.

Before graduating to big car racing that included two shots at Indy, Frank McGurk was a die-hard Chevy hot rodder at the dry lakes. Here, Frank accepts a small track trophy. In 1948, he began manufacturing superior speed equipment for Chevrolet and GMC. By the mid-1950s, Frank had built more hot modified Chevy Six engines for more winners than any other builder. (Circa 1930s)

CLAY SMITH
Mr. Horsepower

It is quite a paradox of motorsports that many great mechanics and designers go unheralded simply because they don't crave national publicity. Clay Smith, the wrench-turning legend, fits that category. Smokey Yunick once said, "Clay Smith was the world's smartest mechanic, but competitors probably used more expletive names for him when they were trounced by Clay Smith-prepped cars."

As the consummate All-American backyard mechanic, the redheaded Clay Smith's fame came first as an engine builder and then as a camshaft designer. Clay's reputation arose from the rough and tumble of midgets, dirt track sprint cars, and track roadsters to record-breaking Bonneville, stocks, and Indy cars.

Mechanical wizardry came naturally for Clay, who relied on a gutsy "sixth sense" to solve complex problems. It was said that Clay could tell how well an engine was running by pressing his loins against the body of the car.

As early as the late 1930s, Clay was literally hand-grinding camshafts—sometimes fine-tuning each cam lobe differently to maximize the volumetric efficiency of a particular engine.

Known by his peers as "America's Genius Mechanic," Clay Smith modestly accepts a prestigious achievement award. (Circa 1950)

The comical "Mr. Horsepower" woodpecker decal was originally drawn as a characterization of Clay Smith as a redheaded, cigar-smoking fun-loving person.

The Founding Forties

In the late 1940s, Clay set up shop in the Los Angeles area. Although he specialized in camshafts, his racing engines soon were setting world records.

Fleeting national prominence came to Clay when he teamed with Bill Stroppe to run Art Hall's winning hydroplane in a 1947 regatta. Their high-revving Ford inline six had won, despite factory engineers saying that it would be impossible. (Clay's correction of the oil starvation and vibration problems of the engine saved it from extinction.) Afterward, Clay and Bill Stroppe were hired on the spot to do special project work for Ford Motor Company.

Their next triumph came when the Mercury they "fine-tuned" won the 1950 Mobil Gas Economy Run. The partners again made history by putting together teams of Lincolns that dominated the large stock class until the Pan-American races ended in 1954.

During his short lifetime, Clay also wrenched for dozens of top Indianapolis entries and was chief mechanic for Troy Ruttman's 1952 Indy winner.

Who knows what else Clay Smith could have achieved if his life had not been cut short, at the brink of greatness, by an out-of-control race car that struck him at the DuQuoin (Illinois) speedway pits in 1954?

George "Honker" Striegel, who now heads Clay Smith Engineering, stands next to one of Clay's first cam grinders. In the 1960s, Honker set numerous drag records with his A-Gas Henry J.

Don Francisco

While Don Francisco happily tuned his classic car, he reminisced about when he had set records on the salt with the Burke-Francisco belly tank lakester in 1949. Don was the first technical editor of *Hot Rod Magazine*. Other early HRM staffers such as Racer Brown, Ray Brock, Tex Smith, Bill Burke, and Jim McFarland contributed to the growth of hot rodding as builders and/or drivers. Jim McFarland continues to serve as an industry consultant, as does Gray Baskerville, "Good Ol' Dad," the senior editor of *Hot Rod Magazine*.

ROBERT "PETE" PETERSEN
Big-Time Rodding Publisher

Probably the biggest single contributor to furthering the hot rodding sport has been Robert "Pete" Petersen, who founded *Hot Rod Magazine* in 1948. The cover of the first issue boldly boasted the "world's most complete hot rod coverage." It was a wild promise that the fledgling magazine had a hard time keeping because of its scant circulation limited to the Los Angeles area. However, by the late 1960s, the magazine exceeded a million copies and was one of the most popular automotive magazines in the world.

Robert Petersen was born in 1926 to immigrant Scandinavian parents. During high school, he began showing an entrepreneurial bent by trading gas for used tires that he would resell for a profit.

Petersen worked as a free-lance photographer and as a $35-a-week junior publicist for MGM. Studio layoffs forced Petersen and others to hype "Mad Man" Muntz, a used car dealer who later became known for his Muntz Jet sports car.

After promoting Los Angeles' first hot rod show in 1948, Petersen saw the need for a hot rod magazine and was determined to produce one. The task was not easy. Thanks to $1,000 worth of credit, he and partner Bob Lindsay had 10,000 copies of *Hot Rod Magazine* printed. Although the initial run was sold out at 25¢ a copy, primarily at hot rod gatherings and even on street corners, the first effort was far from successful because the magazine had no subscribers, no advertisers, and no money.

By the mid-1950s, circulation of *Hot Rod Magazine* had grown to approximately 400,000, with an abundance of paying advertisers who found the magazine effective (as it is today) for introducing new racing products. For early speed equipment makers, it became "our store front to the nation." Petersen admits to the naiveté of early advertisers.

"We sold most ads to small businesses who didn't know any more about guaranteed circulation than we did," said Petersen. "We'd show them the magazine ad, collect for it, and then rush back and pay the printer."

In 1948, Pete relaxes in his first office.

Pete (left) with partner Bob Lindsay (right) worked in this tiny office with only a desk, a typewriter, a telephone, and a lot of imagination.

The Founding Forties

The current publishing empire now produces numerous other automotive magazines, including *Motor Trend*, *Car Craft*, and *Rod & Custom*. In 1960, combined sales exceeded $10 million.

The Petersen Automotive Museum, a $40 million world-class showcase of auto history, was started by Petersen and his wife with their contribution of $15 million. The museum features famous and record-making race cars, hot rods, custom cars, and exotic dream cars. It is a wonderful tribute to the automobile and its technology, and to Robert Petersen himself, a deserving hot rod pioneer.

The first cover of *Hot Rod Magazine* (January 1948) had a photo of Eddie Hulse, who had barely beaten Randy Shinns' Class C roadster record of 129.40 mph.

Pete proudly poses in 1968 with 20 years of issues of *Hot Rod Magazine*.

Pete is shown taking pictures at the dry lakes in 1948.

PAUL AND CARL SCHIEFER
The Famous Father-and-Son Team

Paul's little Texaco station became a speed shop, which is where it all began.

Early West Coast hot rodding owes much to enterprising men who turned a racing hobby into a lucrative business by making products that bettered the emerging sport. As a true pioneer of the performance industry, Paul Schiefer, a founding member of SEMA, acquired fame for the development and production of safer lightweight flywheel and clutch assemblies for racing.

After serving in the Navy during World War II, Paul resumed his quest for speed with a flathead Ford V8 powered "T" that became his "rolling laboratory." His experiments eventually produced the first ribbed-type, lightweight cast aluminum flywheel that would not warp or distort under severe operating conditions. As engine horsepower increased, so did the danger of an exploding flywheel. Therefore, Paul developed a process whereby the entire contact area on the aluminum flywheel and pressure plate was coated with a patented copper/steel spray to provide the ultimate in coefficient of friction and wear resistance.

Next developed was the first bonded bronze clutch facing and a new forged aluminum flywheel called the "Albro." In 1956, Bud Bragdon joined Schiefer to mass-produce the first guaranteed blowup-proof clutches. These became "must-have" items for serious drag racers.

As Hemi Chryslers and 265/283 Chevys came into top demand with their higher horsepower, the bottom had fallen out of the flathead market, and a Harman & Collins and Schiefer merger was imminent. Together they were able to put cams, magnetos, clutches, and flywheels all under one umbrella. Top teams were soon using Schiefer/Harman & Collins equipment.

Paul Schiefer holds the first forged racing flywheel that was guaranteed explosion-proof. (Circa 1960)

This test car was one of four race-prepared vehicles used to evaluate Schiefer products.

The Founding Forties

Pit crew prepares Big Daddy Garlits' dragster at the Indy Nationals. Carl (front) began driving the Top Fueler in 1965. Don Garlits is on the left.

Records fell with Schiefer-sponsored cars driven by Carl Schiefer, Roland Leong, and Don Prudhomme.

Behind the wheel of the Schiefer and Hamilton Gas Dragster, Paul's son Carl clinched the coveted 1964 AHRA World Championship. In 1965, when Don Garlits was recuperating from a back injury, Carl drove "Big Daddy's" Top Fueler. During this time, Carl also headed Schiefer Manufacturing marketing operations, including advertising, sales, OEM, wholesale, and sponsorship programs. His early hands-on experience was the bedrock for his provocative sixth sense in proficient advertising and promotion.

In the early 1970s, Schiefer was sold to the Hurst Corporation, and Carl formed Schiefer Agency, representing baseball, basketball, and football celebrities. In time, he began operating Schiefer Media Marketing, producer of television productions in the motorsports field.

Although Paul, a 1969 SEMA Hall of Fame recipient, has gone on to a better life, his son Carl continues to carry the great strength and perseverance of his father. Paul Schiefer's heart was rather similar to his powerful Rev-Lock clutch—always there for you. Like his products, Paul was tough and steadfast. His son, to this day, proudly lives out the Schiefer name all across the global network of television.

Carl Schiefer drove TV Tommy Ivo's two-engine gas dragster in 1964.

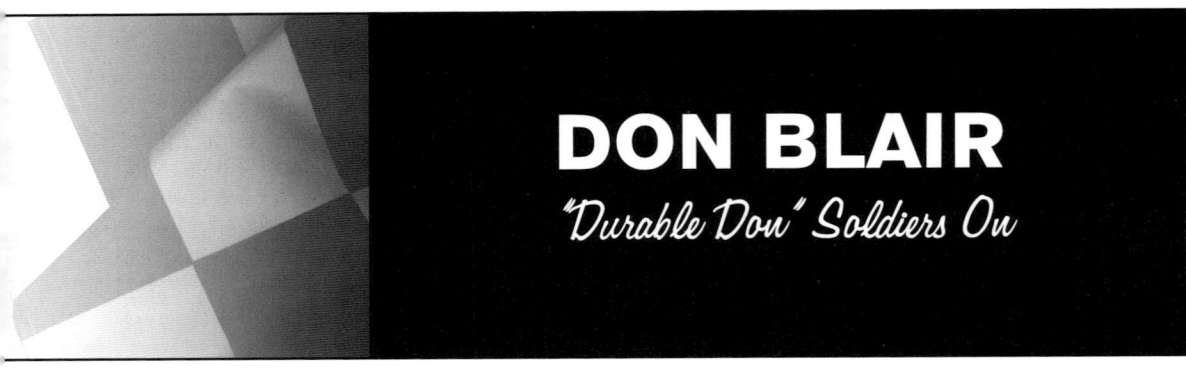

DON BLAIR
"Durable Don" Soldiers On

If you ask an old-time rodder such as Don Blair an inane question such as, "What first caught your interest in cars?," you would probably receive a curt but frank reply such as, "To make them go faster, Stupid!"

Similar to other renaissance men, Don learned speed tricks the noisy way. As a dry lakes participant, Don had one of the fastest roadsters after World War II. Then, as a member of the Pasadena Roadster Club, Don remained active in both early time trials and reliability races. Later, Don ran one of the first blown Chevy funny cars to many match race wins. In addition to sprint car racing, Don won the California Roadster Association (CRA) championship with 1974 Chevy power.

Reflecting on his half century of hands-on experience as a driver and mechanic, Don said, "Over the years, racing has contributed greatly to the change and improvement of the average automobile in terms of efficiency, performance, durability, and safer handling—and it will continue to do so."

Blair's Speed Shop, established in 1945, became one of the most popular hot rod emporiums in California. In 1974, Don sold the business, but later started Blair's High Performance, which specialized in engine building. This may bring Don the record for being the world's longest-practicing speed merchant.

Driver Don Blair makes his way to a quick 141-mph run after setting a two-way 130.27-mph average dry lakes record in 1945.

The Founding Forties

Blair's Speed Shop, started in 1945, was well known in the Pasadena area for new and used speed equipment.

Don Blair waits his turn at an SCTA meet on a cold desert morning in 1946 at El Mirage.

Two carburetors atop an old Roots-type Mercedes Benz supercharger fed Don Blair's flathead Ford V8 modified, making it one of the fastest in its post-World War II era.

FRED C. OFFENHAUSER
A Credit to His Namesake

Born with the greatest name in racing was probably a mixed blessing for young Fred C. Offenhauser, from whom much would be expected. "Like his uncle who made the famous Offy engine, my father, Fred, was driven to excellence," says Tay Offenhauser. "It was in his blood."

Before graduating from high school in 1935, Fred worked in his uncle's Los Angeles plant, where he was groomed in racing engine technology and the promise that the business would be his someday. The tall, spirited youngster soon discovered that the brilliant elder Offenhauser was a tough Prussian taskmaster who earlier had been a one-time associate of the great Harry Miller.

Although the Offy engine remained the preferred engine of the racing fraternity in the 1940s, the senior Fred Offenhauser surprised everyone by selling out to Meyer and Drake. Deeply disappointed, young Fred left the Offy enterprise to pursue other mechanical interests, which included souping up his rumble-seated Ford street racer.

In 1947, after service in the U.S. Navy during World War II, Fred partnered with Fran Hernandez and started a small localized speed equipment business, eventually making high-compression heads and dual intake manifolds for flathead V8 Fords and Mercurys. After Fran went to work for Ford, Fred moved the fledgling company in the early 1950s to its present location in Los Angeles. While making many nationwide sales trips, Fred tuned into the varied needs of competition-minded car buffs.

By 1955, the manifold line was expanded to include the new OHV V8 (and later 4 and V6) engines. Fred, his brother Carl, and well-known engine builder Ollie Morris hit on a new idea for more efficient 360-degree intake manifolding. The result was a full line of quad and dual port 360-degree manifolds that became an industry breakthrough.

After Fred died, the Offenhauser family name went full circle. The company again is producing flathead Ford and Mercury V8 equipment. However, this time, it is for the nostalgia and classic car marketplace.

Fred Offenhauser (right) and Ollie Morris (left) dyno test a Chevy V8 engine with a monstrous experimental induction system. This was a prelude to high-rise manifolds introduced in the 1960s.

The Founding Forties

Foundry patterns for producing an aluminum intake manifold require precision molds, core boxes, and match plates. The tooling shown here is for an Offenhauser Buick quad.

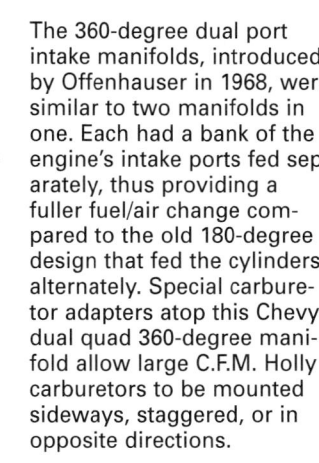

The 360-degree dual port intake manifolds, introduced by Offenhauser in 1968, were similar to two manifolds in one. Each had a bank of the engine's intake ports fed separately, thus providing a fuller fuel/air change compared to the old 180-degree design that fed the cylinders alternately. Special carburetor adapters atop this Chevy dual quad 360-degree manifold allow large C.F.M. Holly carburetors to be mounted sideways, staggered, or in opposite directions.

Jackie Cooper

Former child motion-picture star Jackie Cooper was all smiles as he tooled his mildly customized pre-war Lincoln Continental around the studio lot. Some say its handmade hubcaps with chromed crossbar started the "flipper" trend.

The Birth of an Industry: Zero to $20 Billion a Year

The dynamic growth of the hot rod motorsports hobby spawned an exploding specialty equipment industry that grew from practically nothing in the mid-1940s to the megabillion-dollar giant of today. This is more than a parallel story about fast cars and hot trends; it is the story of American ingenuity challenging and often leading the automotive world in performance technology.

At first, hardscrabble hot rod innovators developed crude parts to fill their own personal agendas. Then, as the sport became more popular, those who saw a marketing opportunity began making parts in enhanced multiples, often in backyards and garages.

Overcoming a peddler's nightmare was the biggest challenge—how to attract distant customers or even get products into their "itchy" hands. At that time, the only effective advertising avenues were men's publications that carried classified ads for soliciting mail orders. Even when car magazines such as *Speed Age* (1947) and *Hot Rod Magazine* (1948) finally appeared, their circulation was pitifully scanty.

Even communication was a problem. Shipping by mail, rail, or truck seemed slower than Pony Express, with coast-to-coast delivery of a product often taking a month to reach the customer. Furthermore, no distribution system had been established. Unfortunately, traditional auto parts houses refused to stock hot rod items because they considered hot rodding a passing fad. No credit system existed, and thus early "mom-and-pop" specialty shops had to order direct, either on a prepaid or C.O.D. basis.

Beginning in 1946, mail-order catalogs such as these made Almquist Engineering the largest supplier of speed equipment and customizing accessories in the world. Later, an associate company, Sparkomatic Corporation, took over the manufacture of all Almquist products, with Almquist as a co-owner. Today, the once-free catalogs sell for $5 to $50 among avid collectors.

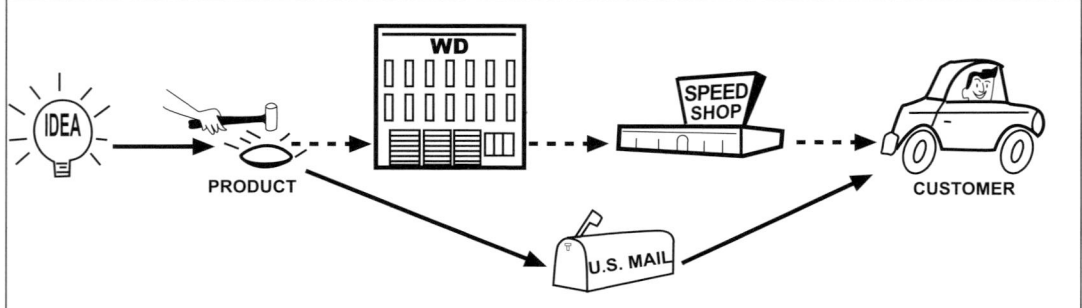

The traditional aftermarket distribution chain shows how early specialty equipment makers sold direct to the customer via mail order, bypassing the manufacturer distributor (WD) and speed shop.

The Founding Forties

As the sport matured, so did the fledgling supply-and-demand ratio for the industry. The bustling market for speed equipment and custom accessories forced more efficient manufacturing and a sophisticated one- and two-step distribution system, often with manufacturer distributors feeding local speed shops.

Even throughout the many changes in the distribution chain, one constant remained. That constant was mail-order marketing, which initially was the most rapid way to introduce a new product. Many would-be speed parts makers parlayed tiny ads in *Hot Rod Magazine*, and a booming business was born.

What was once a core speed equipment aftermarket now has become a polarized and segmented specialty equipment universe, split into hundreds of businesses catering to diverse and often "niche" markets that include street rods, muscle cars, traditional hot rods, low-riders, off-road vehicles, sport utility vehicles, custom cars, vans, and trucks. Today, almost half a million Americans race annually in all forms of organized competition events, including drag racing, stock car, open wheel, sports car, autocross, and off-road. Many other millions of motorists add components to make their cars and trucks more powerful, more efficient, more beautiful, and more fun to drive.

What is next? New vehicle evolution, technological breakthroughs, computer-aided manufacturing, and marketing, coupled with consumer demand and desire, will continue to define and reshape the automotive specialty aftermarket. Will more passing fads develop, such as conversion vans, while street rods continue to grow? Is the next trend (after import cars, pickups, and sport utility vehicles) looming around the corner? Only time and perhaps a crystal ball will tell.

In the early 1950s, the little Almquist "Traveling Speed Shop" was trailered to area tracks where Almquist-equipped springs and drag and stock cars raced regularly.

FREE catalog—Speed, power, custom equipment, from the East's only manufacturer, Almquist Automotive Engineering, Milford, PA.

This tiny classified ad in 1946 started a nationwide business. A $10 ad in *Popular Mechanics* could gross as much as $500 in mail orders. No automotive magazines existed at that time.

BOB ESTES
He Has Seen It All

Bob Estes (left) speaks to Bensen Ford prior to the 1950 Indy race. Famed mechanic Judd Phillips was chief mechanic. Estes' car, powered by an Ardun-equipped Mercury engine, qualified at 124.17 mph—then the fastest stock engine on the speedway.

Few men in America have been involved in so many facets of motorsports competition and with as much style as has Robert Estes. Bob has seen it all—from hot rodding to big-time Indy 500 racing.

While Bob was in high school, he hopped up a beat-up Model T Ford roadster with a Fronty OHV head that sped to 111 mph at Muroc Dry Lake, a record run in 1930.

"Before I could shave, I worked as a grease monkey and a used car peddler," said Bob. "I had gasoline in my veins."

After World War II, Bob acquired a new-car dealership in Inglewood, California, and built a Mercury stock car, in which Bill Taylor won the Pacific Coast Stockton title in 1948. Estes-owned Offys took Midwest AAA and USAC sprint crowns in the 1950s, and Estes cars made six top-ten finishes at Indy.

Bob tried for the Class C World's Land Speed Record in 1952 with a sleek streamliner named the Bob Estes Special. With driver George Hill, it captured the International Class C record run of 229.77 mph at Bonneville.

"In 1951, I hired mileage-wizard Les Viland to prep and to drive a stock '51 Lincoln for an overall Economy Sweepstakes win," said Bob. "The following year, with driver Bill Stroppe, we won the title in a stock Mercury at 24.409 mpg."

Bob Estes competed in four of the Mexican Road Races and took second place in 1952, while Johnny Mantz set a speed record of 115 mph in a Lincoln.

Octogenarian Bob Estes remains a classic car collector, still revved up and on the go as one of "Motorsports' Elder Sportsmen."

Bob Estes (left), Judd Phillips (center), and driver Joe James show off their 1950 Mercury-engine Indy car.

The Founding Forties

Luckily, Don Freeland escaped without serious injury after crashing this earlier Estes Special. Note fuel gushing from the overturned car.

Don Freeland's best showing in three tries at the Indianapolis Speedway was in 1954 when he finished seventh in the Estes Special. A year later, he led the race for 460 miles, but the transmission gave out.

In 1930, Bob Estes exceeded 100 mph at the lakes with this 1925 Model T equipped with a Frontenac OHV and high-speed gearing.

Jack Kelly

Long-time hot rodder Jack Kelly built this dry lakes racer. Before its top was removed, Jack's beautiful 1934 Ford was a *Hot Rod Magazine* cover car in 1959. Recently rebuilt as a roadster for the XF/GMR class, the racer has topped 149.791 mph for a record at El Mirage. The streamlined nose, hood, and belly pan are fiberglass, and the engine is a 1948 Ford with Kong 9:1 aluminum heads. (For SCTA competition in the XF/GMR class, the "XF" means that the engine can be any production Ford or Mercury flathead V8 up to 325 c.i. The "GMR" identifies the car as running on gasoline as a modified roadster.

This belly tank lakes racer is under construction. The fiberglass body made by Jack Kelly was constructed in two pieces and is a reproduction of the original "Scotty's Muffler" tanker of the 1940s.

LES VILAND
America's Fuel Economy Champion

It remains a deep, dark mystery why Iowa-born Les Viland, the top gas mileage guru in the nation, was retired by American Motors so soon after America had suffered its worst fuel shortage.

"Had Les Viland not been forced out [from American Motors] in 1975, his gas-saving savvy might have saved AMC from eventual extinction," stated one Detroit auto writer.

Some time after fun-loving Les bolted a used Frontenac OHV head on his 1926 Ford, Les said, "I beat everyone who'd race me because the old 'Tin Lizzie' would go almost 80."

Now reformed, Les Viland said, "I once was a lead-footed hot rodder, but now I proved to the motoring world that good driving habits save fuel and keep cars—and drivers—safe. When mandatory gasoline rationing began in

Former Michigan governor and AMC president George Romney (right), shown here with Lenore Romney (second from right), and Marjorie Viland (left) and Les Viland (second from left). Romney said, "Les Viland worked for me for over ten years and helped American Motors become Number One in fuel economy in the U.S."

July of 1942, I stretched many more miles out of each gallon of gasoline by observing a dash-mounted vacuum gauge to sharpen my driving technique. Even Indy racers on an allotted amount of fuel have learned that throttling efficiently will maximize their chance of winning."

Amicable Les loved racing, and on the side he worked as a mechanic for Bob Estes, who entered cars in the first Mexican road race. Les also wrenched with Clay Smith at Indy in 1950.

Les tuned and drove an Estes-sponsored stock Lincoln to a record-setting sweepstakes win over all classes of cars in the 1951 Mobil Economy Run. In 1953 while driving a Ford, Les repeated the sweepstakes win over all classes.

1968 marked the sixth time in seven years that American Motors cars won in the famed Mobil Economy Run contest. This Rambler posted the best mileage of all cars in its class, averaging 24 mpg for 2,272 miles from Anaheim, California, to Indianapolis, Indiana. Les Viland is on right; Les Scott is on the left.

The Founding Forties

As an engineer for American Motors, Les helped develop the first fuel-efficient automobile (Rambler). He also drove on to win a dozen class championships in the famous fuel-economy contests run by Mobil. In addition, Les took top MPG wins for seven years at the NASCAR–Union Oil Performance Trials at Daytona Speedway.

Les' technological knowledge provided the first use of high-temperature thermostats (105°F) and the first isothermal production intake manifold that allowed for leaner air-fuel mixture and cleaner emissions. He also invented a gas-sipping carburetor, which exceeded 51 mph in a NASCAR test with a Rambler Six.

In 1976, at age 65, Les began a second career with Detroit Testing Laboratory, researching the use of electrolysis-produced hydrogen as an alternative fuel. Despite encouraging mileage increases of up to 26%, the project was cancelled in 1996. According to a prediction by Les, "Hydrogen still has promise as an energy source for the future."

TV variety-show host Ed Sullivan presents Les Viland with the AAA Trophy for winning the Mobil Economy Run in 1951. Les averaged 25.488 mpg from his heavy Lincoln on the rugged 840-mile course. What was his secret? "I drove smoothly (feather-footing) to maintain manifold vacuum as high as possible. Even in today's [1951] traffic, the average motorist could save one gallon in ten and still reach his destination in time by using his right foot efficiently," Les said.

Benson Ford (right) congratulated 1951 Mobil Economy Run contestants (from left to right) Clay Smith, Bill Stroppe, and Les Viland. Les drove the winning Lincoln and received a new Lincoln as a gift.

BOB AND DICK PIERSON
The Brothers and Their Historic Coupe

As youths, Dick Pierson (right) and his brother Bob (left) show off their hot rod at a Russetta exhibit in 1950.

If longevity is any measure of the basic appeal of an American street rod, then the Pierson brothers' 1934 Ford coupe is the all-time champion. Before it became a famous hot rod, the 1934 Ford coupe had several owners. The last owner probably was glad to get rid of the car by selling it to Dick Pierson for $25.

In converting the rusty coupe into a streetable hot rod, Bob and Dick Pierson accidentally accomplished an enduring profile that was repeatable in other hot rods, even in the Plymouth Prowler 60 years later. The Pierson coupe sports a low, wedge-shaped silhouette with a hood tapering down to a rounded grill.

"With a cheap welding outfit, a hacksaw, and a hammer for rough, cut-and-fit work, we finished our dream project within five weeks without blueprints," Bob Pierson explained. "After spending $200, we ran over 140 mph in 1949 at a Russetta meet. Then, in 1950, we were 'loaded for bear.' We won three out of eleven classes with a top speed of 153.06 mph in '2D.'"

After the Pierson brothers whipped the competition at Russetta, some questioned the accuracy of the clocks.

"Quite possibly the SCTA invited us to run as a guest to prove our clocks were 'screwy,'" said Bob Pierson. "Well, the upshot was that because Russetta's tracks were 1/4 mile long and SCTA's were slightly shorter, we ran 2 mph faster at SCTA."

Later, when the SCTA revised its regulations to allow hardtops to run, the Pierson brothers beat everyone except the So-Cal streamliner. That ended the clock controversy.

In 1951, the Pierson brothers sold the coupe for $600. During its hot rod life, the car has had numerous engines and modifications, including more than 200 holes filled and ground for restoration. According to Bruce Meyer, the last restorer of the car, more than $100,000 has gone into restoring the vehicle to its current show-car beauty. This is a real tribute to the lasting vision of the Pierson brothers.

The famed dry lakes coupe built by Dick and Bob Pierson set a Class D record of 140.32 mph in 1949. A year later, the car turned 153.06 mph. The car features a channeled body, with top chopped five inches, and racy hand-formed nose. Handling was improved with reworked spring rake and 50/50 action Ford shocks. Zephyr gears in the transmission turned 3.27 to 1 gears in the 40 Ford rear end (locked by brazing the spiders together). Tires were 7:50 X 16 Indy skins. Weight of the car was 1700 pounds.

This sleek 1936 Ford coupe owned by Bob Pierson did 148 mph after it was the cover car for *Hot Rod Magazine* in 1948.

The Founding Forties

GENE WINFIELD
Car Customizing: Heaven on Earth

Gene Winfield, a customizing genius, "smoothed" and "leaded" his road to fame with talent, innovation, and one-of-a-kind designs. He produced scores of impeccably restyled and customized cars that won trophies and graced magazine covers worldwide. However, Gene's talents did not stop there, for he went on to create imaginative cars for television shows, commercials, and motion pictures.

In his high school in Modesto, California, Gene made his own speed equipment for his first roadster.

"I welded up tubing to make a dual carburetor manifold with equalizing tubes on a '34 Ford flathead," said Gene. "I even beat out Lee Chapel, who had a slow day with his Ardun overhead streamliner."

After serving in the Navy in World War II, in 1947 Gene converted a chicken house behind his mother's home into his own shop, naming it Windy's Custom Shop.

Gene's busy California shop may not look like much from the outside, but magnificent rods and customs are created inside the unimpressive building. These vehicles include bizarre and exotic cars created for the movie and television industries. A $2-million prototype car was once made here for Chrysler.

Gene (left) and his friends are ready to tow car #113 to a dry lakes meet. By switching bodies, the car could run as either a coupe or a roadster.

"The going rate for custom work when I first started was only $5 per hour, but it is now over $50 per hour," said Gene, whose color-designed 1956 Merc was dubbed the Jade Idol and eventually earmarked him for national recognition. "Initially, I charged $150 to $200 for a good custom paint job. But now a Winfield multi-blend, consisting of six to eight tones, can run $7,000."

When Gene was a consultant for AMT, the model kit company, he built a car for "The Man from U.N.C.L.E." television series and created a shuttlecraft called Galileo 7 for "Star Trek." In "Blade Runner," starring Harrison Ford, Gene's engineering knowledge was used to create 25 futuristic vehicles. Gene's expertise also was used in other television series, such as "Mission Impossible," "Batman," and "Bewitched." One show featured a car called the "Reactor," built totally of aluminum.

Nothing is impossible for Gene. His clean living and independent spirit glow in the limelight, in the same way as his stunning repertoire of vehicles.

The famous Strip Star built by Gene in the early 1960s had a sharp all-aluminum body. It ran more than 140 mph before going on show tours with the International Championship Auto Shows (ICAS).

TED HALIBRAND
Mag Wheels Daddy

The name "quick-change" comes from the ability to change the final gear ratio by merely removing the rear cover plate and switching the straight-cut two-gear set.

At one time, the term "mag wheel" was almost a synonym for Halibrand, who was the first to produce light-weight magnesium wheels for the racing world. The mag wheel was invented in 1946 when Ted Halibrand was looking for a way to give his midget race car more zip. While working as an aircraft mechanic during World War II, Ted's experience with light alloy materials prompted him to make a set of experimental wheels cast from magnesium. They were an immediate success, and Ted formed Halibrand Engineering Company, which eventually dominated the racing wheel industry. For the next 17 years, every Indy car winner proudly "wore" Halibrands.

For more than four decades, Halibrand also has become famous for its quick-change rear end units for Model A and V8 Fords, which reduced the time to change gear ratios from

Ted Halibrand shows his new "mag wheel" with knock-off hubs to Lou Moore, owner of a winning Indy car.

Halibrand's original style of magnesium wheels, shown in this ad, are still marketed today.

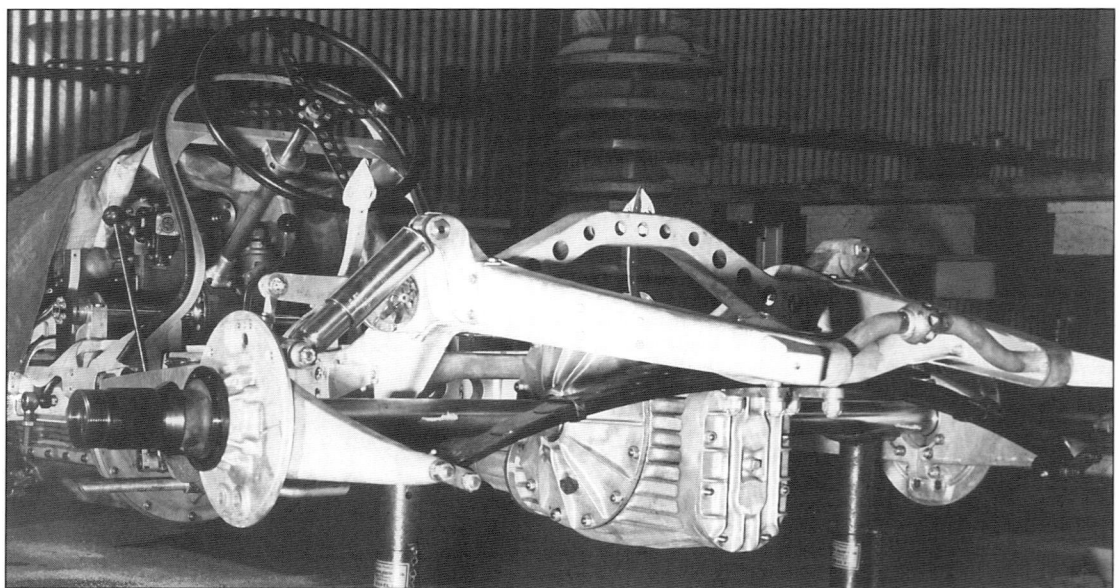

This early dirt track car has a Pat Warren locking rear axle with a five-minute quick change gear box. From the 1930s onward, there were many other makers of gear boxes and associated components, including Stelling, Hough, Green, Roof, and Getz. (Circa 1948)

The Founding Forties

several hours to a few minutes. From 1948 to the 1970s, most hot midgets, sprints, drag cars, Bonneville, and Indy cars ran a Halibrand rear.

Ted sold his company in 1979 and retired. Since then, the ownership of the firm has changed several times and was moved from California to Wellington, Kansas, as a subsidiary of Lamar Corporation. There, the same high-quality Halibrand Q/C and "mags" continue to be produced for competition and for the fast-growing, nostalgic street rod market, despite the demise of Ted Halibrand in 1991.

Introduced in 1964, the "Halibrand Shrike" was a rear-engine Indianapolis car composed largely of magnesium parts weighing only 1140 pounds. The sleek racer was driven successfully by A.J. Foyt, Roger McClusky, and Eddie Sachs (seated).

Gary Cooper

Like today, the rich and famous of the old days loved classy custom cars. Here, Western movie star Gary Cooper admires Coachcraft's customized Lincoln *sans* the car's Carson top. Derham in Rosemont, Pennsylvania, was the first coachbuilder in the United States and once made customs for kings and presidents, including a flashy 21.5-foot Cadillac convertible for the Sheik of Kuwait.

ED "ISKY" ISKENDERIAN
The Legendary Camfather

Isky, a half century later, has more about which to smile.

"They threw away the mold after the Lord made Isky," said a friend of Ed Iskenderian, the congenial pioneer cam grinder who is one of the folk heroes in the industry.

After his middle-class parents lost their vineyard to a severe frost in Tulare, California (where Ed was born in 1921), the family moved to Los Angeles. After graduating from Polytech High School, Ed took a shine to mechanics and subsequently qualified as an apprentice machinist. Later, he borrowed several hundred dollars to open his own shop.

In 1939, after Ed built a T-bodied, V8-powered Ford roadster, he ordered a camshaft from Ed Winfield, the master cam grinder who then had difficulty supplying the demand.

"If racing cams are so hard to get, why don't I grind cams?" asked Ed. "I have some machine shop experience, and I like the hot rod game." Ed had purchased a secondhand cylindrical grinder—to which he built a cam grinding attachment that has since produced more than $1 million worth of cams.

It was a bootstrap beginning, with machining and shipping done in a two-car garage behind Ed's father's house. The first employee was Norris Baronian, who began part time by rough grinding about five camshafts per day, while Ed would finish them at night. As business grew, other talented rodders such as Ted Frye ignited sales.

"We didn't have champs to brag about, so we advertised 'Winners use Isky Cams,'" recalls Ed. "At first, the California boys wouldn't buy my cams because they considered me a fellow racer who didn't know anything. However, the early Eastern stock car racers were open-minded, and they would try any cam. In fact, the NASCAR boys kept it a close-mouthed secret that my cams' broad power curve made passing quicker."

Ed's ads in *Speed Age* and *Hot Rod Magazine* eventually touted both big and small winners in an East-versus-West competitive theme, which subsequently encouraged the Bakersfield, California, Smokers Club to post a $1,500 challenge to Easterner Don Garlits' 180-mph record.

In the 1950s and 1960s, Ed gave his new camshaft profiles trick names such as Polydyne 505 Magnum, five-cycle, and Superleggera 550. At top racing meets, he was one of the first to promote T-shirts made from his own silk-screening outfit. For cheap

Ed Iskenderian was an Army Air Corps B-24 tail gunner and crew chief in World War II.

The Founding Forties

advertising in the early days of drag racing, Ed would lend or give his latest camshaft to a rising star in the sport, which led to contingency awards to drag racing's winning cars that were equipped with Isky cams.

Under a gentleman's agreement, a corporate sponsorship was entered into with Don Garlits, the "Florida Swamp Rat" who was smashing West Coast dragster fuel records. Soon, similar sponsorships with other top names followed.

By 1966, the Isky plant was producing 50 camshafts a day and even more during the racing season. New competitors such as Crane and Crower were expanding to keep pace with the fast-growing hot rod industry. However, as the race for camshaft market supremacy continued, Ed kept coming up with fresh product ideas to maintain his lead.

When the gas shortage in the 1970s occurred, Ed introduced his famous "Mileamore" camshafts with shortened timing that helped improve mpg and low-end torque but sacrificed top rpm.

"Its name was taken from the story of the Mile-or-More bird who ate hot chili peppers and had to fly backwards to cool off his rear end," Ed stated, smiling. "You could hear the bird screaming for a mile or more."

Like most other speed equipment manufacturers, Ed attributes much of his success to the growth of drag racing, which "opened a brand new sport that almost anyone could participate in and enjoy."

"Another thing that helped me grow," recalls Ed, "was the result of hanging around Edelbrock when I was a nobody. Vic, Sr., who took a liking to me, endorsed my cams and began pushing them to his distributors."

This familiar caricature of Isky, "the Legendary Camfather," adds to his image.

Among Isky's firsts are hardfaced cam lobes, hydraulic racing camshafts, anti-pump-up and chilled-iron lifters, and coordinated cam and valve spring assembly kits, which were especially helpful for the beginners. Currently, Isky's catalog consists of 90 pages of camshaft and valve train components.

The Isky operation is now a modern 75,000-square-foot facility on a city-block-long property in Gardena, California. Ed oversees the operation that consists of more than 65 employees who fall under the direct supervision of his sons, Richard and Ron. More than 100 new cam profiles and dozens of new valve spring combinations have been developed there recently for oval track and drag racing (blown and Top Fuel

Ed's first hot rod was this snappy T-body roadster, which he continues to drive today.

Hot Rod Pioneers

classes), thereby maintaining Isky's status as one of the world's largest manufacturers of racing and performance fuel economy camshafts.

Ed Iskenderian, who was president of the Speed Equipment Manufacturers Association during its first two years, has been inducted into the SEMA Hall of Fame and has become a member of Chevrolet's "Legends of Performance," an honorary group of "men whose vision, skill, and perseverance have reshaped the automobile into more than just transportation—and who have elevated motorsports to the high level of prominence it enjoys today."

Ed continues to run the original cam grinder that netted him a fortune.

"TV Tommy Ivo" used Isky cams in many of his cars—including this sleek fuel car.

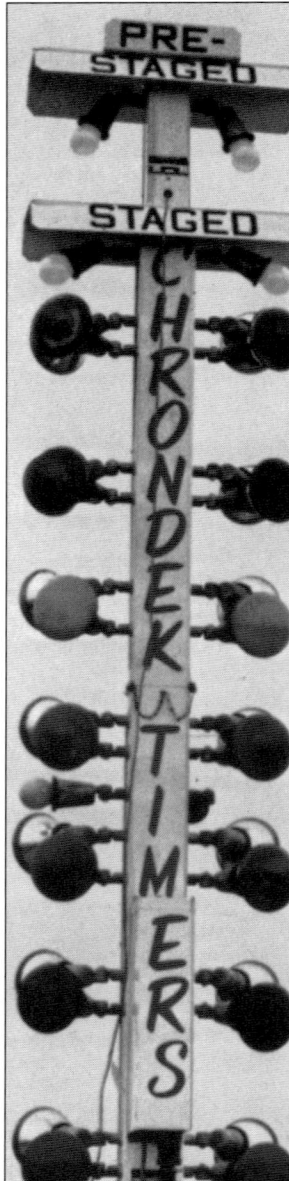

The 'Christmas Tree' Starting Line System

Pre-Staged Indicator Lights: Top amber indicator light warns drivers they are approaching the starting line "staged" position.

Staged Indicator Lights: Signals drivers that they are on the starting line ready to race. A race cannot be started until both competitors are fully staged.

Three-Amber Start System: For "pro-starts," all three amber floodlights flash simultaneously before the green light comes on. Racers in other categories get a countdown of one amber light at a time until the green light flashes. Many nonsanctioned NHRA-event tracks still use a five-amber light countdown for handicap racing.

Green Light: Signals go! The e.t. timer starts after the car leaves the starting line.

Red Light: Signals a foul start and driver in that lane is disqualified.

Christmas Trees and Timing Devices
Keeping Drag Racing Honest

In the sport that now measures success or failure in thousandths of a second, early timing methods were primitive by comparison and even laughable! Sometimes timing was checked on hand-held stop watches, gas station bell-ringing hoses, or tripwires hooked to relays or clocks.

As drag racing became more sophisticated, the concept of separately clocking top speed (mph at the finish line) and elapsed time (from starting line to finish) was firmly established. Often, first-time quarter-mile contestants were surprised to learn that the car with the fastest top speed did not always reach the end of the strip first. (Eventually recordbooks showed both top speed and elapsed time as independent variables, with the latter the criterion used to declare the winner.

Clever San Diego watchmaker J. Otto Crocker responded to the call for more accuracy in early lakes time trials by developing an almost foolproof precision millisecond timing system, consisting of photocell units and light beams that triggered electronic clocks. Otto's next brainchild in the 1940s was a combination timer-recorder that printed times on a continuous tape. Even after Otto became chief timer for the National Speed Trials at Bonneville in 1949, he continued to refine timing equipment for the fledgling sport.

J. Otto Crocker adjusts one of his first timing clocks in the 1940s. The clock had a large sweep-hand for easy reading and was calibrated to 1/100th of a second. Later models had three clocks activated also by lights and photocell pickup units at the start and end of a run. Timing equipment eventually became precise enough to split a second into 1,000 parts. (Courtesy of Bob Ruffi)

"The best way to watch a speeding vehicle is to hold your eyes motionless and turn your head. That way, you see every detail of the run," stated Otto.

By 1954, O.V. "Ollie" Riley, who invented the Chrondek electronic timing system (with accuracy within 0.001 of a second), was producing affordable timing systems for drag strips. Riley's next development, with the help of Lou Bonds, was the "Christmas Tree" countdown starting device (shown here at the far left), which eliminated human flag starters.

After debuting at the 1963 NHRA Nationals, the "Tree" was hailed as a major advance for fair competition; however, a few criticized it for being confusing. Carping drivers, who did not really understand how the starting line system worked, frequently "red-lighted" (fouled).

Because starting time reflexes (reaction times) are vital in drag racing, a sharp, practiced driver could easily get the jump on a less experienced driver and win the match race. This and the explosive brevity of an awesome, tire-smoking acceleration blast down the 1/8- or 1/4-mile strip adds to the drama, excitement, and fun of drag racing.

In 1926, the AAA Contest Board used this electrical timing instrument to record the Auburn Endurance Trials. By 1947, all AAA-sanctioned championship events of less than 10 miles required an automatic timing apparatus registering to within 1/100 of a second, or a split second recording chronometer used to time competitions of variable distances to within 1/5 of a second. At the end of the 1955 season, the AAA dropped from all types of auto racing.

HARRY WEBER
The Likable Premier Cam Man

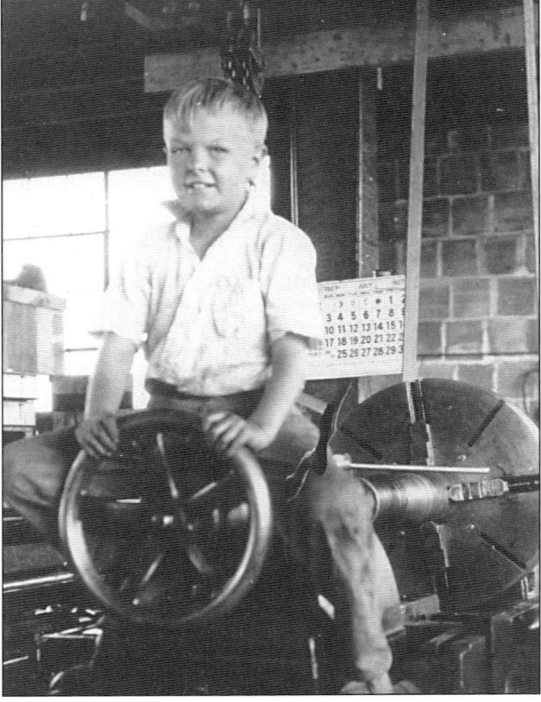

Child labor laws did not exist in 1927 when 10-year-old Harry Weber began working in his father's machine shop. Even at an earlier age, Harry was cutting his teeth on micrometers.

To promote his new Weber products, Harry Weber and his wife Barbara began a "goodwill" tour in the summer of 1950. They crossed the country from California to Milford, Pennsylvania, with their souped-up 1946 Ford coupe loaded with camshafts. The car hit bottom so often that Harry later said, "My fillings were loosened before we had traveled out of California."

A decade earlier, Harry had been racing his hot Deuce roadster on weekends while apprenticing in his father's machine shop. In a small 1,600-square-foot building near Los Angeles in 1946, Harry launched his first Weber Camshaft Company.

"One of my greatest triumphs was converting an old Van Norman crankshaft grinder to a cam grinder—and doing it without any money," recalled Harry. "This was later copied by my peers with many of its features incorporated into present-day, state-of-the-art performance camshaft grinders."

Harry's Weber cam masters were made by a tedious trial-and-error process, which was the custom years ago. Then, flathead Ford and Mercury camshafts were reground to a Semi-Race, 3/4 Race, Full Race, or Super Race profile for a $30 retail price.

Harry Weber checks a new motorcycle racing cam grind.

Harry was a natural at solving problems. After Phil Weiand and Earl Evans presented Harry with their own high-compression cylinder heads, Weber tactfully installed an Evans Head on one side of the V8 block and a Weiand on the other side, to avoid hurting anyone's feelings.

Weber and Almquist collaborated on a much-needed cam grind for aiding heavy modified stock car racers on short Eastern tracks. Each produced an improved cam profile with faster ramp lift and longer duration valve opening. Weber's camshaft was called Weber Super Track #7-4, and Almquist's was called #600 Super Track. Both "short track grinds" were immediate successes. This was

The Founding Forties

the beginning of the early "camshaft advertising war," wherein everyone claimed he had the best and fastest product.

Weber's further product development encompassed lightweight flywheels and heavy-duty, blow-up-proof safety clutches, which became popular sellers.

"We developed the safety clutch after a good friend lost his life when his stock Ford clutch blew up," said Harry. "I felt that a good blow-up-proof clutch would save a lot of lives."

The steady and profitable growth of the Weber business caught the interest of a large Connecticut-based public corporation intent on buying into the burgeoning automotive performance aftermarket.

"To Wall Street investors, the hot rod industry appeared to be a 'sleeping giant,' " said Harry. "As new management was initiated, decisions were often made by individuals who didn't know a carburetor from a cuspidor. When too much emphasis is placed on the 'quarterly bottom line,' any business will lose touch with its lifeblood—the customer's needs and desires."

Despite the fact that the Weber Camshaft Company had tremendous potential, "the mother corporation decided to spit out its stepchild," lamented Harry, who was fired after three months. "The company was practically given away at a board dinner party."

The Weber Camshaft Company was eventually "piecemealed." The clutch portion was sold at a bargain price to a former employee, "Red" Roberts, who now operates under the McLeod name.

Jovial Harry Weber, who had begun to devote his time to his small motorcycle/Weber cam business, caught his last checkered flag when he died on December 24, 1995.

Harry Weber considered it fun to work at his beloved Weber-modified camshaft grinder.

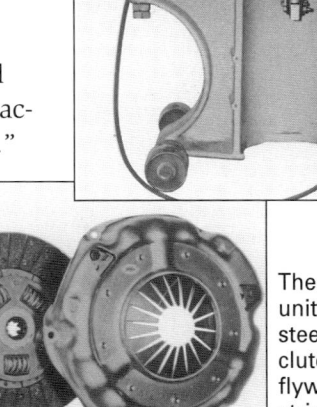

Drag racers were trying a new slipping technique in the mid-1960s. Weber developed a novel clutch "dyno" with a dial indicator to test and calibrate clutch spring pressures because dimensional tolerances often resulted in variation between seemingly identical pressure plates.

The Weber double disc drive unit—with two clutch discs, steel floater plate diaphragm clutch, and a steel billet or cast flywheel—was the hot setup for strip, track, or street in the 1960s, priced under $350.

Early drag racers popularized Weber equipment. Here, Mike Wallace is ready to start a race at the Paradise Mesa drag strip in 1953.

Holy Grounds of Speed

Daytona Beach

Today we would call them the "jet set," but can you imagine two feisty old millionaires quibbling over whose horseless carriage was the fastest? That friendly wager reportedly started the motorist's thirst for speed.

Since then, Florida's Daytona Beach (and nearby Ormond) became known as America's "Birthplace of Speed." In 1905, the world's first "official" drag race debuted on the hard-packed sand when Louis Ross' underdog Stanley Steamer surprisingly stole "top eliminator" by trumping the more powerful, gasoline-engined rivals.

Beginning in 1903 with Ransom E. Olds (of Oldsmobile fame), who was the first to race on the beach in a timed run, more than 80 officially recognized auto speed records were established. The final beach speed trial occurred in 1935 when England's Sir Malcolm Campbell drove his huge Rolls-Royce-powered Bluebird to a record 276 mph.

NASCAR co-founder Bill France, Sr., was a dirt-poor mechanic and stock car racer before he put together the most elaborate and most effective sanctioning body in motorsports. Big Bill (who is six-feet five-inches tall) left a legacy that continues to give superstar status to stock car racing. (Circa 1939)

This photo of Frank Lockhart and his Stutz Black Hawk was taken shortly before he hurtled to his death in a failed attempt to top the world speed record at Daytona Beach in 1928. The sixteen-cylinder streamliner had an estimated potential speed of 240 mph. Previously, Lockhart had set a land speed record of 198 mph for a single-engine car.

Speed Capitals of the World

Eventually, weather and waves eroded the raceway surface of the beach and made land speed record runs too dangerous. Therefore, a new 3.2-mile course was created in Daytona that was part beach and part public highway. The first organized "strictly" stock car race was won there by Ford driver Milt Marion in 1936, signaling another era of speed. When the World War II blackout ended, racing resumed with fervor in 1946. The following year, a sanctioning body was formed by Bill France, along with Bill Tuthill and "Red" Vogt, who proposed the name NASCAR (National Association for Stock Car Automobile Racing). Their consistent set of rules, point system, and guaranteed purses encouraged top drivers and even automakers to participate.

The first winners in 1948 in the NASCAR-sanctioned modified class were Fonty Flock and Red Byron. The latter also won the first Grand National race (today's Winston Cup) over the beach-and-road circuit in 1949. The next year, NASCAR staged its first "Speed Weeks," featuring acceleration tests and speed runs. Once called "a redneck's high holy days," Daytona's "Speed Weeks" event now draws multitudes of hot rodding participants and fans.

In 1959, the totally new Daytona International Speedway hosted the first Daytona 500 race. After three days of review, Lee Petty was declared the winner. In 1960, former moonshine-running veteran Junior Johnson received $20,000 for winning the 500-mile event in a 1959 Chevrolet, reaching almost 150 mph. Today, winners pocket huge amounts of money for a single race and often retire as millionaires.

Today, stock car racing has grown into the largest spectator motorsport in America, and it supports a $2-billion industry.

Daytona International Speedway opened in 1959 and today is a 2.5-mile D-shaped oval track with a 3.56-mile road course. It is one of the fastest racetracks in the world, easily handling speeds of 200 mph and higher on its highly banked turns.

Indianapolis Speedway

If racing were a religion, the Indianapolis Speedway Brickyard would be Mecca. This photo shows the start of the 1946 race. Since then, the place has grown, both in size and in tradition. The triangular-shaped, 2.5-mile track was built in 1909 as a proving ground for automobiles. Today, its original brick surface is covered with asphalt. The Indy 500 program stretches for longer than a month with practice and qualifying, but it ends as the largest single-day sports event in the world—the final act in a dramatic festival of speed. (Tower Photographers, Indianapolis, Indiana)

This photograph shows the first lap of an early Indy 500 race.

These scenes from early Indy races reveal an unusual six-wheel entry and what appears to be unhurried pit activity. Many Indy champs, such as Pete DePaolo and Troy Ruttman, began their careers driving hot rods and roadsters on dirt tracks. (Courtesy of Ray Kuns)

Bonneville Salt Flats

Salt Shakers

Over the years, thousands of amateur speed enthusiasts and their machines have made the annual trek to the desolate wastelands called the Bonneville Salt Flats. In the quest for higher speeds and new records, many have volunteered time and energy to this annual racing event that lasts several weeks. The barren flats first were used as a speed course in 1911 when W.D. Rishell drove his big Packard to a roaring 50 mph. However, it wasn't until the early 1930s that Utah and its foremost racing citizen Ab Jenkins popularized the miles of rock-hard salt beds as a raceway. After breaking numerous world racing records, Jenkins, who later became mayor of Salt Lake City, was named "Utah's First Citizen" for his barnstorming efforts to "sell the salt." Nonetheless, success came slowly.

There, in 1939 and again in 1947, John Cobb of England set sizzling world records of 368.9 and 394.2 mph with the finest racing machine money could buy. In 1949, approximately 200 youthful Americans participated in the first annual National Speed Trials, with home-built cars rarely costing more than $1,000. As the annual pilgrimages continued, speeds and records jumped to greater heights every year, indicating that hot rodders' ingenuity seemed unlimited.

By 1952, seven cars had passed the 200-mph threshold, with the Kenz-Leslie streamliner being the fastest at 244.66 mph. Since then, notables such as Craig Breedlove, Art Arfons, Mickey Thompson, Al Teague, Bob Herda, and the Summers brothers have tested their mettle on the gleaming white salt.

The original course was over a straightaway, with photoelectric timing devices clocking a measured mile. Official records were established by a two-way average within a specific time period. At first, competition included four general categories: Special Construction, Modified, Production, and Vintage. These categories were subdivided further into classes defining engine size, body type, modification, and other criteria.

Although speed and endurance records continue to be broken and set at the Bonneville Salt Flats, natural erosion and the mining of the salt beds continue to diminish the fragile landmark. It is hoped that a "save the salt" project (a salt deposition process that continually renews the surface) will be successful. Otherwise, the world could lose one of its oldest and historically most important motor racing venues.

Bonneville, called the world's fastest speedway, echoes with the mighty roar of powerful engines pushing sleek cars to new speed records every year. These three streamliners were the pinnacle of early rodding art and were among the fastest in their era. Driver Willy Young is in the #777 Kenz-Leslie car (foreground). Alex Xydias and Dean Batchelor drove the So-Cal Special (center). Lee Chappel's streamliner (background) usually was driven by Sonny Rogers. All set records in the early 1950s.

This old sign marked the entrance to the Bonneville Salt Flats by boasting of records set by Ab Jenkins and John Cobb.

In 1937, Ab Jenkins set a new 224-hour distance record by driving his single-motored Mormon Meteor an average of 157.27 mph for a total of 3,774 miles. All three of Jenkins' Meteors were powered with 1200-horsepower Curtiss twelve-cylinder airplane engines that each gulped 3.5 gallons of 86 octane gasoline per mile.

SMOKEY YUNICK
Racing's Miracle Mechanic

Smokey expected perfection from himself and other technicians. Here he painstakingly measures cylinder head volume.

Henry "Smokey" Yunick, who was born in 1923, is the colorful owner of the "Best Damn Garage in Town." In his Daytona Beach shop in Florida, he wears cowboy boots and a ten-gallon hat and peppers the air with cuss words and earthy observations.

"Some long races are so boring," remarks Smokey. "It's like watching two grasshoppers having sex."

Racing fans love Smokey, but competitors fear him because of his dominance in the winners' circles. Over the years, he has built red-hot engines and cars for road racers, drag strip, Indy, and circle track competition. He was a "biggie" for Hudson, Ford, and General Motors in early NASCAR successes.

"When I was in a motorcycle race, the damn old Indian Scout, which I had paid $15 for, began spewing a lot of oil smoke," said Yunick. "The announcer, who could not remember my name, began calling me 'Smokey.'" Yunick later named his son Smokey.

After flying bombers in World War II, Smokey Yunick opened a small garage in Daytona and teamed with stock car racer Marshall Teague to build a stock car for the Hudson Motor Company's driver, Herb Thomas.

"After secretly connecting a line from under the carburetor to a bottle of nitrous oxide under the seat, I instructed the driver to pull up the hidden handle just prior to qualifying and to push it back down after taking the checkered flag. That nitrous oxide really got the S.O.B. going," said Smokey, about the Hornet engine trick. This was a typical example of how Smokey would bend the rules, which ultimately led to many feuds with NASCAR.

Smokey's run-ins with auto makers are epic.

"I've seen many factory mistakes, but Chevy's initial error in computing compression ratio in 1955 really shook me up," said Smokey. "They said 9:1 was the absolute maximum effective compression. It was hard to agree

This sign in front of Smokey's huge Daytona shop brazenly challenges all comers.

The Founding Forties

with that when I had been running 12:1 successfully. Surprisingly, many didn't understand gasoline—other than it smelled bad and burned. Also, they didn't have an accurate flow bench, so they copied my homemader, but it cost them a small fortune."

In 1955, when Smokey was the "answer-man" for *Popular Science* magazine, his cars began beating the nation's top stock cars. The top brass at GM wanted to hire him, but he declined.

"I was more interested in a one-shot deal," said Smokey. "Relocation would interfere with my drinking and social life."

"Fireball" Roberts waits while Smokey (head under hood) fine tunes his storming stocker Pontiac that did 150+ mph at Daytona. (Circa 1950)

Finally General Motors CEO Ed Cole offered Smokey $10,000 to build a Chevrolet stock car racer equipped with a Chevrolet V8 engine.

"After its first big NASCAR record, I got more respect, and Ed Cole made me another offer of $10,000. This time, Ed wanted a 24-hour record of 100 mph plus. We ended up making two cars and set the record. Then, Cole talked me into running a Chevrolet racing facility in Daytona. Everything, at first, was on the QT," said Smokey, who relished being in the unofficial headquarters for high-performance Chevys.

Smokey worked closely with former Indy winners Mauri Rose and Zora Arkus-Duntov to race-ready the first "official" V8 engine in a Corvette. Then, Smokey helped put Buick and Pontiac on racing's top playing field. In a 1961 Daytona race, Smokey's Pontiac, with "Fireball" Roberts behind the wheel, averaged 152.529 mph, then the fastest 500 miles run by any kind of car. After Smokey's association with Chevrolet became strained in 1957, he contracted with Ford Motor Company for $40,000 per year.

Smokey's involvement with Ford resulted in many records and stock car wins for new Ford products.

Smokey also worked with Pete DePaola, an Indy winner and manager of the Ford racing team. The Ford group worked well together until politics pressured car makers to drop out of racing.

Smokey was a chief mechanic for former hot rodder Jim Rathmann, who drove an Offy for the 1960 Indianapolis 500 victory. Later, Smokey entered one of the first fiberglass-bodied racers at Indy. It hit the wall doing 140 mph, but its light weight and durability eventually made it universally acceptable in racing.

"If you develop something good, it takes Detroit forever to utilize it. It's like peeing against a 100-mph wind," recalls Smokey about the initial rejection of his variable-ratio power-steering invention, which was later incorporated in the modern automobile.

Smokey's latest invention, the Safety Wall, promises to reduce race track injuries and absorb impact energy from a speeding race car.

"If one life can be saved, it's worth the small fortune I've already put into the development of the Safety Wall," said Smokey.

Jim McFarland, a popular auto writer, once stated, "Where there's fire, there's Smokey!"

Smokey's cars became top contenders in all forms of racing. His Turbo Chevy Eagle race car proved that the little stock-block Chevy was faster down the straightaway than the most costly turbo-Fords and turbo-Offys. Later, the engine was nicknamed the "Smokenhauser."

FRANK DOMINIANNI
The Road Racing Hot Rodder

In the not-so-distant past, sports car racers and hot rodders were similar to oil and water—they would not mix well together. Sports car road racers, the gentlemen of the sport, wore tweed jackets and caps. Hot rodders, the grit of the earth, wore T-shirts and blue jeans. Sports car drivers in their imported road racers often frowned on the no-nonsense mentality of the hot rodders in their crude homemade cars. Although hot rodders often would beat the highfalutin sports car racers, old-time hot rodder Frank Dominianni tried to amicably improve that negative, competitive rivalry while racing Corvettes in the 1950s.

After serving as a World War II combat infantryman, Frank studied aircraft mechanics. Eventually making speed his specialty, as a driver and builder of cars, Frank opened in 1947 one of the first speed shops in the East. He called his enterprise "Hi-Speed Power Equipment," a firm that continues to operate today on Long Island. The tiny business grew into one of the first warehouse distributors (WDs) in the nation. During the 1950s and 1960s, Frank's speed emporium was noted for quick supercharger installations and engine swaps. Also, Frank helped Willy Frick popularize the Ford-Cadillac and Cad-Allard engine swap combinations.

Hop-up parts shown for a hot street flathead in 1951 included exhaust headers, dual straight-through mufflers, lighter flywheel, high-compression cylinder heads, multiple carburetor manifold, aluminum three-ring pistons, semi-race camshaft, and related valve train. Frank Dominianni points out the beefed-up center main bearing on this Ford 59A block.

When Corvette introduced fuel injectors in its 1955 models, the Rochester system had a design weakness that Frank helped correct. For the cranking signal valve, which often was faulty on 1963 and earlier models, Frank substituted an automatic transmission solenoid.

In 1964, Frank won the SCCA "B Production" National Championships, driving a 1962 Corvette. The experts considered Frank's winning as quite a feat because the solid axle of the car made cornering extremely difficult.

"It's like dirt tracking," said Frank. "There is no way to steer a corner. The trick is to turn the car sideways to slow it down and get it around."

As one of the most feared Corvette road racers during the 1960s, Frank's unique "foot-to-the-metal, stab-and-steer" driving technique helped fill his shop with awards and trophies. Beating Carroll Shelby's Cobras at Bridge Hampton brought personal satisfaction and national accolades to Frank in 1971. Frank's 1962 Corvette crossed the finish line by only half a car length, proving that top driving skills are more important than expensive top-notch machinery.

One of the first speed shops in the East, Frank's Hi-Speed Power Equipment is still going strong.

TOM MEDLEY
Stroker McGurk's Father

Tom Medley, who did the early photo shoots for *Hot Rod Magazine* and blaring Bonneville movies with a sound bite taken from a microphone attached to a tanker, gained fame as the creator of the cartoon character Stroker McGurk in the early years of *Hot Rod Magazine*. Everyone loved Stroker, who was the hot rodders' alter ego because he was always in dire predicaments similar to those of hot rodders trying to wring more speed from their rods.

As editor of *Rod & Custom*, Tom, who had already visited hot rodders' working environments that were often "little hole-in-the-wall shops" and backyard garages, noticed the contagious spirit of hot rodding spreading nationwide when Wally Parks started the NHRA.

"At that time," said Tom, "we began to make the sport a lot more visible."

Mega-brain Tom was born in 1920 in Oregon. By the late 1930s, he had heard about the speed secrets and the lore filtering on the "highway of information" from Seattle to Los Angeles.

"In our minds, there was not a lot of difference between a custom and a rod," said Tom, about the men who made cars their lifeblood. "Many years later, we separated the two building ideas. But I'll tell you one thing. Those early guys were real craftsmen. They did it the hard way, without all the fancy stuff that they have around today."

After serving in World War II, Tom, who was in the Battle of the Bulge, majored in advertising design at the Los Angeles Art Center in Southern California. In his spare time, he would hang out at Don Blair's Speed Shop in Pasadena.

"I started drawing 'toons and putting them up on Blair's bulletin board," explained Tom. "One day, Pete (Robert E. Petersen of Petersen Publishing Company) came into Don's shop peddling *Hot Rod Magazine*, saw my artwork, and liked it."

Tom later concentrated on hyping street rodding for *Rod & Custom*, which helped create the National Street Rod Association (NSRA).

Tom worked for Petersen Publishing for 37 years and retired in 1986 to resume his car hobby. He will go down in history for masterminding hot rodders' mentor, Stroker McGurk, who was "the first cool guy to use a drag chute." Tom Medley, who skipped elementary school on Fridays because he had hitchhiked the preceding night to watch midget racing in Portland, has always been similar to science fiction—bordering on reality and way ahead of the times.

Tom Medley (left) and Rex Burnett (right) at a 1996 reunion of old-timers. A half century ago, Rex began drawing cutaway sketches for *Hot Rod Magazine*.

Tom Medley had the bright idea that parachutes could effectively stop cars on the drag strip. He expressed this concept through this Stroker McGurk cartoon, which appeared in *Hot Rod Magazine*. The idea caught on.

BILL STROPPE
Dubbed "Wild Bill" by Reporters

When the postwar tiny boat "Miss Art Hall" swept the 225 class at Detroit's 1947 Regatta, the top management at Ford Motor Company took note of Bill Stroppe and Clay Smith. Not only did their little boat show up the mighty unlimited hydroplanes, but it also floored the factory big shots when a serious flaw in the Ford 6 engine was corrected. "Miss Art Hall" was a hit, and Stroppe and Smith were heroes.

After driving a Mercury to win the sweepstakes in the 1950 Mobil Gas Economy Run, Bill became team manager for Ford Motor Company entries in the rugged Mexican Road Races, where Lincolns dominated the large stock class. However, it was the 1950 Mobil Gas Economy Run that eventually brought Bill Stroppe to national prominence. While trying to reduce road friction for the run that spanned 751.3 miles from Los Angeles to the Grand Canyon, Bill fitted Art Hall's Mercury with symmetrical Firestone tires containing the same rubber compound as Indy tires. An official AAA observer, Les Viland, an automotive engineer, and Clay rode with driver Bill for the sweepstakes' win, with the best overall gas mileage

Bill Stroppe teamed with hard-driving Parnelli Jones in the "Big Oly" Ford Bronco to win the 1973 Baja-500 road race.

of 26.524 mpg. In 1952, Bill's Mercury won again, averaging 24.409 mpg.

After splitting with Clay Smith, Bill worked quietly for Ford's West Coast team under the name of Bill Stroppe and Associates. His crew prepared Mercury cars for stock car drivers Troy Ruttman, Sam Hanks, Marshall Teague, and Tim and Fonty Flock.

Although Bill and Clay won major races, Bill was almost put out of business when car makers dropped out of racing. However, when the 1963 anti-racing pact ended, Stroppe received

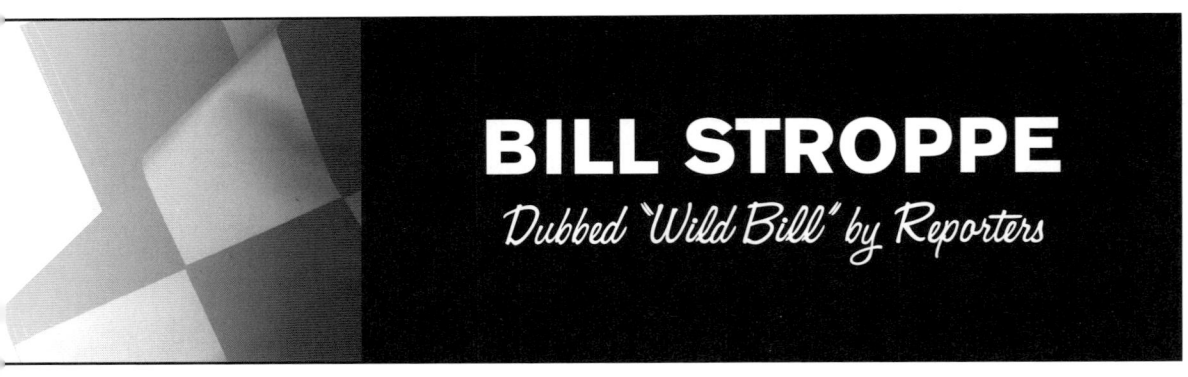

Bill Stroppe, holding a trophy, is being congratulated by sponsor Art Hall for winning the 225-power boat class at the Henry Ford Regatta in 1947.

The Founding Forties

the green light from Ford and continued building stock car racers for the USAC and NASCAR circuits. Top drivers included Roger Ward, Bobby Unser, and Parnelli Jones, who won eight major races including the 1964 USAC Championship in Stroppe's Mercury Marauder.

When Baja-type off-road racing peaked in the 1960s, Ford provided Bill and Parnelli Jones with new Bronco off-road vehicles for desert races. Co-driver Parnelli remembers when Bill's "while-you-drive" urinal released a powerful surprise for the unsuspecting user. Derived from a former battery's acid container, the receptacle worked fine until warm moisture reactivated the dried acid and triggered pungent fumes and subsequent pain for the user.

In the 1970s, as off-road racing hit the big time, Bill's team expanded to include pickup truck drivers Ak Miller, Ray Brock, and Walker Evans. Bill Stroppe died in 1995. During his lifetime, he had teamed with the best and shared his expertise with actor James Garner, musician Ray Coniff, Mexican President Lopez Mateos, and former President Richard Nixon.

The Mermaid was a streamlined 1957 Mercury with a highly modified Stroppe engine. Driver Art Chrisman did 159.91 mph at Daytona Beach in 1957.

TV star Larry Wilcox drove a Stroppe-prepared Mercury during record runs at Bonneville.

An "Exhausting" Business

Despite thundering exhaust sounds heard a mile away, old-time speed freaks would gut their mufflers in a basso attempt to wring more power from their jalopies. To minimize exhaust back pressure even more, early race cars builders shaped headers from steel tubing, often with bends from old metal beds.

Soon after crafty speed shops began custom building dual exhaust systems for street rods, enterprising Sandy Belond was first to aggressively market a Ford V8 dual header system with mufflers ($45 in 1949). By 1954, Belond's bootstrap operation became so successful that it attracted a buyout by OEM "Muffler King" John Gorlick, who soon folded the business because he did not know how to "woo hot rodders." It's ironic that Gorlick once said with a snide chuckle, "I'm going to put all you little hot rod muffler bastards out of business."

Later, Bob Hedman, who had been Belond's partner, became the first mass producer of headers in the United States. When Detroit followed with factory dual exhaust systems, back pressure remained problematic because of contorted cast iron manifolds and restrictive mufflers. One quick remedy was to install a scavenger-type header system that utilized the European principle of "pressure wave" tuning to better evacuate the exhaust. For drag racing, headers made by Hedman, Hooker, Thorley, Stahl, and others showed dramatic increases in horsepower when "tuned" to the engine.

Dual header exhaust systems such as the one shown here (right) eliminate the back pressure of the restrictive stock system by carrying exhaust gases out of two separate systems, thereby boosting horsepower.

Once the fastest stock car in America, this early supercharged eight-cylinder Duesenberg exhibits a Siamesed exhaust header system that was state-of-the-art in the 1930s. In theory, the exhaust flow in one cylinder helped to draw from the next cylinder in sequence. A muffler cutout, located under the running board with a free exhaust pipe, discharged the exhaust near the rear wheel.

The Founding Forties

The evolution of the popular "hot rod muffler" was much less scientific. In the 1930s, a visionary character known as "Old Man" Smith began making a nonrestrictive muffler by stuffing trashed steel shavings between a perforated center tube and an outside casing to allow for free straight-through passage of exhaust gases. The result was the famous steel-pack muffler called "Smithy," which soon became almost a generic name for a throaty exhaust purr.

Later, muffler makers such as Belond, Porter, Douglass, and Almquist used catchy trademarked monikers such as Deeptone, Mello-Tone, Hollywood, and TwinTone. They became dominant players in the fledgling specialty exhaust industry. For many rodders, dual mufflers were hot stuff and offered perhaps a psychological as well as an aesthetic appeal because cars really appeared and sounded fast. Then, a "split" manifold dual kit developed by Almquist was the only inexpensive way to achieve dual exhaust benefits for a four-, six-, or eight-cylinder in-line engine.

In the 1950s, many states legislated against "noisy" exhausts in automobiles, although trucks continued to shatter eardrums. This "unfair mandate" put some manufacturers out of business; however, the Almquist Silencer, a miniature muffler that clamped on the tailpipe, allowed the driver to adjust the exhaust tone from the dashboard for quiet city driving.

Today, for all forms of racing, more and more sanctioning bodies and tracks require competition mufflers that decrease exhaust noise to an acceptable decibel level—usually 95 maximum.

Once popular for dragsters, upward sweeping zoom headers such as the one shown here directed exhaust over the top of the tires, with the intent of blowing away dirt and rubber bits and thereby improving traction.

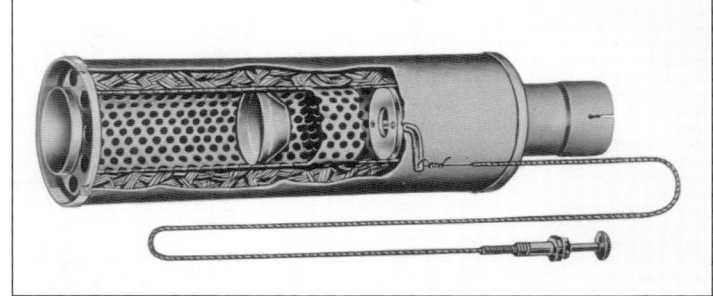

This Almquist Exhaust Silencer with dash control helped "legalize" loud mufflers.

Identification with winning racers has always been an effective way to advertise products. Here, early header maker Sandy Belond sits in the car in which Sam Hanks won the 1957 Indy 500. The following year, Sandy's Belond Equa-Flow Special also won, with Jimmy Bryan driving.

DICK MARTIN
Triumph Over Tragedy

When the open-car craze hit Portland, Oregon, in the late 1940s, Richard Martin was in the thick of it. In a competition T-bodied Model A with a souped-up V8 flathead, Dick raced at Northwest speedways and won almost half the races he entered. Then Jack McGrath and Manuel Ayulo blew off Dick's cherished Portland record. After shortening his roadster, Dick captured the 1948 Oregon Roadster Racing Association championship, eventually taking home several thousand dollars and some trophies.

As a teenager, Dick craftily drafted a letterhead for a bogus speed shop, thus acquiring a dealership for Porter Mufflers and Thickstun manifolds, which made him a "walking speed shop." After Dick acquired other dealerships, including Belond and Edelbrock, Dick's booming business began specializing in exhaust equipment. By 1957, Dick had installation shops in Seattle, Washington, and Portland, Oregon. He had become the largest exhaust systems distributor west of the Mississippi. Later, as a warehouse distributor (WD), he marketed a full line of speed equipment.

At Bonneville for 1952, Dick's 1933 Ford Coupe ran 121 mph for a class record.

"It was a gorgeous red-channeled, three-window coupe that I sold for only $525," reminisced Dick, who was adept in quick engine swaps. "I wish I had it now."

In the mid-1950s, Dick competed in an automatic transmission car and won the Columbia Class A Competition Coupe Championship.

"It was in 1958 on a strip run, after I had just set a new record, that all hell broke loose," said Dick. "The Buick's dynaflow transmission exploded and tore off my right leg. That ended my drag racing career."

After recuperating, Dick began skiing the slopes of Sun Valley with his own personally made outrigger skis. At that time, only six amputee skiers existed in the United States, and the Portland Junior Chamber of Commerce, which ran the ski school at Mt. Hood, took on Dick as an instructor. Dick had designed some special equipment for physically challenged students and for patients at regional hospitals, and he started amputee skiing on a national level. After the national races for amputees commenced in 1964, Dick became a national champion.

"Despite my misfortune, the muffler business made me a millionaire, and I owe a lot to it," remarked Dick, who was inducted into the SEMA Hall of Fame in 1993.

Dick Martin poses with his flathead Ford "speed shop" pickup in 1947.

In the first roadster race held at the Portland Speedway in Oregon, Dick Martin finished fifth and made $150 in 1946.

Adolph F. Braun
Inventor of the Four-Barrel Carburetor

The forerunner of the modern muscle car was the 1941 Buick with a straight-8, overhead-valve 165-hp engine that must have seemed like rocket science compared to the flatheads of the day. Its compound (dual) carburetion setup gave Rochester carburetor engineer Adolph Braun an idea for a resounding carburetor breakthrough.

"Why not let the engine itself control carburetion for better efficiency by simply putting two carburetors in one, with each half having two barrels?" reasoned Braun. "The second half would cut in only when needed. At higher engine speeds, the force of incoming air would progressively open the counterweighted valves, thus delivering fuel more efficiently."

After secretly making his two-in-one carburetor prototype in his home workshop, former shop teacher Braun, who had worked with Buick since 1928, taunted corporate bureaucrats by referring to his efficient invention as "the hot rod carburetor you guys should have made years ago!"

Braun's new four-barrel carburetor was introduced on 1952 Buick Roadmaster models and boosted the horsepower of the big car approximately 10%. Later, it set a new standard for carburetion performance. Among rodders, the "big-mouth" four-barrel carburetor was a quick way to boost power via special adapters on other big engines.

Adolph, nicknamed "Brownie," was a good-hearted, hard-drinking man, who coaxed me to become a member of the Society of Automotive Engineers. After informing him that I had a marine engineer's license but not a college degree, he blurted, "Damn it! Many of you rodders have more ideas and have accomplished more creative things than a dozen of my engineers!"

Brownie retired to Milford, and our friendship grew. However, when he was asked to assist me on my fuel pressure regulator invention, he informed me that his retirement contract with General Motors would not allow him to "work on anything automotive."

When Adolph Braun invented the four-barrel carburetor, he did not realize that he would become the "hot rodders' hero." Here, Braun (right) explains the automatic operation of his double carburetor prototype to fellow engineer Verner Mathews (left).

JOE HUNT
He Sparked Winners

Although much of Joe Hunt's professional career was as a TWA flight engineer, he is most remembered as Indy's "Magneto-Man." Before and after becoming a certified aviation mechanic, Joe zealously indulged in auto racing as fan, driver, crew, and sponsor. With driver Johnny Parsons, Joe ran a Dodge-powered sprint car and took a wild shot at Indy in 1940 but failed to qualify.

After realizing that magnetos are the most reliable way to fire a high-winding racing machine, Joe opened a small sideline magneto shop in 1949 and converted Bendix aviation magnetos to automotive use with new cam drives and adapters. After Johnny Parsons won the 1950 Indy classic with Joe's modified magneto, a switch was made to Vertex Scintilla. Joe's magnetos then became standard ignition on many of the country's top winners for almost two decades—from midgets to drag racers as well as boats and motorcycles. However, Joe's cars didn't fare too well because during 30 years of sponsoring cars, there was only one Pike's Peak hill climb win and only two Indy qualifiers.

In 1981, Joe's son Tommy, a USAC vice president, took over the business. Joe Hunt died from cancer in 1985, but he was a lucky man because he knew that his magnetos fired more winners than any other ignition in history.

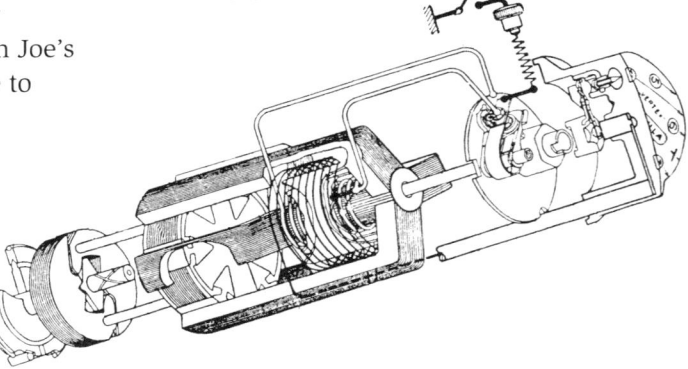

The Vertex magneto lit the fire in racing cars around the world. Its basic design remains unchanged from this 1930s model.

The 1954 Pike's Peak winner, Keith Andrews, is congratulated by Joe Hunt (second from left) and his pit crew.

Joe Hunt (standing) was part of the pit crew for this early big car driven by handsome Johnny Parsons, an Indy 500 winner who touted numerous commercials in the 1950s and 1960s.

ERNIE HASHIM
First Ran A Blown Dragster

In the early days of street and drag racing, when homemade parts and "hope-it-works" experimentation were the norm, Ernie Hashim was a typical rodder who tried some zany ideas—and some even worked. Ernie installed a high-output GMC Roots blower in a flathead Ford V8 coupe, and that screamer clocked a whopping 143 mph in 1948. In an era when dragster speeds higher than 150 mph were considered unattainable, Ernie's Chrysler-powered dragster became one of the first to break that barrier. To prove themselves to all the "doubting Thomases," Mickey Thompson invited Ernie and driver Bill Replogle to the Lions Drag Strip, where Ernie made a record run of 153.91 mph and was awarded five $100 bills by Mickey.

"I was a California native, born in 1924," said Ernie. "I raced a Winfield-equipped Model B roadster and worked during high school for Minor T. 'Fat' DeBolt, who was Eddie Rickenbacker's riding mechanic in the Indy 500 races. He also taught me how to be a machinist."

In those days, bearings had to be "hand-poured" and pistons "machine-fitted" to each bore.

After returning from World War II, Ernie worked for Fat again and raced at the dry lakes. After a two-year stint at Lord's Speed Shop, Ernie opened his own shop in 1954. A few years later, he worked on race tire testing with M & H Tire Company before becoming the West Coast distributor for M & H Tires.

After induction into the International Drag Racing Hall of Fame in 1992, Ernie "drummed-up" a "memory room" at Hashim Automotive in Bakersfield, "where racing buffs can relish in memorabilia."

Racer Brown reported that this Hashim-Waters-Radon Chrysler-powered 1932 Ford high boy was "drunk with horsepower and could be heard for five miles." It did a sizzling 189.27 mph for a new class record at Bonneville in 1955.

The first to use a blower in a dragster, Ernie Hashim (standing) poses with driver Bill Replogle. The car did a record 153.91 mph in 1956. The GMC 4-71 blower is driven off the nose of the crankshaft on the Chrysler 354 c.i. engine.

CLIFF COLLINS AND KENNY HARMAN
Harman & Collins, Inc.

In 1945, a pool of more than 13 million Ford V8s existed, with more added each year for the next eight years. This was a gold-mine opportunity for speed equipment makers such as the Harman & Collins firm that produced Ford and Mercury performance camshafts almost nonstop for a decade.

When partners Clifford Collins and Sidney Kenneth "Kenny" Harman began their joint venture in the back of a filling station, they shoehorned a used cylindrical grinder with a simple cam-grinding attachment. Despite their homemade rig, their master cams produced profiles superior to most racing cams of the day.

On the family farm in Arizona, Cliff Collins' fascination with machinery was tweaked. When his teachers told him he was not smart enough to attend college, Collins "fixed their wagon."

"I shut them up good by graduating as a mechanical engineer," Collins said. "Afterwards, I moved to L.A. in 1936, and wrenched and raced sprint cars before working as an engine designer."

Kenny Harman apprenticed with George Riley, making racing engines and ground cams as a sideline before teaming with Collins. In 1951, Kenny split. During their working relationship, it was a trade secret that Harman & Collins made camshafts for Clay Smith.

When the auto industry switched to overhead valve V8 engines, wear often was a problem with reground camshafts. Harman & Collins got the jump on other cam grinders by hard-facing cam lobes to increase durability. At the close of the 1950s, it was apparent that roller tappet camshafts would be the new standard. However, Harman & Collins' first run of 8,000 rollers was a disaster because of a heat treating mishap that produced severe cracking.

At its peak, Harman & Collins employed 30 people and grossed more than $600,000 per year; however, the company began biting off more than it could chew. After a merger and a new marketing method failed to rekindle sales, Harman & Collins, Inc., became defunct in 1961.

Typical Harman & Collins ad from a 1950 *Hot Rod Magazine*.

In 1950, H&C advertised their totally new "cast billet" camshaft as completely interchangeable with a stock cam. The larger core stock heel diameter with flame-hardened lobes promised longer life and quieter, more positive valve action even with radical grinds. Note comparison with a typical reground cam profile. Special grinds for Ford or Mercury V8s cost $40 to $50.

EDDIE EDMUNDS
The Manifold Man

Eddie Edmunds once said that he enjoyed hobnobbing with the Hollywood movie crowd and being in the fast lane. In his Los Angeles shop, he would enthusiastically install Edmunds' equipment and completely reworked engines into the cars of affluent motorists and movie stars.

Some true-blooded Oregonians—such as Edmunds, who turned from amateur racing to designing speed equipment—claim that hot rodding began in the Northwest and that many speed industry pioneers came from Portland. After failing to find a suitable aluminum foundry, Eddie moved to Los Angeles where, in the late 1940s, he began producing high-compression cylinder heads (mostly 8:1 ratios) and dual intake manifolds for numerous makes of engines. Because Eddie was always in a hurry, he rarely made detailed working drawings for new designs. Instead, he hoped for the best and depended on pattern makers to translate his ideas into wooden molds suitable for casting.

By the early 1950s, Edmunds had the largest product line of manifolds. However, production runs were small, and quality often suffered because the foundry never had time to resolve production flaws. Nonetheless, the exteriors of the castings were always smoothly buffed. Unintended internal roughness from the sand cores probably aided fuel atomization at highway speeds, so customers were satisfied.

Dual intake manifold prices ranged from $40 to $80 for Kaiser, Hudson, Nash, Packard, Studebaker, and most Chrysler, Ford, and GM products.

Eddie lost a lot of Ford and Mercury manifold business to other makers because his compromise design did not

Edmunds' dual intake manifolds and high-compression cylinder heads were cast in aluminum and polished to a sparkling chrome-like luster. The manifold shown here is for a six-cylinder Plymouth, Dodge, and Chrysler.

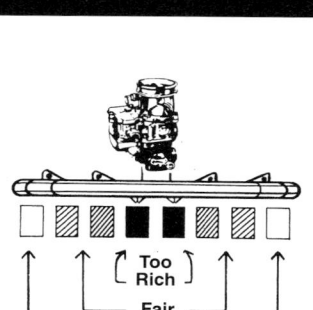 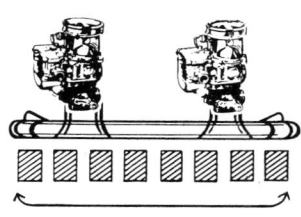

Edmunds advertised that "Dual carburetion provides more uniform fuel distribution to all cylinders."

provide equidistant carburetor spacing necessary for better mixture distribution. Why didn't Edmunds produce a more competitive model? The answer remains a mystery.

Overexpansion and insufficient operating capital plagued Eddie. As a result, product delivery became so poor that his reputation suffered, and dealers discontinued marketing Edmunds custom equipment.

Eventually Aaron Fenton took over sales and production of the Edmunds product line. Within a short time, Fenton dropped the Edmunds products because of incompatibility with his rapidly growing mix of Fenton products. The Edmunds name in the industry came to a close in the late 1950s.

BOB MORTON
Expert Engine Builder

Bob Morton at age 18 is proud of his hot Mercury engine that won the streamliner class at a 1948 Russetta meet in a Tattersfield belly tank. The vehicle clocked 136.57 mph for a new record that year. (Circa 1949)

"Our golden age of roadster racing was 1947 through 1950. We often were the 'test bench' for speed equipment makers who supplied parts for free if we showed promise. Back then, we had no cash," said Robert Morton, who teamed with George Rubio and others to set numerous land speed records.

While working after school for Ansen Automotive, Bob learned the art of engine porting from Jack Andrews and Lou Senter.

"Later on, those Ansen engines would tear up the stock car circuit that was just beginning in the Southeast," recalled Bob, who became an expert engine builder prior to the Korean War.

This faded photo of the first meet in Santa Ana in 1950 shows Morton's 1932 Ford V8 coupe (left) about to win at 118 mph.

Morton and Rubio's Ansen Special took third place at 134 mph in the roadster class at Bonneville in 1949. The same year, George Rubio did 144 mph at a Mojave meet.

Bob's Mercury-powered roadster made history by beating the hottest bikes in the SCTA's first "Motorcycle vs. Hot Rod" contest that was held at an inactive Navy airstrip near Santa Ana in 1950. All four classes of roadsters (including Larry Shinokas' "Chopstick Special" 25-T Mercury) administered a shellacking to the bikes, which included hopped-up Harleys.

George Rubio in his "daily driver" 1929 roadster mounted on a 1932 frame. When stripped for weekend racing with a 1947 Mercury engine, the car did an all-time top speed of 151 mph in the C Roadster Class at the first 1950 meet of the SCTA at El Mirage, with Bob Robinson driving.

ROGER WHIPP
Fanning the Sparks of Motorsports

Like many returning G.I.'s, Roger Whipp was caught in the youthful "hot rodding craze," which the media claimed was producing a country of rambunctious motorheads. Whipp, a co-founder of the Shasta Roadster Club in 1948, remembers when abandoned airport runways served as ideal quarter-mile drag strips for clandestine drag racing.

While working as an apprentice machinist, Roger built and drove circle track and drag racing cars that captured many class records. His "Redhead" Bonneville streamliner, his fastest car in 1966, clocked more than 311 mph.

Over the years, Roger retained his passion for motorsports and fanned that same emotional spark for hundreds of youths at his hometown raceway in Redding, California.

"As one of the first tracks sanctioned by the NHRA, it still is a popular place for kids to let off steam and do it legally," said Roger, a compassionate track organizer and city official.

This 331 V8 Chrysler-powered Kurtis did 124.5 mph at Bonneville in 1953. Shasta Club members wait their turn for a run down the salt.

Jim Johnson (left) and crew prepare Roger Whipp's modified roadster for a 200+ mph run at Bonneville Speed Week 1961. Rex Clark is kneeling; Bob McGrath is watching.

Named the "World's Fastest Hot Rod" in 1966, Roger Whipp's "Redhead" streamliner won the Class B Streamliner record at more than 311 mph that year. Bob Herda's four-wheel-drive B streamliner is shown in the background. The two Chrysler V8-powered cars traded records through the 1960s.

Whipp's rear-engine competition roadster had a destroked Model A for power that was air-cooled to save weight. Its best time in 1958 at Bonneville was 114 mph.

In 1955, the starter waves off two roadsters at California's Redding Drag Strip, one of the oldest continuously operating strips in the nation.

Hot Rod Pioneers

Speed Equipment

What does "speed equipment" bring to mind? More than 35 years ago, *Popular Hot Rodding* magazine said it best with these words:

"Speed Equipment" is a vast, glittering, powerful-looking array of strange and wonderful items that are the heart and soul of hot rodding. With it the hot rodding enthusiast can accept the challenge to explore the unknown and the untried.

Speed Equipment is a magical name. It means many different things to many different people. It is sometimes extended to cover almost every kind of specialized automotive extras. Yet, if the term is applied in its strictest sense, speed equipment is performance-producing equipment.

Before World War II, the limited amount of speed equipment made in America was used for souping up production engines, mainly for racing. However, by 1950, more than one hundred small firms produced speed parts, mostly for Ford and Mercury V8 flatheads. A decade later, the grassroots innovators had created a multimillion-dollar industry, about which the late Griff Borgeson wrote:

Hot rodding would be nowhere today if it wasn't for speed equipment makers who have spent vast amounts of time and money developing new equipment, thus furthering the sport.

Major speed parts for a hot Ford or Mercury flathead in the early 1950s were a dual intake manifold, high-compression head, reground camshaft, dual-ignition kit, and a dual-header exhaust system.

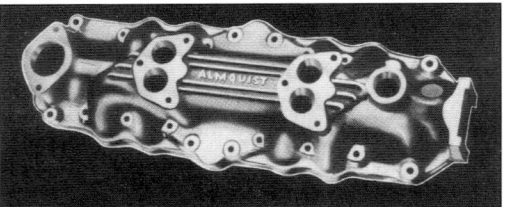

Multiple carburetor manifolds have always been popular because they were a quick way to add horsepower by equalizing fuel distribution to all cylinders. This Almquist dual-intake manifold with 180-degree porting also had a removable oil filter cap, enabling it to fit all flathead V8 Fords and Mercurys from 1932 to 1952.

Many early racing intake manifolds such as this Almquist open-plenum design had short runners that positioned the carburetor venturis directly over the intake ports. The fins were there for cooling and appearance.

The six-cylinder engines also were under-carbureted for high rpm use. Therefore, a dual- or triple-carburetor intake manifold was essential for equalized fuel mixture delivery. The basic hop-up parts for a Chevrolet Six were exhaust headers, straight-through mufflers, camshaft, and "hot" ignition. The engine pictured is a Ford Six with Knudsen triple-carburetor intake manifold, headers, and Spalding ignition.

The "Heart" of Rodding

Then and now, the motor enthusiast rather than the hardcore racer was the typical parts buyer. He was well aware that auto makers' mass-production compromises, which adversely affect performance, could be remedied easily. Factory weak spots usually were the first thing the serious speed tuner considered.

The average gearhead, bent on making a "good car better," usually asked, "What kind of performance, and how much?" However, tastes varied. Some men wanted extra power for safer passing ability, without sacrificing fuel economy; others could not be satisfied unless their machines could beat the competition at local drag strips.

Throughout the years, the basic paths for boosting engine performance have remained surprisingly elementary: induction of fuel, efficiency in utilizing the fuel's "canned" energy, and thorough exhaust scavenging. All this came with a choice of bolt-ons and internal speed parts for improving all venues.

Speed equipment has evolved tremendously with each decade, and recently the words "performance equipment" were substituted because of their sophisticated lingo that sounded much "safer" to the uninitiated public. Although most special-performance products offered today are functionally the same in principle, their technology, metallurgy, and production methods are greatly improved, in spite of the price escalations caused by inflation. For example, a special intake manifold that sold for only $35 to $50 years ago now would cost a motor enthusiast at least tenfold for installation on a late-model car. Pictured here are some popular "go-fast" goodies from the past.

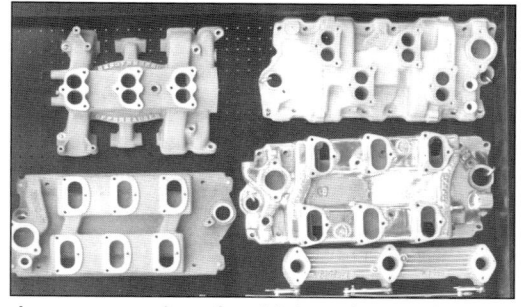

A new generation of multiple-carburetor intake manifolds included a bewildering variety of models for street, dragstrip, or track. The dual-plane, 180-degree manifolds have a multi-level plenum that separates adjacent intake pulses, and they are preferred for street use. The single-plane, 360-degree model feeds all cylinders through a single plenum and is best for top-end performance. Racing manifolds also came in a "log" design, where the carburetors would feed into common chambers close to the intake ports. Ram-type manifolds (not shown) have long runners between carburetors and ports.

Called the "Poor Man's Hop-Up," this Almquist Y-manifold adapted dual carburetion to virtually any six- or eight-cylinder car. The slip-rod linkage could be adjusted so the second carburetor cut in at any desired speed for faster acceleration (similar to a four-barrel carburetor). Its long runners may have produced a "ram tuning" benefit because many who won drag races with this "Mickey Mouse" setup shocked the unbelievers.

Boosting the engine compression ratio is one of the oldest hop-up tricks. Because each point added to the static compression ratio increased horsepower and fuel economy 4% or more, the desired result was accomplished easily on old model engines by either milling the head or installing high-dome pistons or aluminum high-compression cylinder heads. Almquist also supplied special thin copper head gaskets that increased compression significantly on most cars, at a cost of $6.95 to $9.95.

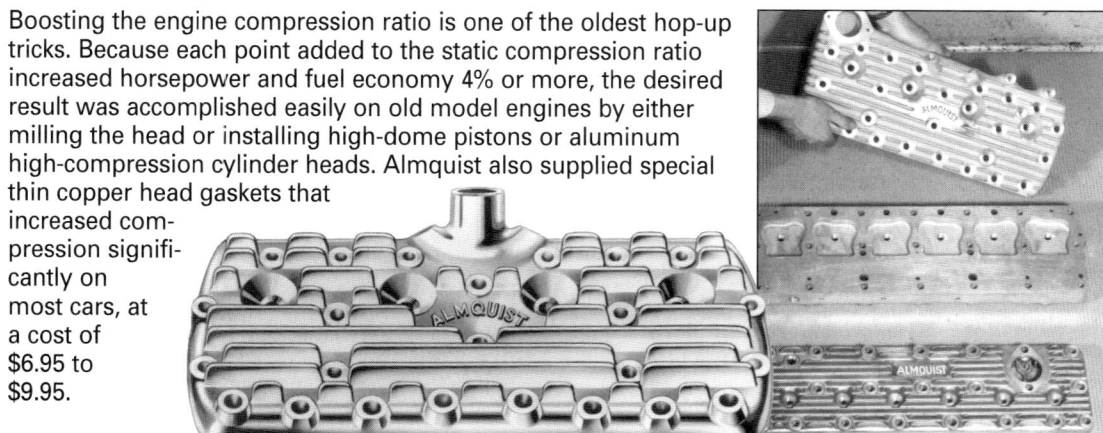

Hot Rod Pioneers

Choice of the appropriate camshaft was always a big speed secret. Because of the advertising "hoopla" common to the camshaft industry, it was and continues to be difficult to separate fact from fiction. We all claimed our camshafts were the "world's fastest," although many of the grinds were similar. Early camshafts that came in four basic grinds—semi-race, race, full-race, and super-race—gave way to "mushroom" and roller-tappet cam kits that worked best when coordinated with other full-race modifications. A slightly higher lift rate and four or five degrees more duration usually worked well for a street machine, whereas the entire valve train had to be "tuned" to the cam lift curve for OHV engines used for competition.

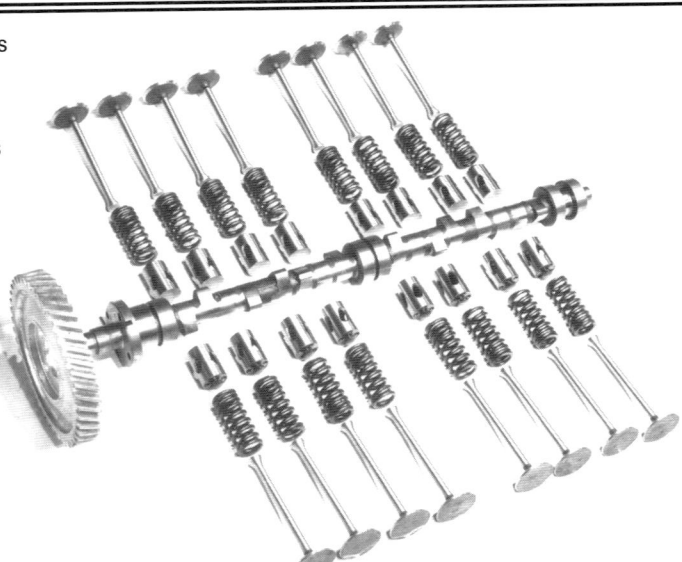

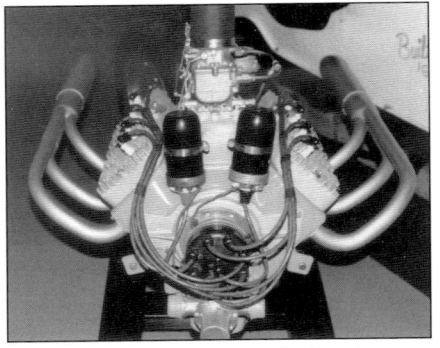

Stock ignitions were marginal, at best, for high performance in the old days. Therefore, Kong, Spalding, Mallory, and others made conversion kits or dual coil distributors to ensure full power at high revs. Magnetos always were the choice of top racers. However, how much "oomph" did hotter ignition really add? The truth was: not much. Without it, though, the engine could not develop the output other modifications were able to give it.

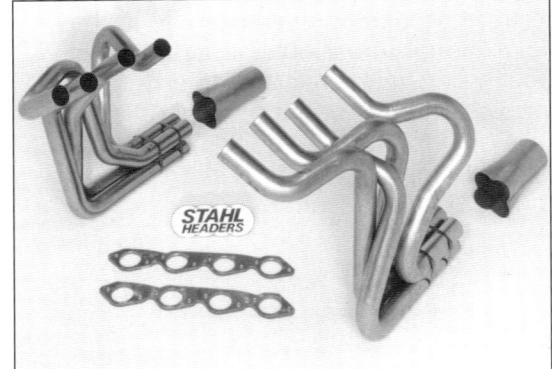

Similar to many other automotive innovations, fuel injection was first used by the racing world. By 1950, Stuart Hilborn's fuel injection system was replacing the carburetor on big-time Indy cars. The Enderle injection system, shown here, delivered 98% nitro in this 1960s-era Ford dragster engine. Although electronic fuel injection ensures more efficient fuel delivery at high speeds, especially in turns, many sanctioning bodies such as NASCAR continue to buck its superior technology.

Dual exhausts and "tuned" headers to reduce back pressure and provide free breathing always were a top priority, even with early rodders. Popular even today, this spaghetti-like header system developed in the mid-1960s featured equal-length primary pipes—one for each exhaust port—extending into large tapered collectors. The runner configuration and collection size were critical in accomplishing maximum scavenging effect and exhaust flow. One maker provided an adjustable collector that allowed the primary pipes to be lengthened for more torque and low-rpm power. Dyno tests showed dramatic horsepower increases in all cars with significant differences among brands.

Boring and stroking to increase piston displacement, either together or separately, added greatly to low-end torque, especially when accompanied with full-race modifications. Matched extra-strength stroker kits with special pistons, forged rods, billet crankshafts (often increasing stroke up to one inch), and related heavy-duty components were available for racing.

The Founding Forties

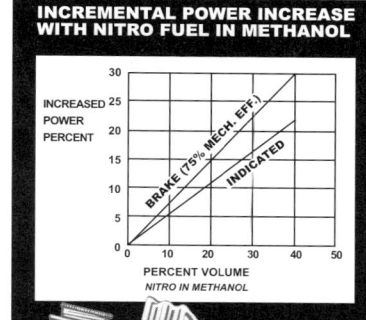

By the mid-1950s, the secret chemical formula was out, and racers who switched from straight alcohol to nitromethane blends claimed the new oxygen-burning "hot" fuel almost doubled engine power. Although dangerous and tricky to use (e.g., detonation and high pressures often blew up engines), top drag machines could use up to 100% nitro because the engine revved full throttle only about four to five seconds at a time. However, gulping as much as a gallon per run made nitro expensive, even for the pros. Later, sophisticated nitrous oxide injection systems hit the drag racing scene, promising even more "bang for the buck" for the "nitro junkies."

As an alternative to other extensive engine modifications, the mighty supercharger by 1960 gave the drag racer all the volumetric efficiency he or she could handle and often the winner's edge. The GMC Roots-type blowers gave both flatheads and OHV engines such as this Ardun conversion a hefty increase in horsepower. The newer Weiand 8-71 blower, similar to the 6-71 pictured here, for big-block Chevy engines delivers a 5 to 8 psi pressure boost and has two-lobe, straight-cut rotors machined to such tight clearances that Teflon sealing ribs are not needed.

By the 1960s, racers were capturing elusive free horsepower by utilizing "sonic-ram tuning," ram tubes and/or high-riser intake manifolds with long passages built up pressure waves and improved low- to mid-range torque.

Floor shifter conversions were America's fastest-selling performance product in the 1960s. Hurst and Almquist were the biggest producers, each with patented designs for positive and faster shifting through the gears. The popular triple pattern shifter shown here was designed by Ed Almquist. It offered a universal mounting kit so that it fit all three-speed manual transmissions.

"A product that Ed Almquist should have patented was a cool air induction kit which he developed and called Air Ram. This concept was borrowed about 10 years later by Pontiac, and they called it Ram Air!" wrote Albert Drake in *Street Rod Action* magazine. Ed's registered trademarked Air-Ram fed the engine more cooler and denser oxygen-laden air from outside the engine bay. Tests showed power increased 1% or more for every 10-degree drop in inlet air temperature. Hood scoops, ram tubes, and nonrestrictive air cleaners also boosted induction efficiency.

JIM NELSON
The Dragmaster

Jim Nelson has been involved in the sport of drag racing for so long that he refers to himself as the "Dinosaur of Drag Racing." To his credit, headstrong Jim has been a race driver, designer, manufacturer, NHRA inspector, and hot rodder. As the first technical director of NHRA, Jim was with the group's Safety Safari from 1953 onward, working with Wally Parks and the car clubs across the nation in a successful effort to move racing off the streets and into organized competition.

The pace-setting Dragmaster chassis developed by Jim Nelson and Dode Martin was the first "production line" chassis to achieve top NHRA honors. The lightweight space-tube design featured a heli-arc welded tubular steel frame, torsion bar front suspension, friction shocks, and a wishbone radius rod. Dode Martin got low e.t. despite the wheelie.

Not to be confused with his namesake "Jazzy" Nelson, Jim became completely hooked on dry lakes racing after turning 134 mph in 1946 in this Ford V8 flathead-engine 1934 Ford pickup.

Jim Nelson (right) and Dode Martin (left) ponder how to wring more horsepower from their Potvin supercharged Chevrolet dragster engine.

The beautiful national champion Dragliner was driven by both Jim Nelson and Dode Martin. It won its C Dragster Class at 123.65 mph in 11.58 seconds at the 1957 NHRA. The following year, Dode Martin drove the streamliner to a repeat victory at 124.30 mph. The tail section and canopy were fiberglass. The Latham axial-flow supercharger gave a 24 psi boost to the carbureted 1955 Chevy V8 engine that peaked at 750 rpm.

In the mid-1950s, after Jim and "Dode" Martin had won many championships with their advanced-design slingshot "Dragliner," they teamed to form the Dragmaster Company in Carlsbad, California. Their successful, custom-built, lightweight dragster frames made with sturdy chrome moly roll cages were run by top racers such as Mickey Thompson and Pete Robinson. What began as a two-man shop producing rather simple cut-and-fit slingshot chassis designs quickly developed into engineering masterpieces that became pacesetters in safe dragster design.

Their Chevrolet-powered Dragmaster racers won several awards, including "Best and Safest Engineered Car" and the 1957–1960 NHRA Nationals C Dragster titles. In the early 1960s, Jim drove a gas dragster to many top speed and e.t. records. However, his career as a driver almost ended when he

The Founding Forties

Jim Nelson drove this Dragmaster Dart AA/D gas dragster to Top Eliminator honors at the 1962 Winternationals. The big 430 c.i. of the supercharged Dodge engine pushed the car to 170 mph in 8.72 seconds in the quarter mile. The restored car is now in the NHRA Museum in Pomona.

Jim Nelson stands beside the restored Dragmaster Mooneyes car at a 1996 anniversary event.

Brass timing plates, such as this one, were given to qualified racers by the SCTA.

ran off the end of C.J. Hart's Santa Ana drag strip and broke his back. C.J. forgot to warn Jim about the ditch.

Despite discontinuing the manufacture of dragster chassis in 1965 because of liability issues, the business grew to include a line of street roadsters and a complete engine building facility with 28 employees. Jim and Dode Martin, partners for 35 years, built more than 100 dragsters and almost 50 street roadsters.

Over the years, Jim and Dode earned approximately 400 trophies, including the 4-foot-tall NHRA Top Eliminator. Retired at age 70, Jim Nelson still loves to compete today as a mighty marathon runner. Occasionally, he drives the Mooneyes-restored 1959 Dragmaster dragster on exhibition runs at the NHRA hot rod reunions and Super Chevy races.

Dragmaster cars such as this example were used in films during the early 1960s. Here, actor James Darren of the movie "The Lively Set" is seated in this Dragmaster car. Actor Doug McClure (standing) also starred in the film. (Universal Pictures)

BRUCE CROWER
He Turned a Kit into a Fortune

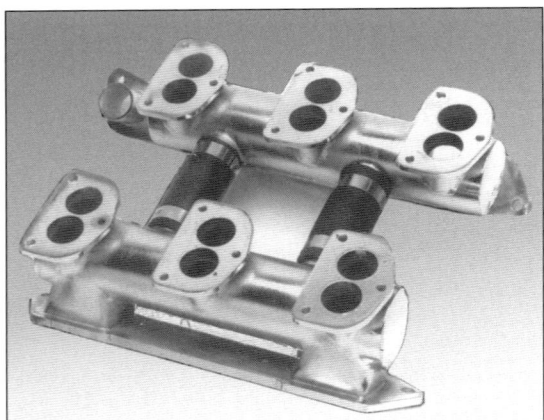

The "U-Fab" do-it-yourself [weld-it-yourself] manifold designed to hold four, six, or eight Stromberg carburetors started the multimillon-dollar enterprise.

Bruce Crower is a dreamer who became a doer. His initial tiny operation evolved into a multimillion-dollar manufacturing enterprise, producing a variety of go-fast equipment for street, strip, and circle track auto racing.

After building and racing flathead Fords and Mercurys, Bruce "shook them up" at Bonneville in the early 1950s when he made a 157-mph record run in an old Hemi-powered Hudson.

Bruce is credited with being the first to top-mount a GMC blower. He had cobbled up his own intake manifold and a pulley system cast in coffee cans by using metal from old pistons. The homemade intake/blower combination worked so well that it put Bruce in business.

With his last $300, Bruce took out a small ad for a do-it-yourself intake manifold kit in *Hot Rod Magazine*. The "U-Fab" kit sold by the thousands and was such an overnight success that Bruce looked for another marketplace void. This time, it was a double disc clutch, similar to one found in a Fiat. The result was a "Crowerglide" centrifugal clutch,

Bruce Crower (left) in his first shop near Phoenix, Arizona, in 1949.

Bruce Crower (center), shown here checking spark plugs prior to a run in 1952, was inducted into the Drag Racing Hall of Fame 40 years later.

The Founding Forties

which quickly became the most popular clutch in drag racing and tractor pulling.

In the early 1960s, Bruce saw a weakness in camshaft technology and developed new drag racing cam profiles that began "taking home the bacon." Bruce's breakthrough came when he began building roller bearings into the camshaft to reduce cam bearing friction.

Bruce also innovated an ultralight crankshaft design, which allowed for faster acceleration by reducing the rotating counterweight mass of the crank. Strong "I-beam" billet connecting rods also quickly became an industry standard.

Today, Crower Cams & Equipment Company is headquartered in Chula Vista, California, and employs more than 200 people. The state-of-the-art CAD and CNC facility allows Bruce to roll with the ever-changing demands of the industry.

Bruce set a 157-mph record in this Hemi-powered Hudson at Bonneville in 1954.

This forerunner of the "slipper clutch" is prevalent in racing today. The early "Crowerglide" centrifugal clutch had multiple discs engaged by a normal throw-out bearing through double-acting arms.

After building a downsized 209 CID stock block Chevy that produced more than 850 hp at Indy, Bruce in 1971 designed outside port heads for the 355 Chevy to improve breathing efficiency. Only 13 sets were made because many racing organizations outlawed them. They delivered an enormous horsepower boost up to 10,000 rpm.

THE NIFTY FIFTIES

The middle of the twentieth century brought another "Golden Age" for the automobile, with Americans enjoying a prosperity that was unlike any known before or since World War II. As hot rodding exploded throughout the "Fabulous Fifties," Detroit recognized a trend and responded with the sporty fiberglass-bodied Corvette in 1953 and a completely redesigned 1955 Chevrolet that had 3,825 new parts, including a new V8 engine. The starting price was $1,835.

Once shunned as rebellious teenage entertainment, drag racing on a quarter-mile strip shifted into high gear and soon evolved into a sophisticated sport, attracting millions of fans only a few years after its 1950 jumpstart. To overcome an "outlaw image," practicing hot rodders throughout the country began forming clubs—many with safety themes. The first official drag strip meet was held in 1951. Soon afterward, the National Hot Rod Association (NHRA) formed and became a major sanctioning body for organized drag racing.

Stock car racing, which began as recreation for moonshiners running liquor across the South, was gaining momentum and soon would become one of the biggest spectator sports in the United States. NASCAR, the National Association for Stock Car Auto Racing, formed in 1948 and would become its ruler.

Bobby socks and the jitterbug were hip; rumble seats were passé. By mid-decade, cruising was a teen's Saturday evening rite of passage. Rock 'n' roll was the new sound. The big black car was king of the road, and Detroit added more chromed glitz to it. However, customizers kept removing the trim to achieve a voguish, custom-car appearance. "Don't fix it—just add chrome," defined the car makers' "bigger-is-best" mentality. Harley Earl's tail fins, introduced on 1948 Cadillacs, continued, but the new GM hardtop with its two-color paint could not upstage hot rodders' "candy-colored" customs.

In 1951, Chrysler introduced its heavy-breathing, 331 c.i.d. Hemi-head OHV V8 engine, which eventually toppled the dominance of the flathead Ford V8 in most racing classes.

In 1954, Ford introduced a new OHV V8 engine, while GM produced its fifty-millionth car. Oil companies introduced more types of premium-grade gasoline. Competition among the oil companies escalated to high drama, pushing credit cards, tiger tails, and glassware. Regular gasoline cost less than 25 cents per gallon.

Despite a $1 hourly minimum wage, 1955 was a boom year for Detroit as it churned out almost eight million cars, some of which are the classics of today.

Customizing evolved into an art form, with Californians doing flamboyant custom show cars, upstaging the rest of the country's milder restyling. "Fuzzy dice" replaced fox tails. Headers and twin pipes with rumbling exhaust sounds were the rage, despite attempts to ban the loud mufflers that cost $6 and higher. Dress-up accessories and bolt-on speed equipment were becoming big business. White would continue as customers' top choice in color. It is interesting to note that light-colored cars statistically are involved in fewer accidents.

Every two Americans owned a car as the motor-vehicle population doubled in the five years following World War II. However, the 1957 Edsel, once described as "an Oldsmobile sucking a lemon," failed within three years after being introduced and became the biggest industrial blunder in history.

For some individuals, horsepower was similar to money—something of which a person could never have enough. That began the "engine-swapping" craze, where big V8 engines were shoehorned into small cars.

Injected 283 Chevys boosted "one horsepower per cubic inch," a ballyhoo soon topped by Chrysler. Aftermarket speed equipment added enormous power increases to both.

Drag racing boomed. At 200 legal drag strips, more than one million officially timed runs were made in 1958, with the NHRA accounting for half the total in the nation.

In 1958, Don Garlits became the first dragster to break the 180-mph mark, in a NASCAR event in Montgomery, New York. However, people on the West Coast did not want to believe it.

AK MILLER
His T-Roadster Beat the $25,000 Ferraris

By the end of the 1940s, there were probably many hundreds of thousands of hot rod devotees scattered throughout America. Unfortunately, some "hell raisers" were giving the sport a bad image. Therefore, hot rod visionaries Akton "Ak" Miller, Wally Parks, and others upped the ante by forming the NHRA in 1951 to promote a more positive image of street rodders and drag racers. As co-founder of NHRA, Ak worked tirelessly, without pay, for years.

Life for Akton Miller started in Denmark in 1920. Five years later, he and his family immigrated to the United States. Ak's lifelong love affair with speed tinkering began as a pre-teen at the Nixon Grocery Store in Whittier, California, where his employer was Hannah Nixon, the mother of U.S. President Richard Nixon.

In 1934, Ak and his brother modified a 1928 Chevrolet roadster and powered it with a four-cylinder Chevy V8 engine for their first dry lakes run. At age 14 and without a helmet or safety belt, Ak sped the car to 94 mph.

In World War II, Ak was inducted into the military service. After fighting in the Battle of the Bulge, Ak spent six months recuperating in a British hospital, hoping that the nightmare would soon end.

Like many of the other old lake racers who returned form the war, Ak became involved in hot rodding. Before he knew it, Ak was elected president of the Southern California Timing Association (SCTA) and served for three years. Because the salt beds at El Mirage and Muroc had deteriorated, the returning G.I.'s had to seek other places to run. Bonneville, Utah, had seven miles of salt surface; thus, it became the ideal place for speed trials. After becoming a proficient builder, Ak captured a new Bonneville record for B- and C-Class in a rear-engine Mercury-powered roadster, which was constructed in his backyard garage. The unusual straight-away machine was called "Minnie's Missile," and its aerodynamic nose helped Ak set a new Bonneville record of 168 mph in 1951.

For the Mexican Road Race, Ak built a rugged Olds-powered Model T roadster that was nicknamed "El Caballo" (the Iron Horse) by Mexican fans. Although pitted against factory teams with their $25,000 Porsches and Ferraris, it was the only U.S. car to finish in its class, coming in eighth in 1953 and placing a remarkable fifth in 1954.

While reminiscing, Ak said, "The Pan-American road race was a hell of a race! I think our little buggy finally showed the Europeans what $1,500 and American hot rod ingenuity could do!"

Ak Miller (seated) drove this $1500 "rag-tag" T-roadster to fifth place in the 1954 Mexican Road Race. Ray Brock (center) was his teammate.

The Nifty Fifties

With his hot rod knowledge, Ak built a sports car with a Devin fiberglass body. With Ray Brock as his pit crew, Ak won his first Pike's Peak Hill Climb in 1963. Since then, he won nine times.

In 1962, Ak was hired by Ford Motor Company to race Fords in the Mobil Economy Run. He became Ford's official "answer man" on performance in monthly magazine columns titled "Flak by Ak."

At the first Baja 1000, Ak and Ray raced their trusty Ford 6 pickup to three wins in Class 7 in 1963.

When Ak's bones began to rebel against the rough off-road terrain, he quit racing at age 57.

"But I suddenly realized that I was not yet in the 200-MPH Club at Bonneville," Ak said. "So I hopped in my little Crosley with a 265-c.i. Chevy engine and did 225 mph to break the Austin's record."

By age 72, Ak was doing 225 mph. As a true hot rodding pioneer, Ak Miller's fingerprints literally are found on every segment of the sport, simply for the pure joy of it.

Ak's home-built 1927 T-roadster with rear engine capped three records.

Ak's rear-engine Missile did a record 168 mph at Bonneville. Its independent wheel suspension has a tubular axle hinged in the center. Rear U-joints and shafts are converted Ford parts. Torsion bars are used for springing.

At Bonneville in 1963, Ak set the E/SR record at 176 mph in his 289 Cobra. The headlight domes and comical front snout improved aerodynamics.

DAVE MARQUEZ
Goodwill Ambassador

This Baja champion two-seat off-road buggy built by Dave and his crew features a sculptured fiberglass body by Sam Foose. For Class Two (2180 cc), a big Volkswagen engine by Sandmaster was used. Class 10 (1200 cc) had a smaller Volkswagen engine modified by Precision Bug Works. Both used Sig Erson cams and dual port heads.

When the eyes of the nation were on the emerging new sport, early drag racers faced the challenge of improving their public image. Racers such as Dave Marquez and dedicated drag strip organizers helped change media bias from sheer loathing to utter respect.

"Forty-some years ago, they used to throw rocks at us, but now people welcome us," said former racer Dave Marquez, one of the two surviving fuel champions from the first (NHRA) race held at the historic Kansas landmark, which is now named Great Bend Motorplex.

"Years ago, the Midwest wasn't the only place hot rodders were misunderstood," said Dave. "Even people in our native California would look at us like we were freaks riding around in our open roadsters."

Beside setting records, including three NHRA national championships, and building and driving cars that won more than 200 trophies, Dave aided racing teams in achieving model standards for sportsmanship and appearance. He encouraged fellow club members to "clean up their act." As a member of the Motor Monarchs of Venturi—one of the first clubs to select a coordinated color scheme for all cars and uniforms—Dave was the club member carrying the "big stick."

"Our impeccable fluorescent orange and white machines and matching uniforms made us one of the most popular clubs in the nation," Dave recalls.

After devoting time to his Mexican food products company in the 1960s, Dave again began chasing a checkered flag when he acquired a "Baja Bug" from a bankrupt car dealer.

Dave campaigned in off-road races for several years and then took to the 1994 winner's circle for the BRA BAJA 300. A year later, he built the two-seater Pancho Villa Express, which took fourth place at the MINT 400 and later won the famous SCORE BAJA 500.

National NHRA drag champion in the 1955 fuel roadster class, Dave Marquez's 1932 high boy was chosen "America's Most Beautiful Competition Roadster" at a 1956 show. Motor Monarchs Club members could remove the body in two minutes. Eric Rickman's cover photo appeared in *Hot Rod Magazine*.

The Nifty Fifties

This 1971 Corvette was sold to actress Farah Fawcett by Dave Marquez after it won eight "best-in-class" car shows. Sam Foose's customizing included hood and side scoops, rear spoiler, and "Christmas tree" rear lights—red outside for brakes, inner green for accelerating, and amber for coasting.

Winning driver Dave Marquez sits in #880 at the NHRA inaugural national championship meet in 1955 in Kansas. The crew left to right, is John Davison, Howard Clarkson, Robert Olinger, and Ben Martinez, whose brother Ed was a co-builder.

Dave Marquez, shown seated in his 1932 fuel highboy, set more than 20 records and gathered over 200 trophies. Wally Parks said "Since the early '50s, Dave has been an able ambassador for many forms of racing, both in America and Mexico."

Flatheads Forever

The Hot Rodder's Favorite—Hot Rodding's Icon Engine

Henry Ford's flathead V8 will go down in history as being the most prolific engine in the world. From 1932 to 1953, that engine propelled everything from farm machinery to Indy race cars and, of course, it was the classic hot rod mill.

Originally rated at 65 hp, improvements and a duplex carburetor increased horsepower to 85. The postwar "59-A" block flatheads of 100 hp could be easily souped up to reach 185 hp for street—while a full race version could produce 250 hp or more with supercharging and special fuels. Some reportedly even dynoed at more than 300 hp.

Until the mid-1950s, the venerable Ford and Mercury flathead remained the backyard mechanic's engine of choice. Virtually any degree of performance improvement (from mild to red hot) could be accomplished with simple modification and additional readily available hop-up goodies. Typical modifications made to the popular 1959 A block Ford V8 engine included the following:

- **Increasing piston displacement** to boost power and torque commonly involved boring to 3-5/16" or 3-3/8" and increasing the stroke 1/4". Special aluminum alloy pistons were used with stroked crankshafts.

- **Increasing the compression ratio** would add 5% to 15% more horsepower and was accomplished by either shaving the head, installing thinner head gaskets, or adding special aluminum heads that increased compression 7-1/2 to 1 or higher. Optional pistons designed to "pop up" out of the block also boosted compression.

- **Improving volumetric efficiency** included "porting" and "relieving" the block. Valve heads were undercut with intake ports enlarged and polished. Heavier valve springs prevented "floating."

- **A reground camshaft,** considered the "heart" of the engine, produced a bhp gain of 15% to 20% by giving the valves higher lift and a longer opening for competition. Many racers preferred "mushroom" or roller-tappet setups for even more horsepower.

- **Dual carburetor manifolds** were "standard equipment" for most hot rods, and they boosted power by 10%, and more when combined with other modifications. Manifolds were also available for three or four carburetors, crowned with either chrome air cleaners or velocity stacks.

- **Tubular exhaust headers** freed power by replacing the restrictive stock system. Many cars had dual mufflers with removable "Lakes" plugs that minimized back pressure for competition.

- For **improved ignition** at high rpm, a hotter coil and dual spring points were popular for road use. Either magnetos or reworked distributors, with double breaker points for two coils, were preferred for racing.

- **A lightweight flywheel** gave snappier acceleration. This was accomplished by either turning down the stock 39-pound flywheel to about 18 to 22 pounds or by installing an aluminum-alloy flywheel (under 12 pounds), usually with a heavy-duty clutch.

- **Other hop-up tricks** included static and dynamic balancing, installing heavy-duty (harder) bearings with slightly increased clearances, and boosting oil pressure to 60 psi for hot road or 80 psi for competition (either racing pump or stretched spring). Overheating was corrected by special thermostats or installing "restrictors" (with a 3/8-inch center hole) in the radiator hose to slow the coolant flow.

Ed Almquist works on a flathead V8 engine equipped with Almquist racing equipment. The four-carburetor manifold was a prototype. (Circa 1950)

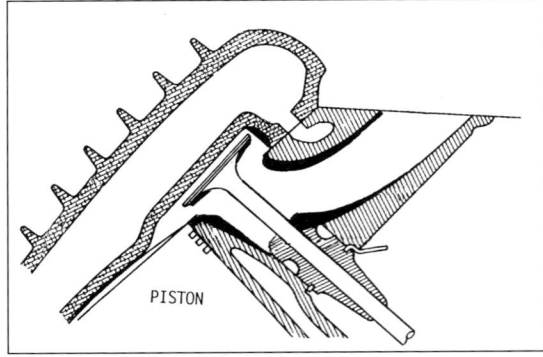

The shaded areas show where excess metal is removed and polished in a typical porting and relieving job. The underside of the intake valve and a portion of the valve guide also are cut away to increase gas flow.

The Nifty Fifties

The most popular modification gambits and their approximate costs in the 1950s:

Medium Road Engine—$200
1. Dual intake manifold
2. Aluminum 8:1 heads
3. Semi-race or 3/4 grind camshaft
4. Adjustable tappets
5. Dual exhaust system
6. Dual-point ignition
7. Miscellaneous
   Porting and relieving with addition of a aluminum or chopped flywheel was also popular.

Hot Road or Track Engine—$500
1. Triple carburetor manifold
2. Aluminum 8-1/2:1 heads
   (Higher compression ratio if special fuel is used for racing)
3. Track cam grind
4. Adjustable tappets
5. Header exhaust system
6. Dual coil ignition
7. Aluminum flywheel
8. Porting and relieving
9. 3-3/8" bore
10. Late Merc 4 "stroke" crankshaft
11. Racing type pistons
12. Special hard bearings
13. High-pressure oil pump
14. Cooling system improved
15. Miscellaneous

Super Racing Engine—$750 and Higher
1. Triple-or four-carburetor manifold
2. Aluminum 9:1 heads
   (Higher compression if special fuel is used for competition; overhead valve heads would add $500 to the above estimate)
3. Super race or track or "mushroom" grind cam
4. Racing style headers
5. Bore and stroke to maximum
6. Port and relieve
7. Racing three-ring pistons
8. Special hard bearings
9. Increased clearances
10. 60–80 pounds oil pressure
11. Magneto ignition
12. Aluminum flywheel
13. Heavy-duty clutch
14. Miscellaneous

Bruce Crower's full-race flathead dynoed at 240 hp. It was one of the first 3/8" stroker flatheads using the 1949 Merc crank offset ground to use the earlier smaller rods. Its 1-3/4" diameter valves and mushroom cam permitted generous breathing room.

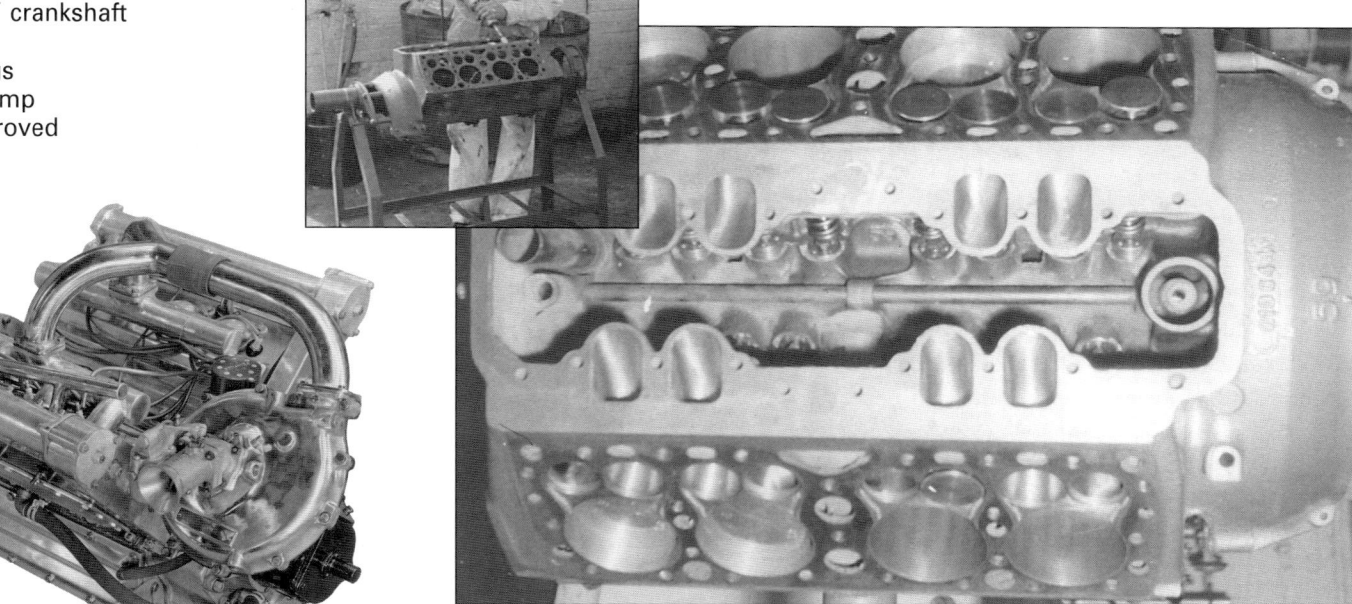

This supercharged Riley V8 overhead valve conversion was made briefly for Ford and Mercury V8s by George Riley, producer of the famed two- and four-port Riley head for Model A's. Except for the Ford crankshaft, connecting rods, and pistons, Riley made the balance of the powerful engine.

Porting and relieving boosted horsepower approximately 5% by improving engine breathing. Ports were enlarged and polished and the block "relieved" by grinding a 3/16" channel between each valve port and cylinder. Inset (above, left) shows Tom Sparks with rotary grinder. (Circa 1952)

HOWARD JOHANSEN
The Quintessential Rodder

Howard Johansen was born in 1909 in Shelby, Nebraska. Similar to many hot rodders, he loved to follow the road less traveled. Rather than sticking to standard modifying procedures, talented Howard Johansen thrived on trying bold new ideas and radical theories.

After driving on Midwest dirt ovals, Howard became involved in the California dry lakes and drag racing scene while fielding many top cars, including the first successful twin-engine gas dragster. Howard's twin dragster was driven by Glen Ward and later by Jack Chrisman, who captured the first Top Eliminator title at the 1961 Winternationals.

Among Howard's many innovations was a chain drive hookup to the GMC blower. He also produced forged aluminum connecting rods, camshafts, and other engine components. The LA speed shop, the forerunner of the original Howard Camshaft Company where Howard built his own cam grinder from an old lathe and tool post grinder, is now run by his sons.

Howard was the genius behind numerous top cars from the 1950s through the 1970s. He died in 1988, but he will be remembered as one of the most resourceful hot rodders of his time.

The "Twin Bear" gas dragster had two supercharged Chevy engines side by side, with the left engine mounted backwards and geared to the right engine that connected via a single driveshaft to the rear axle. The firing order was reversed on the left engine. With it, Jack Chrisman did 176 mph/8.99 e.t. for the top win at the 1961 NHRA Winternationals.

One of Howard's interesting design experiments was an odd twin-tanker lakes-type-racer that utilized two aircraft wing tanks for its body. A direct drive Mercury flathead was in one, and the driver's cockpit was in the other. Wheels were enclosed with a free-turning rear axle. The car hit 147 mph at Bonneville in 1949.

CHET HERBERT
Triumph over Adversity

Adversity causes some men to break—and others to break records. Chet Herbert was one who chose the latter.

While in his twenties, Chet was stricken with polio, which cut short his plans for racing motorcycles in 1949. Undaunted, he got his buddy Al Keyes to ride his hopped-up Harley motorcycle ("The Beast") at the emerging drag races in Santa Ana, California. From the first time out, they dominated the meets for several years with an eventual top speed record of 135 mph.

Determined to succeed in spite of the obstacles that confronted him, Chet parlayed his abundant self-taught technical savvy and tiny savings into a small camshaft grinding business in 1950 that became a forerunner in the industry. He adapted the Harley-Davidson motorcycle-type roller design to create the first roller cams for performance engines. Ten years later, Chet said, "We've been making racing cams with compatible roller-tappets for years, saying their reduced friction would boost power far more than solid lifter setups. The same grinders that panned them are now in the roller business." By the mid-1950s, most Bonneville speed records running with overhead valve engines had roller-tappet cams.

Throughout the 1950s and 1960s, Chet became involved in serious racing to "prove out" new cam profiles, as well as some of his unique car ideas. A 1952 project was a slinky little Hemi-engined streamliner named "Beast III" that did a record 236 mph at Bonneville. Others included the 180-mph Cagle-Herbert slingshot and the co-sponsored "Goldenrod" streamliner that blew the top off the land speed recordbooks with an incredible average of 409 mph in 1965.

Chet Herbert sits at the cam grinder with which he developed the first power roller cam in the hot rod industry.

The "Rapid Transit" streamliner was another Herbert-cammed record car. Bill Burke drove it 242 mph on the salt. It crashed later at 291 mph.

THE CHRISMAN CLAN
The Good Old Boys

Art Chrisman as he appears today. A half century of hot rodding excellence has earned Chrisman a place in the Motorsports Hall of Fame.

As an original veteran in organized drag racing, Arkansas-born Art Chrisman mastered the trick technique of building record-breaking cars, while his uncle, Jack Chrisman, raced roadsters, stockers, and slingshots before becoming one of drag racing's first touring professionals. Their famous 1930 chopped and channeled coupe with flathead, Ardun, and Chrysler engines held records in several classes in the early 1950s. Before Art restored it, the car appeared regularly on the "Dobie Gillis Show" on TV.

Immediately after World War II, Art and his brother Lloyd joined their father in setting up a shop in Compton, California, which spearheaded the family's racing dynasty as the two brothers competed successfully in dry lakes and drag racing.

The Chrisman Garage was best known for producing top-performing machines, featured on early *Hot Rod Magazine* covers. Art partially built and drove Chet Herbert's "Beast" streamliner to more than 235 mph, thus becoming a charter member of the 200-MPH Club in 1952.

In those days, variety was certainly the spice of life, with some gas and fuel slingshot dragsters showing up the sport with mid-engine "sidewinder" configurations having engines mounted transversely. Although the Jones Sidewinder was difficult to drive, Jack almost won the 1959 Nationals in Detroit with it. In 1961, Jack won the Top Eliminator title at the first NHRA Winternationals when he drove the Howard Cam Special, a two-engined supercharged Chevy dragster, to an 8.99-e.t./176-mph run. The following year, Jack swept the Nationals with a Top Eliminator win driving a Pontiac-engined AA/Dragster.

The handsome Chrisman #25 car had many lives. It was an oval track racer in the 1930s, a dry lakes competitor in the 1940s, and later a dragster. The car was powered by a Rajo-T, a Cragar Model A, a flathead Ford V8, and an Ardun-Mercury before switching to a Chrysler Hemi. With the car lengthened to improve traction, Art was one of the first to exceed 140 mph at the drags in 1953. He also made the first run at the first NHRA National Championship Drag Races at Great Bend in Kansas in 1955.

In 1962, Autolite Spark Plugs took on Art as a racing representative. After a decade of instructing mechanics on the technique of tune-ups and "reading spark plugs," Art ran a dynamometer service for Ed Pink that developed into a test facility for SEMA, as well as for "gray market" preparation of import cars for state emission certification. In the early 1980s, Art and his son Mike began restoring and turning out show-car-quality street rods, one of which was the "America's Most Beautiful Roadster" winner of 1994.

Over the years, both Art and Jack raced—sometimes winning, sometimes losing—against top gassers and fuelers, even at friendly feuds such as the "East vs. West" competition that fired-up the fans.

Jack Chrisman (left), Dee Keaton (center), and Gene Mooneyham (right), with Jack's Comet that debuted as an exhibition stocker. The "Super Cyclone" precluded "funny cars" but was so drastically modified (with gutted interior, fiberglass front end, plastic windows, and a hot GMC-blown, 427 c.i.d. V8 engine) that NHRA classified it as a fuel dragster.

The Hustler I, built by the Chrismans and wealthy Frank Cannon, was named "Best Engineered" at the 1958 NHRA Nationals. A year later, Art became the A/Dragster Champion while driving the GMC-blown, injected Chrysler-powered machine. Initially, the car had a Thunderbird engine.

ZORA ARKUS-DUNTOV
Patron Saint of the Corvette

Zora Arkus-Duntov was a good friend to hot rodders. Like his double surname, he related to two legends—the Ardun OHV Ford and the Corvette sports car.

Born in Belgium in 1909 to Russian parents, Zora was educated in Germany as a mechanical engineer. During the 1930s, he raced motorcycles and worked on supercharged sports car engines. As war clouds loomed, Zora fled Nazi Germany for Paris. To earn extra money, young Zora smuggled gold into Belgium for wealthy French people. After briefly serving in the French Air Force during World War II, Zora immigrated to the United States in 1940.

With his brother Yuri, Zora formed a company in New York City to do machine work for the military. Also, during the war, the brothers began improving the inherently poor breathing characteristics of the Ford V8. The resulting Ardun (a derivative of ARkus-DUNtov) overhead-valve conversions with aluminum hemispherized heads easily upped the stock 85-hp Ford flathead to more than 175 hp when in full-racing form. However, the well-designed Ardun OHV conversion kit sold poorly at $359 in 1949. (Speed equipment probably was more salable then because it cost less to accomplish equivalent horsepower increases. Later, Zora wrote to me and stated, "I can't compete with you hot rod guys.") After weak sales, the disheartened brothers sold their New York shop. Zora then went to England, where he spent two years improving the brawny Allard sports car.

After returning to America and attending a 1953 GM car show, Zora fell in love with the newly introduced Corvette. Zora wrote a letter to GM's Ed Cole, asking for employment and brazenly expounding his ideas for improving the anemic six-cylinder Corvette. Zora was hired by GM, and the rest is history. Despite his enormous engineering wizardry, Zora first had to battle with manage-

For the 1946 Indy 500, Zora Arkus-Duntov drove a rear-drive six-cylinder 274 c.i. Talbot but failed to qualify.

The Nifty Fifties

ment to push his "Vette" ideas. In 1953, Duntov delivered his 1953 "manifesto memo" to the heads of GM, coaxing them to become "tuned in" to the hot rod youth market. (The Chevy V8 engines soon rivaled the best in the world.)

Believing more horsepower was needed to ensure speed attempts at Daytona, Zora produced his Duntov high-lift camshaft, which jumped the engine output the extra 30 hp needed at 6500 rpm. By 1956, Smokey Yunick, Paul King, and Mauri Rose had installed the first "official" Chevrolet V8 engine in a Corvette, which was test driven by Zora at the Sebring road racing course. Mauri enjoyed heckling Duntov, "Arkus, why don't you open it up? You're driving like a little old lady." To promote the Chevrolet's performance image, Zora raced at Le Mans, and in 1956 he established records at Pike's Peak and Daytona Beach.

In 1992, Zora drove the one millionth Corvette off the assembly line at Bowling Green, Kentucky—a fitting tribute to the genius who helped the Corvette become America's ultimate sports car. Zora Arkus-Duntov died in April 1996, at age 86. He is honorably entombed at Bowling Green's National Corvette Museum.

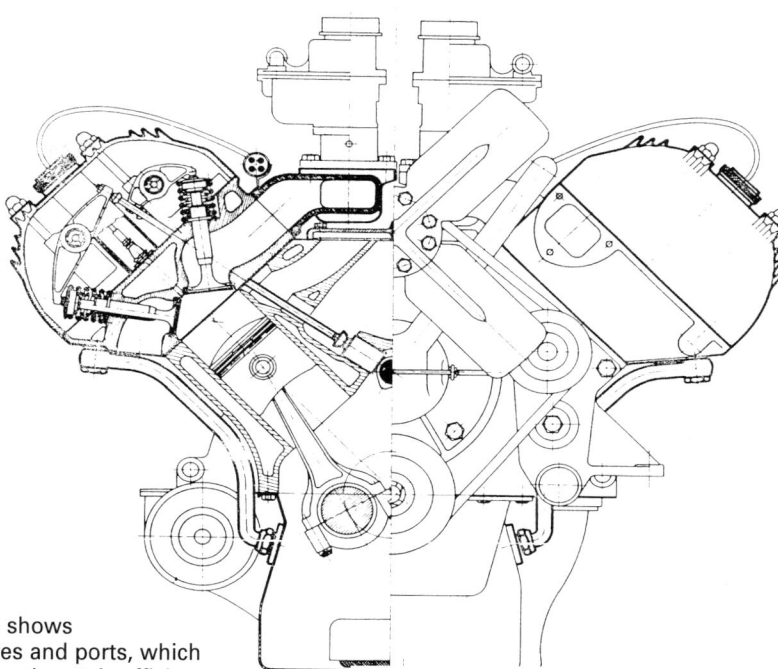

Sectional drawing shows Ardun's huge valves and ports, which allow 50% greater volumetric efficiency at high speeds.

Said to be inspired by American hot rods, the 1950 Allard J-Type roadster was offered in New York with a Ford engine for $3,383 or less engine for $3,000. Duntov did development work on later models using Ardun, Cadillac, and Chrysler engines.

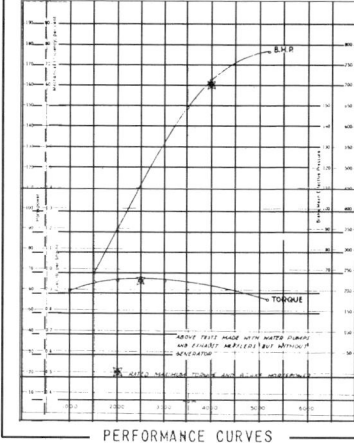

Performance curves on Ardun-equipped Ford engine show 225 ft/lb torque (bottom line) at 2500 rpm and 175 hp (top line) at 5200 rpm.

Hot Rod Pioneers

The Stock Car Racing Boom

The fiery tongues of folklore have stock car racing firmly rooted in the Deep South, with "tobacco-spitting" moonshiners outrunning the "revenuers" on twisty mountain roads during the dark of night. However, on Sunday afternoons in farmyards and cornfields, those hillbillies would legitimately pit their souped-up sedans against each other. Without realizing it, the "rebels" had revived an old sport because stock car racing really began soon after 1900 when auto makers discovered that winning races sold cars, as they do today. Competition tapered off after World War I but resumed again in 1927 when AAA inaugurated a loosely supervised stock car racing program. Then, stripped stock machines, with names such as Auburn, Duesenberg, Studebaker, Marmon, and Stutz, often barnstormed the county fairs.

After World War II, "stripped stocks" and even track roadsters evolved into strictly stock, semi-stock, and modified stock classes, whose popularity soon overshadowed all other types of racing. Spectators who jammed the stadiums could easily identify with cars that resembled their daily transportation. The showmanship and "slam-bang" action appeared to be a "simulated suicide," with frequent spin-outs and crack-ups that thrilled the fans.

Before Bill France, Sr., became the czar of organized stock car racing, he raced this Ford coupe on the sands of Daytona Beach. As a visionary post-war promoter, Big Bill (who was six feet, six inches tall) foresaw the need for an association to standardize rules, establish a championship point system, guarantee purses, and unify the fledgling sport. In 1947, Bill and some fellow enthusiasts created NASCAR, which is now the major nationwide stock car sanctioning body. The Winston Cup Grand National Program that began in 1971 is now the top auto racing series. (Photo courtesy of NASCAR)

The New York State Stock Car Championship in 1950 was won by "The Flying Milkman," Russ Dodd. Dodd was a "dark horse" competitor who had never won a main event. Here, Ed Almquist (right) is presenting a trophy to Russ (left). At that time, more than half the cars at the old dirt track in Middletown, New York, ran Almquist racing equipment.

From Rum-Running to a Major Motorsport

Before he began drag racing seriously, Jay Moss was a frequent winner in this hot V8 Ford coupe, which was the most popular stocker model in the early 1950s. The body was gutted to lighten the car, and the suspension was reworked to keep the top-heavy cars from toppling. Special features included roll bars, glass removed or taped, and Nerf bars for protection against the inevitable battering.

Wally Campbell of Trenton, New Jersey, was the 1951 NASCAR champion in the modified class and ran his Ford as the Almquist Racing Equipment Special. Before that, Wally was a trophy-collecting competitor in the midget, roadster, and sprint car circuits.

In the Northeast, the scene was "crash for cash"—especially on short tracks such as Hinchcliffe Stadium in Paterson, New Jersey, which was originally designed for midget racing. The television media soon sought stock car racing mayhem for exciting programming. Billed as jalopies or hard tops, early modified stocks zipped around the one-fifth mile oval in fifteen seconds or less. With as many as twenty cars on the track barely inches apart, all trying to squeeze through corners only wide enough for a few, spectacular crashes were inevitable. Miraculously, injuries were rare and usually not serious.

Southern stock car races often were zany events with colorful personalities, such as "hell-raising" top drivers like Curtis Turner who, it was rumored, was a "high-rolling partygoer" who made and lost fortunes almost overnight. Other popular characters who are now long gone included Lee Petty, "Fireball" Roberts, Marshall Teague, Cotton Owen, Red Byron, Buck Baker, Herb Thomas, and the Flock brothers. (Many racing daredevils were my steady customers. Frantic phone calls, sometimes with raucous barroom background noise, often would come after a race—even at two o'clock in the morning—requesting rush shipments of pistons and other expendable engine parts. The hitch was the impossible delivery times to speedways somewhere in the Deep South.)

A new era began in late 1947 when the all-encompassing National Association of Stock Car Automobile Racing (NASCAR) was formed. The following year, almost a million people reportedly paid an average of $2 each to witness the rapidly growing sport on Eastern tracks alone.

By the mid-1950s, automobile makers saw motorsports as a low-cost marketing tool and began supporting "big-buck" race teams. Then, manufacturers advertised victories with such hoopla that it brought criticism from Washington, DC, forcing NASCAR to adopt a 1957 rule requiring "honest" advertising. As a result, car factories *en masse* withdrew from both track and drag racing.

In the early 1950s, drafting aids were attempted; within a decade, aerodynamics began in a big way. Smokey Yunick remembers, "By 1967, cars were so butchered, chopped, and changed that you couldn't tell what they were anymore. NASCAR had templates of new cars made, but if they had enforced the template rule, then only two out of sixty cars would have passed."

Hot Rod Pioneers

From 1947 to today, escalation in horsepower and cost of racing have been astounding. In 1947, 200 hp and 5000 rpm were considered big. By 1970, the level had risen to 650 hp with more than 7000 rpm. In 1995, Smokey Yunick said, "Now Winston Cup small-block engines are lighter (with aluminum cylinder heads) and put out well over two horsepower per cubic inch, normally aspirated on good but not great gasoline blends."

The ever-popular early "modifieds" (usually pre-war Fords) could easily be competitive on both track and strip, with only a few hundred dollars in parts. By the mid-1960s, one of the most successful modifieds sold for $4,000. Now a turnkey-modified car costs more than ten times that amount, and a good Winston Cup racer can cost around $100,000. Of course, these are not really stock anymore; rather, they are carefully engineered handcrafted "racing specials" with fiberglass bodies that vaguely resemble the original vehicle.

The race car of yesterday usually was towed by a passenger car loaded with tools and spare parts. That has been replaced by a tractor-trailer equipped with air-conditioned living quarters, a machine shop, extra engines, and often two ready-to-run race cars. Instead of a driver and a lone mechanic, a big-time moneyed team today might have a dozen well-honed professional technicians trained to use elaborate computers to tune handling to each track. Top crews now change wheels, refuel, make adjustments, clean the windshield, and hand the driver a beverage in less than twenty seconds.

The strides made in engine and chassis technology have been so rapid that ever-changing rules, such as carburetor restrictor plates, roof flaps, lower air dams, and taller rear spoilers for more downforce have had to be made to control top speed. Otherwise, cars literally would fly on the super-speedways.

"Compared to yesterday's wild and wooly cats," said former champion Junior Johnson, "it's hard to comprehend the changes in racing today."

A late 1990s survey discovered that 26% of Americans are fans of all types of auto racing. This helps to explain how a humble stock car lineage would grow to become one of the world's most popular motorsports and over a two-billion-dollars-a-year industry.

As stock car racing continued to grow nationwide, especially at the grass-roots level (which is still affordable), the darlings of dirt ovals in the mid-1960s usually were Chevy V8-powered cars with modern chassis and suspensions, but often with old or pre-war style bodies. Today, most modifieds do not resemble any kind of car—"a la World of Outlaws."

Sam Packard's fast Studebaker was the only station wagon to run in a big NASCAR race in the 1950s. Sam, who later was a NASCAR inspector, used the car to advertise his Rhode Island speed shop as he helped introduce both stock car and drag racing in New England.

OTTO RYSSMAN
'Slingshot' Pioneer

Unfortunately, safety mandates came too late for an unsuspecting spectator and Otto Ryssman. Otto will never forget the 1955 tragedy at the Santa Ana drag strip, when his clutch and pressure plate blew up and ricocheted into the audience, killing a spectator. For the next seven years, Otto and promoter C.J. Hart were in and out of court. Unknown to Hart, the drivers of the strip had not been legally covered. Although the suit was later dropped, it drained Otto financially.

In the early 1950s, Otto and his partner Tiny Conkle built a revolutionary slingshot dragster.

"We installed the engine rearward with a one-foot drive shaft to get more weight on the rear wheels for stability," said Otto. "Everywhere we'd go, we'd set a record. The dragster would leap the line like a slingshot, as the style was later named. It really should be considered the first slingshot because it had a longer wheelbase than Dick Kraft's 'Bug' that is hailed as the original as it now sits in Garlits' museum."

Was this the world's first slingshot dragster? Many claim Otto's flathead was the forerunner of the slingshot dragster. It was often used to test fuels and Chuck Potvin's camshafts. (Circa 1950)

At Bonneville in 1952, Otto drove the Post Special streamliner to a two-way average speed of 222.57 mph. In a later record attempt, a tire blew, the car crashed, and Otto escaped with a cut leg and a concussion.

Otto is one of the five charter members of the Bonneville 200-MPH Club and was elected as its president in 1957.

Otto Ryssman was one of the first recipients of the recently established California Hot Rod Reunion Speed Pioneer Award, a deserving honor for another true hot rod pioneer.

The Post Special streamliner driven by Otto Ryssman was built by Harold Post. The car was later sold to engine-builder Doug Hartelt and cam grinder Chuck Potvin, who replaced the flathead with a 300 c.i. Chrysler engine.

TOM SPARKS
Wrenched for the Greats

What could be a more fiery given name for a race car driver than Tom Sparks?

"With a name like Sparks, I was expected to really heat things up," said Tom Sparks, a midget car racer in the Los Angeles area after World War II. "When I was young, I saved my paper route money and bought a 1932 Ford three-window coupe off a used car lot for $35. I taught myself to drive but smashed into a moving car one block from home. The front end was really bashed, so I removed the hood and fenders. That was the beginning of my first hot rod."

"Finally," Sparks continues, "after three years of driving midgets, success wasn't coming my way. So, one day, Bud Meyer said to me, 'Sparks, why don't you stop wasting your time driving and come work for me? I'll make a real mechanic out of you!'"

The next week, Sparks went to work for Eddie Meyer Engineering. Two years later, he joined with engine builder Ray Brown to learn about more speed tricks. Eventually, Sparks partnered with Ted Bonney in a company specializing in automotive performance. Later, he built "camera cars" and customs for Hollywood's Paramount Studios, which was located across the street.

Although Tom Sparks is now semi-retired, he builds a couple Ford flathead engines every year and meticulously grooms his 40 classic cars, many of which are used in movie and TV cameo appearances.

At 80 years old, Tom builds fast flatheads for the fun of it.

Tom Sparks is ready for a run at Pomona drag strip. Note the high position of Tom's foot on the gas pedal. The "Kill Switch" is on the shift knob. (Circa 1953)

This 1937 Willys coupe is one of the first hot Willys drag cars. A blown flathead helped it reach 133 mph in 10 seconds in 1954.

The Nifty Fifties

Night races were popular for midget racing from 1945 to 1951. Both Tom Sparks and Eddie Meyer would chase mighty Offys with their tiny but muscular V8 60 Fords.

The 1954 four-car team of Tom Sparks (left) and Ted Bonney (right) earned them bragging rights and customers for "Sparks & Bonney Automotive."

C.J. "PAPPY" HART
Founder of the First Drag Strip

Pappy Hart was a starter in one of the first races on the strip.

Impromptu acceleration contests have been held since cars had wheels. However, organized drag racing, as we know it today, really began after World War II. Although lesser-known acceleration contests and events sanctioned by the SCTA had run previously, the first commercial drag strip in the world was located at the Orange County Airport near Santa Ana. On June 19, 1950, the founder of the strip, Cloyce J. "Pappy" Hart, and partners Creighton Hunter and Frank Stilwell staged the first organized drag race on a surplus runway.

"It was a surprise hit," said Pappy, who in 1981 received the SEMA Hall of Fame award. "The first night, we had about 55 cars and motorcycles, with 300 spectators who each paid 50 cents. Timing was done with stopwatches and a homemade timing clock. We could only afford to give out small trophies, which we often would buy back at $7 each."

The first races began with rolling starts from as much as 20 feet from the line and with no elapsed times. After some racers began wondering how they were beaten by a car that turned a slower speed, elapsed timing (e.t.) lights were installed. Initially, the crude tim-

At age 70, C.J. "Pappy" Hart helped promote NHRA events.

A match race between two cars at the Santa Ana Drag Strip in 1950. Nine years later, a record 4,000 fans were in attendance at its closing.

The Nifty Fifties

By the mid-1950s, well-organized drag meets were being held throughout the nation.

ing equipment read only up to 149 mph because the experts thought that nobody could exceed that speed in the quarter-mile.

At the "legal" drag racing strip, Pappy's wife Peggy became a feisty driver who trounced male drivers and wounded their egos. The Santa Ana drag strip operated continuously for almost nine years before becoming incorporated into the John Wayne International Airport.

From 1963 to 1973, Pappy was manager of the Lions Association Drag Strip (LADS) in nearby Long Beach. After he was events supervisor for the American Hot Rod Association (AHRA), he joined the staff of the National Hot Rod Association (NHRA).

"I didn't invent drag racing, but I surely helped get it off the streets," said C.J. Hart. Without realizing it, Pappy had helped kick off a new motorsport.

NASCAR began sanctioning drag races in the mid-1950s. Record crowds attended nightly events at Daytona during "Speed Weeks." Later, NASCAR teamed with NHRA in presenting the Winternationals.

MICKEY THOMPSON
A Prince of Speed

Hot rod hero Mickey Thompson, once America's fastest man on wheels, lived a fabulous dream that ended tragically.

Mickey Thompson's saga is similar to a novel: a fierce struggle against improbable odds, a gutsy adventure, and a tragic ending. It would require a separate book to describe Mickey's quantum leap from obscurity to being a "King Fish" in almost every facet of racing.

As perhaps the most innovative influence in hot rodding, Mickey was among the first to develop the slingshot dragster. However, he also was the first with a four-engine racing car and the first with workable, unorthodox engineering concepts. Mickey manufactured speed equipment and racing tires, and he built, sponsored, and drove all types of racers. He was a speedway and car show promoter and founder of SCORE, which was at one time the largest professional off-road sanctioning body.

Born in 1929 and christened Marion Lee Thompson, Mickey worked as a pressman for the *Los Angeles Times* for 10 years after high school. When he began executing his originally bold hot rodding ideas on weekends, Mickey proved he was more than a dreamer.

"We used to scrounge through Ford dealers' trash bins for parts that could be welded, tied, or glued together when Mickey was hopping-up a car," stated his first wife and "Gal Friday," Judy.

After Mickey initially appeared in Bonneville in 1950 with a flathead-powered 1936 coupe, he returned for the next two years with two flatheads in a Bantom car. Mickey built one of the first slingshot dragsters on the West Coast in 1952, which he later fitted with a streamlined body. A year later, Mickey took regional honors with a Pontiac four-cylinder dragster that began as a low-budget experiment. A formidable competitor and hustler,

Among the many cars in Mickey's stable, this Pontiac-powered dragster was one of the most unique. It was equipped with M/T components, including fire-spitting scavenger design headers claimed to increase power by 10%.

Mickey (right) at age 15 had built his first flathead engine for a 1932 roadster he raced at Muroc. The huge hoist shown in the photo later fell off the garage ceiling.

Mickey managed to tune, build, and drive an assortment of cars during the next four years, thus earning 70 international and American land speed records

After one of his failed record attempts at the dry lakes, Firestone refused Mickey's request for better high-speed tires. Gene McMannus, then with Goodyear, became involved. (Later, they established the M/T Tire Company.) Quite by chance, Mickey was introduced to "Bunkie" Knudsen. Knudsen was a top engineer at Pontiac whose company, along with Goodyear and Mobil, helped subsidize the start of the famed Challenger I streamliner. The project moved swiftly with help from George Hill, who designed the aerodynamic body, and mechanic Fritz Voight, who was the West Coast Gas King of the early 1950s. Despite additional aid from other companies, money was short and Mickey had to mortgage his home.

In 1959, Mickey drove his home-built four-wheel-drive Challenger streamliner, equipped with four supercharged Pontiac engines, to the Class A record at 350.5 mph at Bonneville. The following year, Mickey did an astounding 406.6 mph. However, after a drive shaft broke, Mickey could not make the return run, which prevented him from earning an official world land speed record. Although Mickey was broken-hearted, he vowed to come back again, but foul weather dampened further speed assaults that year.

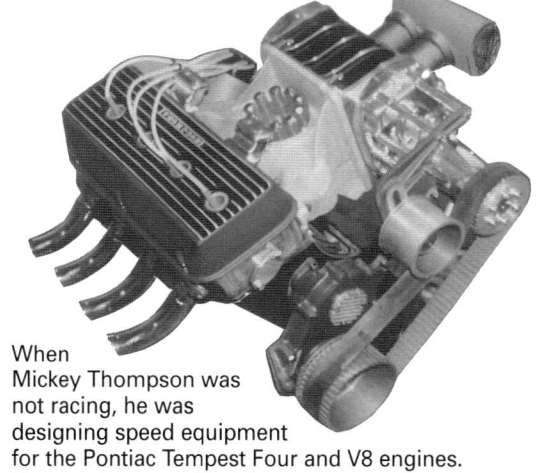

When Mickey Thompson was not racing, he was designing speed equipment for the Pontiac Tempest Four and V8 engines. Output of the modified four-cylinder exceeded 400 hp at 6000 rpm on a small shot of nitro added to gas.

In stretching the speed envelope, Mickey had proven to himself that he was a born engineer even if he lacked formal technical schooling. In his usual confident manner, he established Mickey Thompson Enterprises in 1959. Within a few years, M/T offered a complete line of mainly Pontiac-oriented speed equipment.

For more than three decades, it seemed everything Mickey touched turned to gold. Seasoned drivers such as Jack Chrisman, Terry Smith, Danny Ongais, and Mike Van Stant helped Mickey win

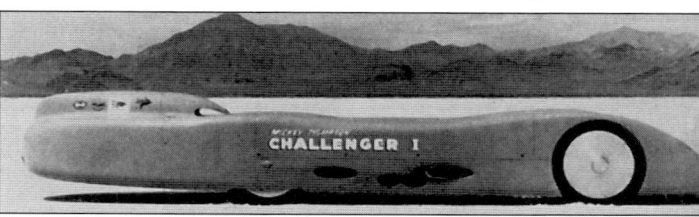

The Challenger I was the last of the conventional land speed record cars in 1960. Rather than rocket or jet engines, it had four Pontiac engines mounted in pairs on a tubular chassis, and it was modified to produce well over 2000 hp. The sleek streamliner was 20 feet 3 inches long, only 30 inches high at the hood, and 39 inches high at the cockpit fairing.

further races and records in major classes of racing. Suddenly, it all ended. In 1988, Mickey and his second wife Trudy were shot to death in the driveway of their home. (The crime was never solved.) Although Mickey's life was cut short, he had crammed it with more racing, more variety, and more accomplishments than most of us would ever dare to attempt.

The Attempt I Class E speed record car was another of Mickey's beautiful streamliners in 1961. The slippery one-piece body covered the trussed tubular frame that was suspended with transverse torsion springs in front but attached rigidly to the rear axle. When run in Class D, its supercharged 180-inch Pontiac Tempest engine was tweaked to deliver approximately 460 hp at 7000 rpm.

In the dramatic Universal Pictures movie "The Lively Set," professional race car driver Mickey Thompson portrays himself, with a cast starring James Darren and Doug McClure.

JOE WOLF
"Rebel-Eating" Yankee Mechanic

JOE WOLF'S HOP-UP PRICES		
Engine	HP Increase	Cost
Cadillac	30	$116
Chevrolet	18	$80
Chrysler	40	$160
Oldsmobile	30	$116
Plymouth	22	$90

It is hard to believe that some people complained that Joe Wolf's prices were too high.

Joe Wolf's love for speed helped him become one of the top wrenches in racing in the 1950s. According to former Indy champ Bill Holland, "Joe was the only Yankee racing mechanic who could take a car down South and consistently beat the rebels at their own game."

In 1956, NASCAR Grand National Champ Tim Flock won the modified sportsman race at 137 mph in an old (1939) Chevrolet equipped with an Olds engine that was modified by Joe. When most speed shops were pushing bolt-on speed equipment, Joe's Pennsylvania Dutch frugality led him to his own brand of mechanical wizardry for souping up customers' road cars. For $116, he could produce a 30-hp increase on early 1950s Olds and Cadillacs. The cylinder head modification included enlarging the ports and boring out around the intake valves for better "breathing." Manifold ports also were enlarged and polished to match. Stiffer valve springs and shortened pushrods for zero valve clearance prevented high-speed valve float common to hydraulic lifters.

Joe's winning stock car racers, speedboats, and hot rods were called "wolves in stock clothing" because his engines often would retain their stock external appearance.

Joe Wolf admires one of many racing trophies his mechanical wizardry helped win.

Why is Frank DeRoy smiling? Duke Nalon (right) was the disgusted driver, and Frank DeRoy (left) was the amused mechanic when the "Wolfe Special" sputtered out of gas while leading in a championship AAA race at Langhorne in 1950.

The Nifty Fifties

Not all stock cars ended up airborne as shown here. Joe's cars all had roll bars and interior reinforcement to protect the driver. (Ed Otto Collection)

After racing this crude roadster with its two-carbureted Model A engine, Paul Reider drove Sportsman stock cars for Joe Wolf.

In an effort to seem trendy, Joe called his Pennsylvania establishment the Hollywood Speed Shop. Fonty Flock drove Joe's flathead stockers to many victories.

CLEM TEBOW AND DON CLARK
C & T's Bold Innovators

This Ardun-powered C & T track roadster was driven by Lyle Dickie at the Hanford, California, half-mile track in 1952. Later, the vehicle was converted to a sprint car that won the California Championship in 1956.

"We were young hot shots who had big aspirations," declared Clem TeBow about himself and his partner, Don Clark. Like other early hot rod teams operating from backyard garages, Clem and Don pioneered new and bold race-winning ideas.

In 1950, the innovative duo opened a tiny machine shop called C & T Automotive, which first specialized in localized engine building. After the pair decided on a "quick-buck" mail-order product, they doodled a one-inch ad in Hot Rod Magazine and received enough orders for flathead Ford Stroker Kits to launch a "fat catalog" of hop-up parts. Eventually, they began making motor mounts and hard-chrome stroked crank assemblies for V8s.

Clem's thirst for hot rodding began with a 75-cent purchase of a farmer's Model T Ford.

"With parts salvaged from the community dump, I got the clunker running," said Clem. "It was my learning tree."

After serving in World War II, Clem and Don spent their spare time tinkering with jalopies in a former chicken coop. There the pair secretly equipped a Ford roadster's 59-A block with an Ardun OHV conversion that was acquired at an unclaimed freight sale.

"Our 'secret weapon' really roared, but it bombed at our first dry lakes run," said Clem. "We replaced the carburetors with a homemade slide valve fuel injection system costing $10 for parts. Then the engine peaked at about 50 more horsepower than we had ever seen on any other flathead."

In 1951, their efforts paid off. After setting a Bonneville Class C Roadster record of 162 mph in 1951 with the Ardun-equipped Mercury

Don Clark (left) and Clem TeBow (right) with an improved Ardun engine on which they boosted horsepower to almost 300 by switching to Chrysler valves, lighter tubular push rods, stiffer springs, a hotter cam, equal length headers, and Hilborn injectors.

The Nifty Fifties

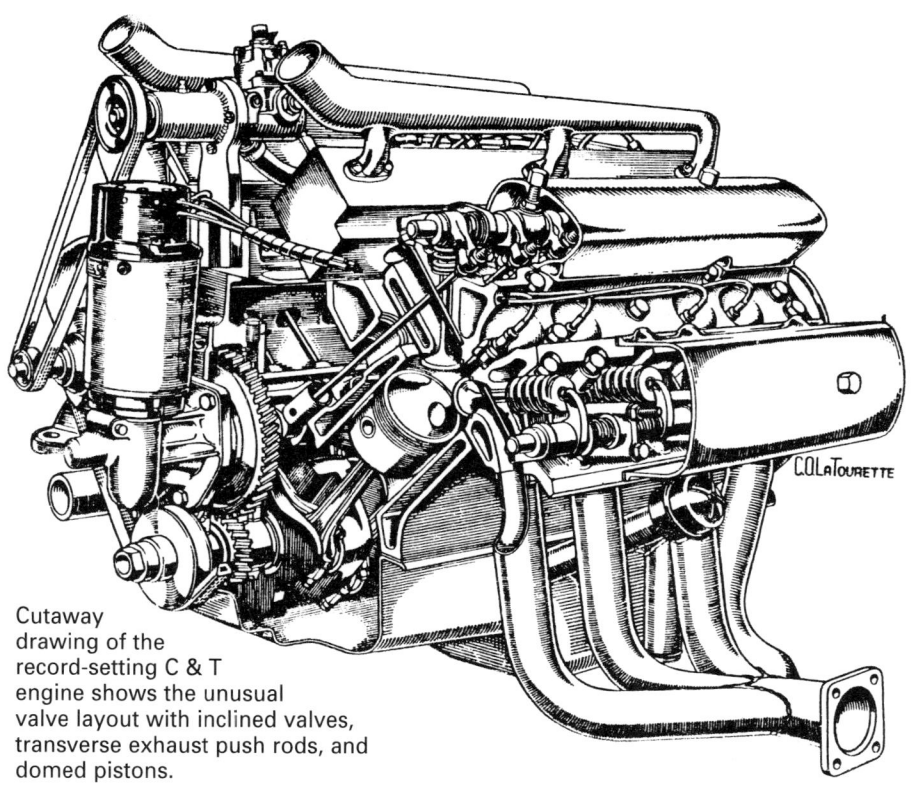

Cutaway drawing of the record-setting C & T engine shows the unusual valve layout with inclined valves, transverse exhaust push rods, and domed pistons.

Partners Don Clark (left) and Clem TeBow (right) pose with the famed City of Burbank streamliner (later renamed the Bob Estes Special). Driven by George Hill, the sleek streamliner captured the Class B record at 230 mph and the prestigious Class C international land speed record of 229 mph at Bonneville in 1952. The 1946 Ford truck engine was converted to C & T's OHV conversion.

engine, the partners decided that linking with OHV conversions was the way to go.

They acquired the head patterns for a new but untried overhead valve conversion for $900 of borrowed cash. By installing the new OHV conversion on a Ford V8 block, the engine revved more than three times its original horsepower. By 1952, a streamliner equipped with the powerhouse C & T engine captured six land speed records.

In 1962, after the pressure of business cut into the fun of racing, Don Clark left the company. Four years later, Clem sold C & T Automotive. Eventually the enterprise, which had grown from a chicken coop to one of the country's largest hot rod suppliers, faded into oblivion.

Albert Drake

Albert Drake's red 1929 A-V8 Ford roadster had a 3/4 race 1949 Ford V8 engine, with Edmunds dual manifold, converted Nash ignition, and milled heads. Now a retired professor, Al writes articles and books about hot rodding. (Circa 1951)

THE BANNISTER BROTHERS
New England's Rodding Heroes

The following was abstracted from a December 27, 1957 article in *The Merrimack Valley Ad* newspaper.

The epic story of Ralph and Fran Bannister begins about a half century ago—just after the two decided to give up street racing and go "legit." Although the youths were charter members of a hot rod club in Sanford, Maine, in 1950, they raced separate cars and won their own share of trophies, approximately 50 in all, but the brothers kept only 32. The story of the missing 18 is their legacy.

The brothers, being mindful that competition was the meat of motorsports, raced for the fun of it, rather than for the money. Both volunteered time and talent by organizing their area's first drag races—inspecting cars, setting up timing equipment, helping opponents, and even loaning parts to another car that eventually beat theirs.

When Ralph fell sick and had to be hospitalized, financial troubles began. After Ralph was released, enormous bills followed. However, a few days later, 35 cars drove up, full of rodders who came to welcome Ralph home. In recognition of all the volunteer work that the brothers had done, Ralph was presented an envelope containing enough money to cover the growing pile of bills. The gesture touched the Bannisters so much that they sent 18 of their best trophies to the New England Timing Association who, in return, named a larger one the "Bannister Brothers Trophy." That trophy was awarded to them in 1956 and 1957. The rule for winning the perpetual trophy was that anyone who won the trophy three times in succession could keep it.

Although Ralph had previously raced modified stock coupes and Fran had run his roadster in Bonneville (1950) and both brothers had raced a dragster competitively at the fourth annual NHRA meet in Oklahoma City, the brothers basked in the glory of being top drivers at local and regional drag race events. They won more than 65 honors before retiring from drag racing in 1958.

When asked about his most memorable incident, Ralph said, "I had a close call on an old airport when testing my dragster. I was about at top speed when a small airplane taxied on a runway that crossed mine. There was no way I could stop, but, fortunately, the plane hit the crossing a mere six feet before I did."

The world of hot rodding is mighty glad the two vehicles did not collide.

Ralph (right) and Fran (left) Bannister pose with their Ford flathead dragster with a 671 GMC blower (inset, above). Note the four-foot trophy for winning the New England Championship at Charlestown, Rhode Island, in 1957. Earlier, Ralph had won the Mobile fastest gas award and sportsmanship trophy at the NHRA 1955 Safari Go in Orange, Massachusetts.

The Nifty Fifties

The first rear-engined dragsters in New England in the early 1950s were built by the Bannister brothers, who raced against the winning No-Mads rear-motored dragster driven by John Sharrigan. Sharrigan won Top Eliminator at the 1955 Regional Championships at Orange, Massachusetts. Both flathead gassers exceeded 107 mph. (Photo by Jack Pennell)

Called "The Thing," Ralph Bannister's mid-engine flathead 1932 Ford made up in speed what it lacked in beauty. It rarely lost a race in 1954.

"Jazzy" Jim Nelson

Colorful "Jazzy" Jim Nelson, pictured with a fast roadster, was a dry lakes competitor even before World War II. He campaigned one of the first Fiat coupe-bodied drag racing machines from 1953 to 1955, when his little, fuel-burning, flathead Ford-engined altered ran in the 9-second, 130+ mph range to beat many big fuel Chrysler dragsters. In 1959, "Jazzy" drove Jocko Johnson's streamliner to a record e.t. of 8.35 seconds.

A Typical Drag Strip Layout

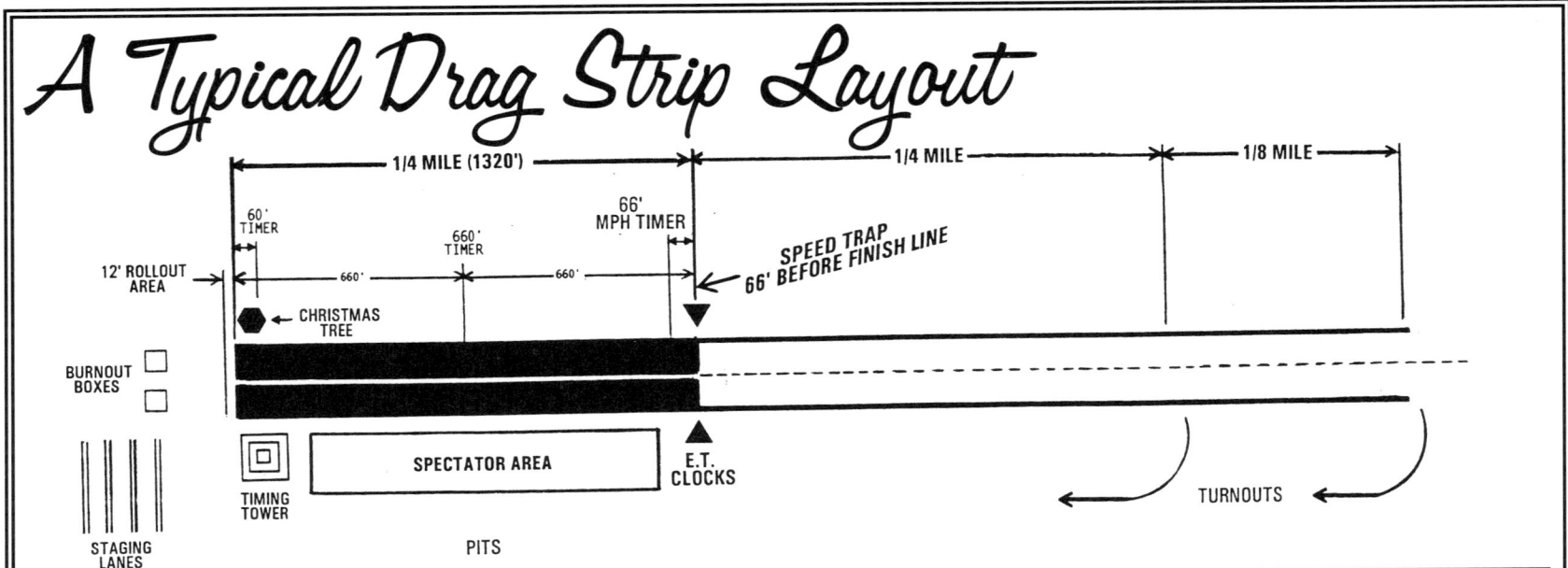

In a typical contemporary 1/4-mile drag strip, the timers measure both e.t. and mph. As the car leaves the starting line, it "breaks the beam" which activates the electronic timer. As the car continues, the timer records the elapsed time until the car "breaks" the finish line beam and stops the timer. Top speed is determined by the car breaking two additional light beams 66 feet prior to and at the finish line. **Previously, mph was calculated in a 132-foot speed trip, divided equally on both sides of the finish line.**

The Modern Moroso Motorsports Park, located in Palm Beach Gardens, Florida, is a multipurpose facility that has a 1/4-mile drag strip and a 2-1/4-mile road course. Annually, it attracts more than a half-million fans to more than 500 racing events and puts $18 million in the local economy.

ERIC "RICK" RICKMAN
Hot Rodding's Top-Dog Photographer

Much credit for the explosive growth of early hot rodding—and later, drag racing—should go to struggling photojournalists who gave national exposure to the sport. One photographer who stood above the rest was Eric Rickman, photo editor of *Hot Rod Magazine* from 1950 to 1970. Before retiring in 1991, Rick had become a team player for *Motorcyclist* magazine and other Petersen publications.

Soon after World War II, Rick began shooting images of midget and roadster races in the San Francisco Bay area. When he was selling action photos at a 1949 auto show, he met Robert E. Petersen, publisher of the new *Hot Rod Magazine*. In May 1950, Rick rolled up his sleeves at the magazine headquarters and converted a broom closet into a tiny, workable darkroom. The in-house photo lab helped *Hot Rod Magazine* grow into the sport's premier flagship. For more than 40 years, Rick documented the evolution of hot rodding and drag racing with vivid photographs that recorded the action of the NHRA Safety Safaris that crisscrossed the nation from 1954 through 1956. When Rick was covering Bonneville, NHRA, and other national events, his stance near the starting line at flywheel alley resulted in impaired hearing. However, he continued to cover sporting events and hot rodders that motored down the salt, track, and roadway.

Rick Rickman has been honored as a charter member of the Oakland Roadster Show, as Bonneville's 200-MPH Club "Man of the Year" award, and by NHRA, ISCA, West Coast Kustoms, and the International Drag Racing Hall of Fame.

Pioneer photographer Eric Rickman always was in the right place at the right time for documenting the evolution of hot rodding and drag racing.

Eric Rickman's enjoys his second childhood.

In the beginning, photo editor Rick helped deliver early issues of *Hot Rod Magazine* to Los Angeles area speed shops in this Model A pickup. (Circa 1950)

ED "BEATNIK" ROTH
Modern Rembrandt of Hot Rodding

The first Beatnik Bandit became a prized 1/25-scale model kit made by Revell.

If success could be measured by a person's contented countenance, then Ed "Beatnik" Roth takes the prize. Besides being a creative car builder, Roth is the Rembrandt of hot rodding. As a "weird artist," Ed inspired Revell, a large model company, to produce scale car models of his "Wild Ones." In the early 1950s, Ed began creating wild candy-colored and multi-faceted paint effects, with pin striping, scalloping, and customizing.

Eclectic Ed has designed and built more than 40 specialty cars. His futuristic show cars and hot rods, which have appeared in movies and on magazine covers, have taken top honors at major auto shows. His original Beatnik Bandit was constructed in 1960 and hailed as the "Hot Rod of the Future." The second Beatnik, a fiberglass-bodied roadster, made its debut at the 1995 SEMA show in Las Vegas.

"The new Beatnik Bandit was roughly executed in a gooey plaster, which was put there handful by handful by my lonesome and applied with a small trowel. Then I began shaping the body with a hand file," explained Ed about this stage that took longer than a month. "Then I covered the whole mess with fiberglass and began shaping it all over again, but this time with an electric grinder. The itch that results from fiberglass dust is gross. It gets into my skin and spreads to my other clothes and bedsheets. I itch like crazy for about six months after I finish a fiberglass body!"

Over the years, Ed has designed and marketed nerfing bars, striping paint kits, fiberglass cycle-type fenders, and a fiberglass-bodied Model T-replica roadster body. He also has traveled extensively throughout the

When I last saw Ed Roth in 1997, he was doing his usual zany signing of autographs for his many fans at the St. Ignace Auto Show in Michigan.

The Nifty Fifties

country, exhibiting his "way-out" cars, his artistically hand-painted T-shirts, and other unique "Roth Things." His trademarked Rat Fink and other grotesque humanoid figures are popular Roth specialties, and his Dumb Stuff Catalog is destined to become a collector's item.

Art is in the eyes of the beholder, and what a pair of colorful eyes Ed Roth has been showing the world!

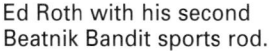

Ed Roth with his second Beatnik Bandit sports rod.

Wild and weird art such as Rat Fink is an Ed Roth trademark.

RAY BROCK
Promoter of the New Frontier

Ray is a native of Blackwell, Oklahoma, and earned a degree in electrical engineering before switching back to his first love. It was not long before his neighbor Don Francisco, a technical editor of *Hot Rod Magazine* who had a nearby engine modification shop, had Ray rolling up his sleeves, greasing his elbows, and delving into speed secrets. Don offered Ray a job as associate editor, which grew to publisher for the magazine.

Ray Brock, a giant in the publishing field, helped elevate the sport of hot rodding and its neo-car culture so much that the nation and Detroit couldn't help but notice. And why shouldn't they? First, hot rodders came out with muscle cars before Detroit copied them and capitalized on the whole business. Today, auto manufacturers continue to look to hot rodders for inspiration.

"Hot rodding is now stronger than ever nationwide," said Ray. "Street rod events attract tens of thousands of entries. Drag racing, nostalgia events, and car shows of all types are popular throughout the country."

This Stroppe-prepared pickup was class winner in Baja races, with Ray Brock sharing the driving with Ak Miller. (Circa 1971)

Ray helped build Ak Miller's T-roadster and was crew chief for the 1953 and 1954 Mexican Road Races. Left to right are Pete Coltren, Ray Brock, Porforio Rubirosa, Ak Miller, Clem TeBow, and Bob Petersen.

"*Hot Rod's* broadening circulation reflected that 80 percent of the U.S. population was east of the Rockies, and so were our subscription and newsstand sales," noted Ray, whose expertise helped the circulation of *Hot Rod Magazine* increase to more than one million. Sales also rose, partly because of Ray's insightful feature articles on automotive performance modifications, analyses of Detroit's new cars, and famous racing personalities, drivers, mechanics, and sponsors.

Over the years, Ray's competitive spirit propelled him into a variety of racing events, including the land speed record attempt with

The Nifty Fifties

Bonneville's first jet car and Daytona's Experimental Class record car. In the 1956 Mobilgas Economy Run, Ray and relief driver Wally Parks captured first place in their class with an Oldsmobile. In the East African Safari, Ray drove a 1964 Comet. Then he drag raced in NHRA Super Stock classes and co-drove in the pickup division with Ak Miller in several class wins in Baja off-road races.

After becoming publisher of *Motor Trend*, Ray became senior vice president for Petersen Publishing Company in 1970. After a 20-year association with Petersen Publishing, Ray left in 1972 and became founder of a new magazine called *Rod Action*, which he later sold.

Hot rodding owes much to men such as Ray Brock, a hot rod participant and technical writer, who helped shape the sport of hot rodding into the national, multifaceted giant that it is today.

Ray Brock was co-designer and crew chief of Bonneville's first jet car. Driven by Dr. Nathan Ostick, shown here, the car failed its 1953 land speed attempt because of soft sand. Best speed was 330 mph.

Don Waite

After a dry lakes run, Don Waite sits in the rear-engined roadster he later streamlined with an old race-car-style grille. Later, with the 1947 Mercury engine highly modified, the C roadster did a record 160 mph at a 1950 SCTA meet. Don set a new two-way average speed of 187.667 mph in a long-nosed DeSoto V8-powered B-modified roadster during the 1953 Bonneville Nationals. Don became a vice president with Edelbrock Co.

Hot Rod Pioneers

Auto Beauty – Shop Glitz

The art of pinstriping is centuries old, perhaps reaching its zenith in the decorative embellishments of fancy carriages, early cars, and fire engines. However, the mid-1950s saw the rebirth of pinstriping, followed by scallops, flames, scrolls, and clever paint schemes. Later, even Pontiac's Firebird had its colorful spread-eagle decal.

"Von Dutch" (Kenneth Howard) was the first customizer to gain national fame for his artistic pinstriping shapes that eventually became standard concepts for many decades. According to folklore, Von Dutch was such a "free spirit" that he might do each side of a vehicle differently.

"After all," Von Dutch once said, "a person could see only one side at a time."

After visiting California in 1952, where he "got the hots" for customizing with paints, Andy Southard (shown here pinstriping the hood louvers on his Deuce) became well known in the East. Many car magazines featured examples of his striping and scalloping. Now an accomplished photojournalist, Andy has written many books on hot rods and custom cars of the 1950s and 1960s, drawing from his collection of 77,000 photographs.

Tricking the Eye with Paint

Other artists such as Dean Jeffries and Ed "Beatnik" Roth followed, each with a unique brand of originality in striping, graphics, and exotic paintwork. Realizing that colorful graphics and sparkling paint finishes are the first things noticed on any automobile (especially show cars), customizers added even more pizzazz with paints in candy color, metal flake, suede, rainbow, multi-tones, and deep glossy clear coats.

Talented purveyors of these unique specialities exist today, along with easily applied pressure-sensitive tapes offering pinstriping, scallops, racing stripes, and other stick-on graphics.

JOAQUIN ARNETT
Co-Founder of the Bean Bandits

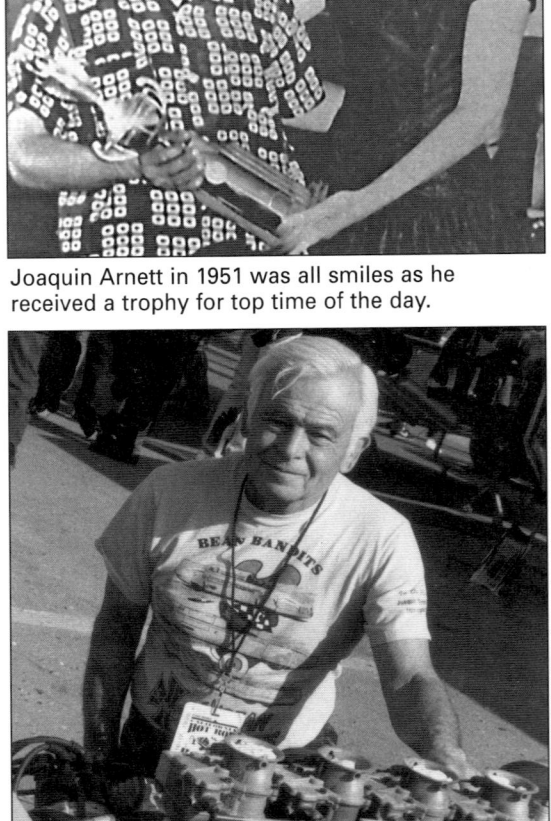

Joaquin Arnett in 1951 was all smiles as he received a trophy for top time of the day.

The legendary Bean Bandits Car Club began in 1951 when Joaquin Arnett and a group of friends voted to "legitimize" their clandestine drag racing at a nearby abandoned airstrip. The club's name reflected the members' Mexican heritage, but soon other ethnic groups in the San Diego area were included.

In 1953, Arnett won the first NHRA Drag Racing Championship at Pomona, California, in a Ford flathead-powered dragster. He and driver Carlos Ramirez topped the country's best rodders, capturing a trophy and a paltry $18 U.S. Savings Bond. Their highest winning time was 132.35 mph with an 11.08 e.t.

From 1951 to 1954, the Bean Bandits Club won more than 250 trophies racing in California, Arizona, Kansas, and Indiana. They traveled on a shoestring, and their oversight in paying the $15 registration fee cost them recognition in the winning record books.

From the start, the innovative Arnett was the club's chief car designer. He and other club members built one of the first twin-engine dragsters in the early 1950s. After competing in the dragster class, the foxy Bean Bandits would switch to the sedan class by bolting on a full-fendered 1929 Ford sedan body. When that race was finished, they could unhook an engine and compete again in a

Members of the Bean Bandits gather around their twin-engine dragster as driver Carlos Ramirez receives a winner's trophy at the Paradise Mesa Drag Strip in 1952.

Joaquin Arnett poses with a flathead, 47 years later and still smiling.

The Nifty Fifties

different class. The unique car won 25 trophies in four classes before the NHRA nixed the swapping idea.

In 1988, Arnett and the Bean Bandits came out of a 30-year retirement to run streamliners at the Bonneville salt flats. With Arnett's eldest son "Sonny" driving, the Bean Bandits set six land speed class records in nine years. After exhilarating triumph came heartbreak, when Sonny was killed in another record attempt in 1995.

Despair did not last long, however. Filled with zeal for life and enthusiasm for racing, Arnett and his reunited Bean Bandits plunged into a new form of competition—nostalgia and antique drag racing.

In 1992, Arnett was inducted into the International Drag Racing Hall of Fame. He was awarded the NHRA Lifetime Achievement Award in 1994 for his contributions to the sport.

Emory Cook, who was Joaquin's brother-in-law, often competed with a borrowed Bean Bandits' Chrysler engine. This Cook-Bedwell slingshot with direct drive had a carbureted, fuel burning Hemi.

Joaquin Arnett raced at Pomona in 1952. A year later, he built a 2/T roadster with a Hemi that was one of the first cars to exceed 150 mph in a drag race. This speed was once considered impossible.

This slippery fiberglass-skinned streamliner is one of four that Arnett and the Bean Bandits built to chase Bonneville land speed records. The first, in 1954, ran more than 130 mph with a Mercury flathead. The next streamliner with an Ardun OHV 1953 Ford reached 202 mph to win the (XX/FS) unblown fuel record in 1991.

Cook-Bedwell

The famed Cook-Bedwell fueler, owned by Emery Cook and Cliff Bedwell, was the first dragster to exceed 160 mph. In 1957, after Emery Cook drove the then-unblown Chrysler to a string of victories, including a record blast of 8.89 seconds and 169 mph, many thought the accelerating speeds of fuelers were becoming unsafe. Therefore, the NHRA initiated its unpopular and controversial fuel ban of 1957–1963.

"HONEST CHARLEY" CARD
Mail-Order Speed Shop

Folksy hot rodder Charles Card earned the name "Honest Charley" in his small Chattanooga restaurant where customers would figure out how much they owed for their meals and then leave their money in a cash box near the front door. In 1948, Charley opened a little speed shop in Tennessee, while keeping his restaurant in full swing. Because most of Charley's business was done by mail—primarily retailing major brands of speed equipment directly to customers—he sold auto products at a discount, and his 8-page catalog ballooned to 200 pages. From that point, the rise of Charley's business was meteoric as it grew from 350 square feet to more than 30,000 square feet by 1964, with several more additions later.

Because of Charley's volume purchasing ability, most manufacturers allowed him extra discounts, which permitted him to chop prices further. As the business prospered, Charley established a small chain of Honest Charley Speed Shops, but the timing was not right and the idea fizzled.

Today, the original Honest Charley enterprise is gone and so is its colorful founder. However, Charley's hot rod heritage continues, along with his comical decals, drawings, and catalogs that have all become collectors' items.

Similar to "Honest Hisself," Charley Card finds *Drag Comics* amusing. At a 1960 auto show, prankster Charley stuck comical "Honest Charley was here" labels on the coattails of unsuspecting guests. At the time, I was his largest competitor; therefore, it was no wonder that people were amused when I walked around with an "Honest" sticker placed where the sun doesn't shine!

Calvin Rice

Calvin Rice, drag racing's first NHRA National champion in 1955 (driving a flathead Mercury at 10.30/143.95), again made history three years later by shattering the FIA world acceleration record that had been held for 21 years by an Auto Union. The average speed of Calvin's Chrysler-powered dragster shown here was 123.56 mph, beating the German car by 6 mph.

Hemi Hopping

When Chrysler introduced its new Firepower V8 engine in 1951, few people realized that the 331 cube hemispherical combustion chambered mill would become the all-time champion on drag strips and oval circuits.

Soon after displacement was increased to 392 and horsepower to 300, the big Hemi hummers began crowding out the old Ford flatheads. When speed merchants tooled-up for Hemi goodies, the fate of the perennial flatheads was sealed forever, and many flathead converts joined the hummers' parade.

Factory production of the mighty Hemis stopped in 1958. However, when Chevrolet and Ford began catching up, especially in the NASCAR circuits, Chrysler resurrected its Hemi V8 in larger sizes. The seven-liter (426 c.i.) Hemi, released in 1963, quickly established itself as the engine to beat on any drag strip or oval track circuit. Unable to meet smog control requirements, production of the 426 was discontinued in 1971. However, the brawny "Elephant Motor" would remain the "biggie" of top fuelers.

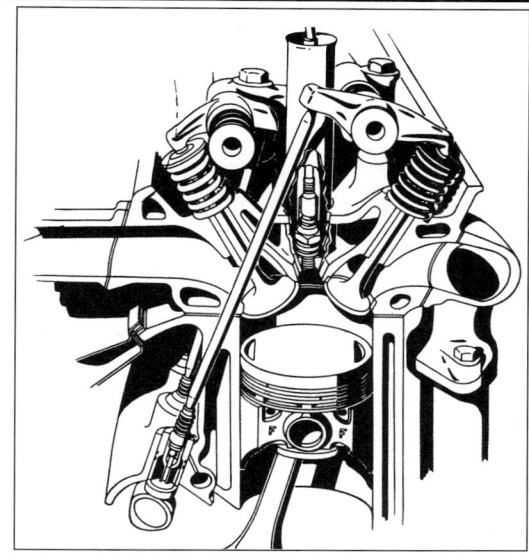

The secret of the Chrysler mill was its unorthodox heads. Unrestricted ports and large valves inclined toward the center of the hemispherically shaped chambers allowed maximized volumetric efficiency. The diagram is a cross section of a 1951 Hemi 5.4 L.

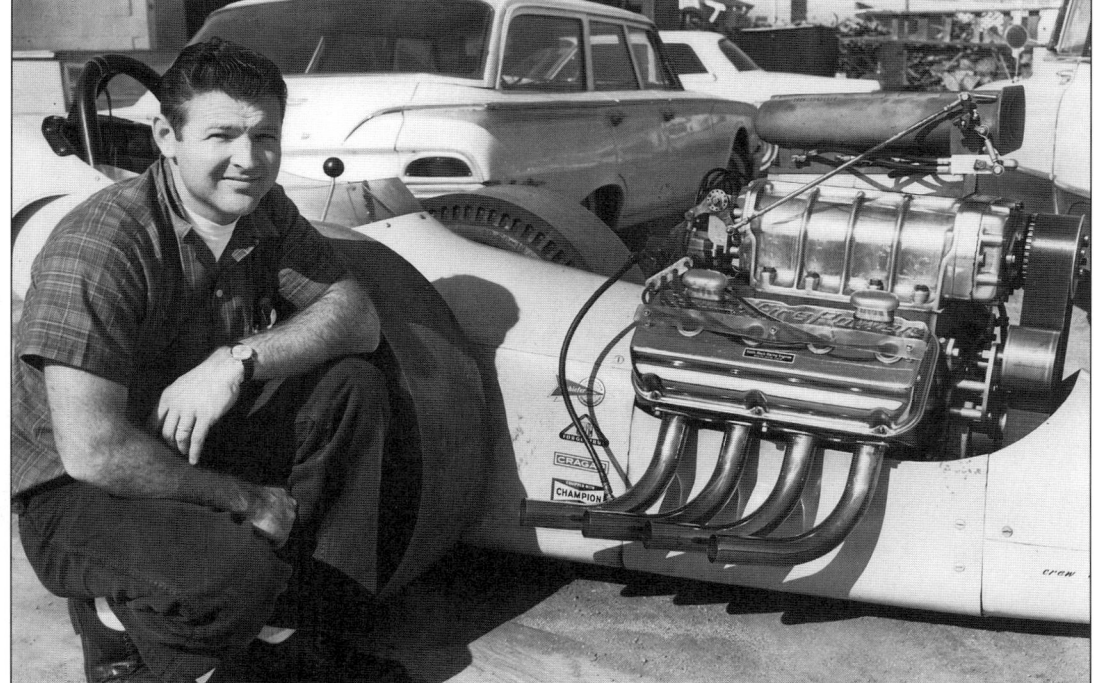

Legendary engine builder Keith Black is shown here with one of his first supercharged Hemi dragsters. Keith's "K-B" motors (and clones) practically dominated top fuel competition during the 1970s.

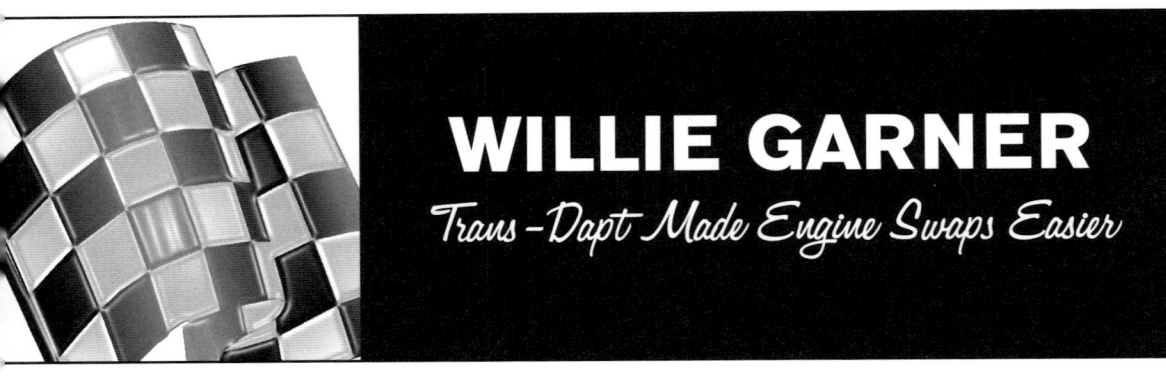

WILLIE GARNER
Trans-Dapt Made Engine Swaps Easier

Crossbreeding of later-model OHV engines into chassis made by other manufacturers reached its heyday during the 1950s, mainly because it was one of the most satisfactory ways to rejuvenate old or tired vehicles.

However, engine swapping was difficult and often impossible. When dry lakes racer Willie Garner saw the need for an easier way to mate the engines with different makes of transmissions, he undertook the monumental challenge of designing special transmission adapters. The resulting Trans-Dapt Corporation that Willie founded in 1959 quickly grew to become the leading manufacturer of engine conversion kits.

At one time, speed merchant "Honest Charley" Card became a silent partner who helped accelerate the development of a complete line of swapping products. As the industry expanded, personable Willie became one of the charter members of SEMA and its third president.

Trans-Dapt would remain Willie Garner's life passion until his death in 1977. Now, the name Trans-Dapt is retired, with Willie's son, Jan Garner, carrying on as sales manager of TD Performance Products, now a multifaceted division of Hedman.

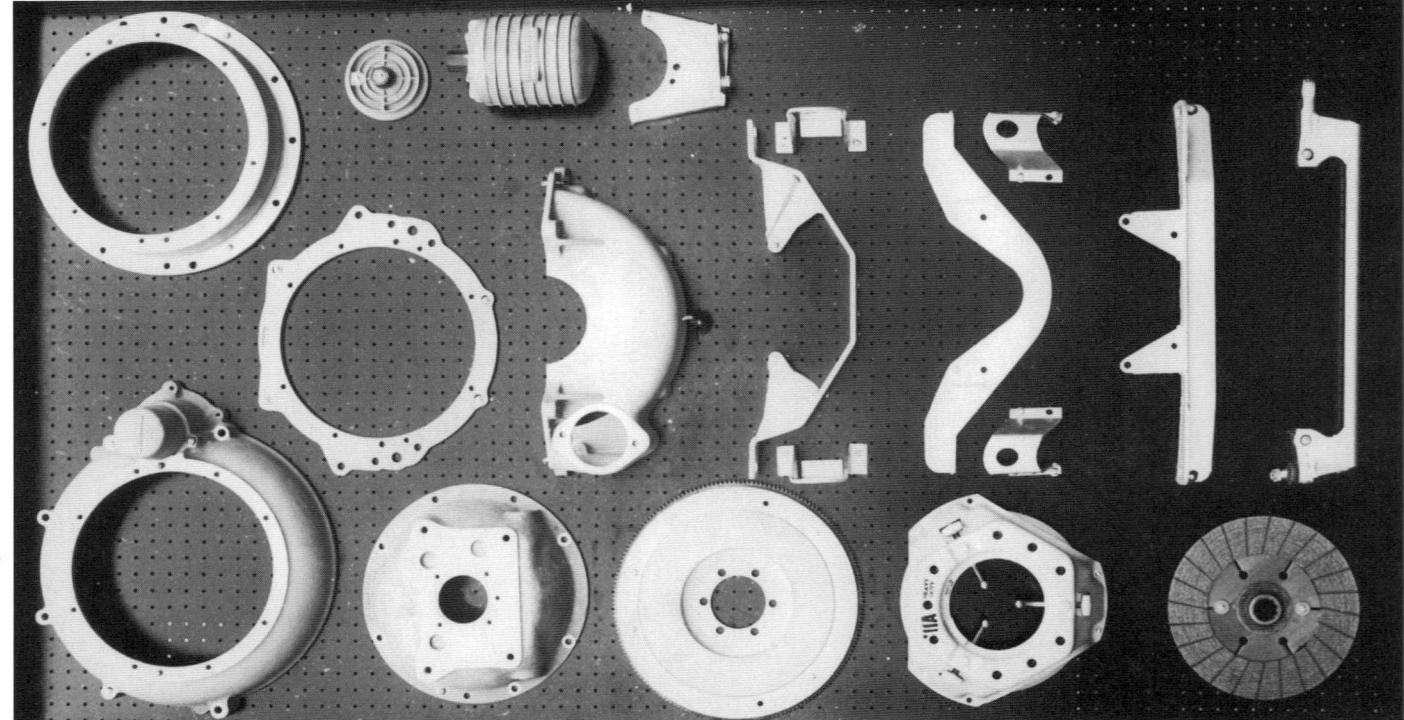

This display board shows typical engine and transmission adapters, motor mount kits, flywheel, clutch disc, pressure plate, and related goodies that were once popular for engine swaps in the 1950s and 1960s. (Photo courtesy of Hank Weidenhammer)

Joe Bailon really "splashed sparkle" in paint treatments when he came up with the "Candy Apple Paints," the envy of all customizers.

"In 1951, I accidentally spilled maroon paint over some gold powder and got a crazy notion for a three-stage formula consisting of a gold base, then a red, followed by a clear coat," recalled Joe. "The first car I 'Candy Appled' was the much publicized Miss Elegance '41 Chevy Coupe."

Since then, Joe has done a myriad of exotic color formulations on more than 1,000 cars that have passed through his custom auto shops. Because Joe's third shop was located in Hollywood, famous movie stars such as Dean Martin, Zsa Zsa Gabor, James Garner, and Sammy Davis, Jr. sought his work.

Now, almost half a century later at the young age of 75, Joe's genius for automobile creativity continues at his shop in Auburn, California, from where another custom styling trend may emerge.

JOE BAILON
Creator of Candy Apple Red

Before: Joe began this laborious restyling project by carefully chopping and lowering the top of the 1941 Chevy.

After: Who would guess that Joe Bailon's smartly customized Miss Elegance was once a 1941 Chevy Coupe? The color of this ground-hugging prizewinner is a deep glossy brownish-maroon.

This wildly decorated 1958 Impala had dozens of scoops. Its toothy chrome grill and nerf bars were made from ordinary electrical conduit pipe.

Hot Rod Pioneers

Drag Racing Dynamics
The Fastest Motorsport in the World

"Drag racing is like sex. People are going to do it anyway, so you might as well organize it and make it safe," commented former Indy winner Wilbur Shaw to the racing officials of the early 1950s. Even before that time (when they were called "acceleration contests"), club-sponsored drag racing, which had been occurring at unused airport runways and country roads, had already become legitimate in many parts of the nation.

Similar to today, the concept of a drag strip race was simple; however, reaching the finish line first was anything but simple. Two cars, matched nose to nose, accelerating in a straight quarter-mile line, called for skillful driving plus the know-how to create the swiftest car in its class—where victory often was measured in thousandths of a second, less than a blink of an eye.

In the early days before drag racing exploded into a national sport, clubs and drag strips in most states played by their own rules, often muddling the sport. Instead of standing starts, some allowed rolling starts, which unrealistically increased top speeds. Reported speeds often were met with suspicion and accusations of phony strip measurements, finagled clockings, and underhanded "track promoters." Also, lack of rule uniformity threatened fair competition, and reported strip "records" quickly were becoming a joke.

In addition to earlier regional and state associations' sanctioning efforts, the strong unification work that began in 1951 by the National Hot Rod Association helped promote safety, standardized rules, and even secured insurance coverage. As "dragging" moved away from its grassroots hobby level, it gradually was recognized as a *bona fide* sport rather than the juvenile delinquent game that the media had once labeled it.

Far from scientific, early competition often was a comedy of trial and error, with no precedent in basics such as weight

When Dick Kraft stripped his roadster to remove weight, he unwittingly created the world's first dragster or "rail." His Bug turned 103 mph at Santa Ana in July 1956. The two lengths of pipe served as a radiator.

At many drag strips, the flathead V8 Ford engine reigned supreme well into the 1950s when the new OHV V8s took over. Here, Ned Gibbons' three-carburetor Ford flathead roadster is being unhitched prior to a class win in Maine. (Photo by Jack Pennell)

This is how drag racing appeared near Boston in the early 1950s. Nick Nicoli's flathead V8-powered car (left) and a Deuce are being flagged for a quarter-mile run on the drag strip. (Photo by Jack Pennell)

reduction, optimum gearing, and improved traction for the quickest acceleration. Racers had to forget about terminal velocity and concentrate on reducing elapsed time. First-time contestants whose cars recorded the highest top speeds were amazed when they lost because of poor initial acceleration. Competitors rarely had perimeter protection such as sturdy roll bars, cages, or explosion shields.

"We drove without safety belts, harnesses, or even approved crash helmets," recalled one old-timer. "Parachutes and fire suits were non-existent. It's a wonder we didn't get killed."

Wally Parks once said, "From the start, the sport encountered—and hurdled—barrier after barrier. Initially, 127 mph was thought to be the ultimate speed for the quarter-mile." After the magical 150-, 200-, 250-, and 300-mph marks fell, pundits stopped guessing. Now a top fuel dragster or funny car accelerates faster from a standing start than a fighter jet and can travel the length of four football fields in less than five seconds.

After the eastern and southern styles of "run what you brung" grew from amateur 100-mph or higher stocker meets to include the pseudo-stock funny car spectaculars where "flip-tops" outran dragsters, Detroit's involvement aided the young sport in its growth, especially after the popular "super stocks" and F/X factory experimental classes officially were established. However, although fast-buck promoters were accused of turning the sport into a three-ring circus with exhibition jets, funny cars, and wheelstanders, contestants often were twisting the rules—so much that veterans such as Don Nicholson, Bill Jenkins, and Ronnie Sox (who probably were no Boy Scouts themselves) began pushing for the enforcement of firm, definitive class regulations. In 1969, when the Pro Stock class finally was established, many top professional teams helped promote its success. Today, Pro-Stock cars that closely resemble their "showroom" counterparts can run the quarter-mile in less than seven seconds with gasoline-carbureted engines that produce more than 1200 horsepower (approximately eight times that of the average automobile).

Well-run amateur club-run drag strips were operating throughout the country by the mid-1950s. This injected Buick-powered slingshot shows advances such as a narrowed rear, zoom headers, and streamlined body.

Hot Rod Pioneers

While the first skinny slingshot dragsters barely outran roadsters and coupes, they would become big time and continue to be rolling test laboratories for furthering new and intrepid ideas. With the advent of nitromethane, the supercharged top fuel dragsters became the "King Kongs" of the drag strip. Today's awesome 5000- to 6000-horsepower top fuelers defy the laws of physics and pull five Gs to leave the starting line with the explosive force of a space shot, going from a standing start to 100 mph in less than a second and reaching the quarter-mile finish line at more than 300 mph.

Gone are the days when a lone contestant could pay a nominal entry fee and run his machine on pocket change. Today, running a modern top fueler or funny car requires deep pockets. Expensive 17-inch rear tires wear out after only six or eight runs, and nitromethane fuel costs $300 for every quarter-mile run. (Ten gallons, at $30 per gallon, are gobbled in each complete pass.) Now, instead of trophies, the payouts (even for nonprofessional events that include contingency sponsors, points, and purses) can exceed one million dollars.

The overall complexity of modern competition classifications and interpretations of rules can be mind-boggling. What had started as a handful of classes running the gamut of "anything goes" has evolved into more than several hundred separate class divisions (grouped into thirteen major NHRA categories), in which tournament-style, two-car elimination matches are run.

Now, as the largest participant motorsport in the world, with million of fans watching a myriad of record-blazing competitors every weekend, drag racing again promises the greatest continued growth of all venues in motorsports.

Who would have thought so short a race would go so far?

The screaming funny cars that evolved from the F/X (Factory Experimentals) have been one of the big professional classes in drag competition. This early ram-tuned Dodge had a radically altered wheelbase and ran in the twelve-second range. By comparison, today's funny cars cover the quarter-mile in less than five seconds, while exceeding 300 mph! Although current bodies resemble production cars, this fiberglass shell hides a top fuel dragster, with the 5000+ hp engine in front of the driver. (Detroit Dragway photo)

From their inception, Stockers represented the majority of all drag strip entries, with Super Stock being one of the most competitive classes. So-called Altereds, which began as a class between street machines and dragsters, were actually light (usually Fiat or Crosley) bodies on slingshot frames that made their debut around the mid-1950s. They were followed by the popular Gas Coupe/Sedan Classes (mainly Willys coupes) that terrorized competition in the big "gasser wars" of the 1960s.

"SPEEDY BILL" SMITH
Founder of Speedway Motors

Call D. William Smith what you will. Call him "Speedy Bill" for his talent, tenacity, and racing attributes. Or call him "Dollar Bill" for his successful real estate holdings and The Smith Museum, which is a treasury of antique and exotic racing engines, cars, and memorabilia in Lincoln, Nebraska.

By 17 years old, Bill was a serious racer.

"When I started in the late '40s," said Bill, "I did it all—from midget and dirt track circle racers to Indy and land speed cars. Basically, anything that had wheels, I raced."

After college, Bill borrowed $300 from his wife Joyce and boldly opened a speed shop. The tiny sideline that began as Speedway Motors in 1952 is now the world's largest racing equipment supplier. The company also manufactures street rod and classic car kits.

With driver Tiny Lund, Bill Smith built one of the first Pontiac late models to be raced in NASCAR in 1956. In that decade and the next, Bill sponsored numerous winning dirt track racers, stock cars, and dragsters.

Bill's many honorary awards include SEMA, National Sprint Car, and National Street Rod Hall of Fame. Despite the devastating news in the 1950s that he had only one remaining year to live, Bill magnificently beat the odds by making his triumphant, entrepreneurial mark in motorsports.

Bill and Joyce Smith's dream turned into a fortune in a cornfield in America's heartland. Their hard work, love for cars, and faith in the Almighty brought the couple success and happiness.

This roaring roadster had a chopped and narrowed Dodge body on a shortened Model A frame. The 1940 Ford flathead was a stock 85 hp. The roll bar made from water pipe helped protect the youthful driver, Bill Smith.

Bill Smith's 1934 Ford coupe ran second in this early stock car race in a Nebraska field. The two cars on the left were driven by Daytona 500 winners Tiny Lund in 1963 and Johnny Beauchamp in 1959.

AARON J. FENTON
A Born Salesman

The 1950s brought fleeting fame and fortune to Aaron Fenton. By the end of the next decade, Fenton's name in the speed equipment industry had become a white dinosaur. Despite his fall from grace, Aaron had an indelible influence on the young hot rod industry, eventually changing the way it did business.

Aaron J. Fenton was a true marketing forerunner who refused to cater to speed shops. Instead, he wooed volume mass merchandisers such as national chain stores, which some perceived as cutthroat competitors. By shaving wholesale prices, Aaron became a thorn in the side of some hard-core manufacturers.

Fenton's moderately priced speed and custom equipment line included cast iron dual exhaust headers, straight-through mufflers, and multiple intake manifolds that he often private-labeled for other firms.

As an amicable and hard-working professional, Aaron Fenton could easily "sweet talk" the big volume purchasers with a promise of quick delivery. At that time, slow delivery had become a common problem in the industry. Therefore, Aaron Fenton quickly seized my Montgomery Ward account when my Almquist dual intake manifolds and cylinder heads that were listed in the 1952 Montgomery Ward catalog were not readily obtainable.

Because Fenton did not wine and dine the "right people," he never received any recognition from the West Coast magazines. Not surprisingly, his health declined from his workaholic "Type-A" personality. Fenton's business, which was prospering to the point of becoming the largest company in the speed business, soon saw its demise.

Aaron Fenton (center) with racing champs Troy Ruttman (left) and Jerry Unser (right), who endorsed the new Fenton dual-triple Chevy V8 manifold. Troy won the 1952 Indy 500 after winning midget and roadster titles. Jerry won the Pikes Peak Run in 1956 and the ASAC Stock Car Championship in 1957.

Lloyd Scott

One of the most unorthodox early drag racing machines was the twin-engined Bustle Bomb. Driven by Lloyd Scott in 1955 at both the big ATAA and NHRA meets, reaching 151 mph. Cleverly built by Scott, Noel Timney, and crew, the front (Olds) engine was positioned in front of the driver, and the Caddy engine was behind the rear axle and reversed. Transmission of power was an ingenious setup whereby the Caddie engine was locked in direct drive; thus, it always was pushing in high gear while the Olds engine picked up speed through the gears. Although fast, the Bustle Bomb proved too complex, expensive, and finicky to continue as a role model for future dragster design.

Road Racing Finally Recognized

Watkins Glen: The Mother Lode

When America's first postwar sports car race was run at Watkins Glen in 1948, the European style of road racing literally was unknown in the United States. At that time, America had no established road courses and few true sports cars. However, after the first road race in 1909 in Portland, Oregon, few drivers in later years would compete where snappy gear changes and supertight left and right turns were the gnarled crux of it all.

A young law student named Cameron Argetsinger was a pioneer member of the Sports Car Club of America (SCCA). Cam's zeal for racing inspired him to start a road course in his hometown of Watkins Glen in picturesque upstate New York. Using both the main street of the town and the countryside to outline a 6.6-mile course, Cam's dream became a reality when 15 assorted cars qualified for the starting line. Among them were owners of small MGs and a few hot rod-type specials. Briggs Cunningham had a hybrid Merc, and Bill Milliken, who had finished sixth the previous year in the Pike's Peak Race, drove a Type 35 Bugatti. Ten cars finished, but an Alfa-Romeo took the flag.

Although later Grand Prix events were successful, an unfortunate spectator fatality had ended racing on the streets of Watkins Glen. Although the race committee began establishing another road course nearby, Bill Milliken, Cam Argetsinger, and others promoted the construction of a safer 2.3-mile course, designed specifically for road racing.

Road racing soon expanded across the country, with world-class events at Sebring, Florida; Bridgehampton, New York; Elkhart Lake, Wisconsin; Riverside and Long Beach, California; and other places where the rigors of road racing tested the mettles of both driver and car.

Often shunned by the elite of sports car racing, the lowly hot rods eventually had their days in the sun at the early Trans-Am and other autocross events. This was especially true when Ak Miller's hot rod beat the imported GT sports cars that cost exorbitant amounts of money.

Many road racing specials that raced during the 1950s were actually home-built hot rods—usually flathead-powered Ford V8s—but often with mediocre suspension and poor handling. Well-known hot rodders who crossed the bar into big-time road competition were Dan Gurney, Bill Stroppe, Troy Ruttman, Ray Brock, and actor Jackie Cooper.

In addition to the popular Trans-Am series, SCCA now conducts a mix of more than 2,000 annual competitions, ranging from highly visible professional series to full slates of amateur classes, rallies, and solo events.

Overall, the hero of America's Grand Prix remains Cam Argetsinger who, with a dedicated group of volunteers, "sparkplugged" its success.

This early road race shows Basil Panzer's Allard J-2 chasing James Seely's Cannon sports car. (Circa 1950)

"BIG DADDY" DON GARLITS
God Is His Co-Pilot

Similar to the old Biblical tradition in which David took on Goliath, Don Garlits took on the super strokers of speed. Time and again, the media brushed him aside and denounced his wins as phonies. However, Garlits, an East Coast guy, continually showed the West Coast boys that he could smoke his ugly little dragster down any strip and make them choke in dismay. Over the years, Garlits, who has sometimes been a paradox of grace and ferocity, became a role model for motorsports.

Born in 1932, Don was raised on a farm in Florida. When his high-school teacher referred to hot rodding as "bringing out the best in mechanical ingenuity," Don listened. Later, while apprenticing at a Chevrolet agency, Don stripped down a Model T and bolted in a mildly hopped-up Ford flathead engine. After the spindly rail job earned Top Eliminator honors at a nearby drag strip, Don was hooked. During 1956, Don and his brother Ed won races all over Florida with the homely little dragster, to which Don added on a used 1954 Chrysler V8 engine and a home-welded six-carburetor manifold. The next year, Don converted the dragster to fuel and set the Florida state record of 144 mph. After beating the famed Cook-Bedwell Special dragster (156.21 mph), Don became determined to break more records.

"I stumbled on another simple speed secret that cured fuel starvation by simply upping hose sizes," said Don. "With nitro fuel, the engine developed 650 to 700 hp, transmitted directly—without gear box—through a 3.63:1 ratio."

Don's Swamp Rat soon made nationwide news at Brooksville, Florida, with the fastest time yet in the quarter-mile (8.79 e.t./176.40 mph).

Don Garlits is shown here with his record-breaking Swamp Rat in 1958. Its total building cost was only $2,500. The 1957 Chrysler engine used Jahn's pistons, Scintilla mag, eight-carburetor log manifold, Isky cam and valve gear, Almquist headers and tachometer, Schiefer double-disc clutch and flywheel, and direct drive with 3.23 gears. Frame rails are 1931 Chevy with rear tread narrowed to 40 inches.

A self-taught mechanic, Don balances a crankshaft on a Stewart-Warner Balancer. Don was at the forefront of numerous technical innovations, including extended wheelbase, cycle front wheels, airfoils, and the first with a winning rear-engine dragster design.

Some of Don Garlits' Racing "Firsts"

- Won 146 national event titles
- Three-time NHRA World Champion
- Ten-time AHRA World Champion
- Four-time IHRA World Champion
- First to exceed 170, 180, 200, 250, 260, and 270 mph

With this stock 1950 Ford, Don won his first trophy at central Florida's popular Lake Wales Drag Strip. Then, after racing the 1927 T-roadster a few times with little success, Don lengthened the frame and installed a Ford V8 flathead to create one of Florida's first slingshot dragsters. (Circa 1953)

"I figured out later that the smug Californians thought it impossible for anyone to outdo them—especially a young hick from swampy ol' Florida," continues Don. "Hearing their dastardly bragging made me want to whip those guys so bad that I could taste it."

By the end of 1957, Don and Ed added five more inches to the wheelbase of the dragster to obtain more leverage. They also switched to a lighter front axle with motorcycle wheels and prayed that the big Chrysler would outpower the West Coast boys at the International Timing Association meet in Chester, South Carolina. (It almost did.) After a big win over Setto Postoian, Don became Top Eliminator in Texas, where he received $450 in appearance money which was big money back then.

A rip-roaring showdown finally came in 1958 at the Freeway drag strip near Houston, Texas, for an East versus West competition for top-cars-only event. Garlits' dented old campaigner, which was not in league with the pristine California cars, belched clouds of smoke from the onset.

"If the jeers about my oil burner were not bad enough," said Don, "hecklers began sneering about my turtlelike starts. After replacing the cam and valve train, I whipped the Californians in three runs and finally got some respect."

Later that year, Don was the first to exceed a top speed of 180 mph with an e.t. of 8.90 seconds. The NHRA certification shamed and silenced the Californians for good, as magazines quickly picked up the scoop. Subsequently, smart race track operators touted the Swamp Rat's moneymaking potential.

In 1964 at Great Meadows, New Jersey, the NHRA clocks timed the Garlits' dragster, which broke two records and was the first to exceed 200 mph with a run of 201.34 mph and an elapsed time of 7.78 seconds.

In 1971, Don was still the top fuel driver and built a rear-engine, front-drive dragster that soon became the standard for the sport and, most likely, saved lives. That same year, Don was honored at the White House.

Don won the 1975 NHRA World Championship with his 250-mph record, which stood for seven years. Eleven years later, "Big Daddy" was again named the NHRA Top Fuel World Champion. In 1987,

Don and his wife Pat enjoy trophies after winning the AHRA National Championship in 1958. Don, who was virtually unknown a few years before that, went on to win 17 world titles.

Hot Rod Pioneers

the Smithsonian Institution accepted Don's Swamp Rat III, the first top fuel dragster to exceed 270 mph in a quarter-mile sprint.

"In the early years," said Don, "my wife and I traveled everywhere in a GMC panel truck, towing the dragster on an open trailer. The tools were stored in the rear of the truck, with a mattress over them for our two young daughters to sleep on. Because we kept such a busy racing schedule, racing on a Friday and driving maybe 500 miles all night to race again on Sunday, the diaper pail, stored in the open dragster seat, was always full. It really is amazing how you can adjust to circumstances."

By the end of his active racing career, Don had acquired armloads of trophies and grossed more than seven million dollars. After winning the 1968 Nationals, Don was named drag racing's Man of the Year, Driver of the Year, and Chassis Builder of the Year. However, as a professional driver for more than 40 years and a drag racing legend, Don had paid a heavy price. His 1959 blower explosion gave him severe burns, his 1965 Great Meadows' crash resulted in a broken back for him, and a 1970 transmission explosion tore off a portion of his foot. At age 60, after almost 10,000 quarter-mile passes, Don was forced to retire due to detached retinas in both eyes.

Whether we refer to him as "Swamp Rat," "Mr. Top Fuel," or "Big Daddy," Don Garlits is the King of Drag Racing. Garlits, the man who once painted a religious cross on his race car, says that his life has been rooted in faith and that God has always been there for him.

In 1959, with a $500 investment, Don opened this small speed shop near Tampa. With business booming, Don opened his "Hi-Performance World" a decade later. After the gas crunch of 1974, Don closed the shop, deciding he had "too many irons in the fire."

It was drag racing at its finest, with almost every top champion competing for seven eliminator crowns at the big NHRA 1968 Springnationals in Englishtown, New Jersey. Here, "Big Daddy" Don Garlits in his stretched dragster is about to capture his fourth NHRA Top Fuel Championship, with a scorching 6.80/222.76.

Retired "Big Daddy" with Swamp Rat after it was restored. The hemi-powered dragster, shown in front of the Don Garlits Museum, also was displayed at the Smithsonian Institution.

The Art of Custom Restyling
It Began Where Car Makers Had Ended!

"It was like turning a sow's ear into a silk purse," Gene Winfield said about the fascinating new hobby and business of automobile customizing, which flourished soon after World War II and suddenly aroused the country's car-hungry youths. Many of these young people were tired of the boxy, "plain Jane" sameness of Detroit cars and their quick obsolescence. They wanted something different, with a sharper and sexier appearance.

Custom shops and individual enthusiasts soon discovered they could greatly improve the appearance of almost any car, especially dated ones, while easily outdoing factory stylists whose designs often were strangled by cost compromises. The early customizers' search for originality and individuality often led to endless experimentation in backyard garages—sometimes spawning new trends that combined aesthetics of form and function that car makers would follow.

In this unique forum, no two custom cars were alike. Tastes differed then as they do now. Some auto buffs preferred the sculpted lines of Italian coachworkers. Some liked the finesse of straight-line streamlining, and others drooled over racy, low-slung open roadsters. There probably are die-hard old-timers who still refuse to drive anything that has fenders.

Mid-century restyling modifications ranged from mild to wild and anything in between, with many requiring expert body-shop work. In the reversal of Detroit's obsession with the gaudy use of exterior body chrome, typical customizers usually removed (shaved) superfluous chrome strips and removed (nosed) hood ornaments and medallions. They utilized molten lead to smooth over holes and body seams in the process called "frenching," which created the prized one-piece sculpted look. Headlamps and tail lights often were frenched or recessed into the fenders.

A full-blown customizing treatment included laborious and costly changes that usually involved chopping the top and/or channeling or sectioning the body to lower the profile. "Z-ing" the frame or reworking the springs or front axle also brought the car closer to the ground. Sometimes the daring and radical show cars of the 1950s pushed the design envelope so far to the edge that even the winners couldn't be driven on streets.

Now that so many contemporary automobiles appear alike, it is a wonder that the magical art of custom restyling has not returned *en masse*, with a new generation of low-riding, late-model custom cars and sport utilities. These can be exhibited proudly beside the many thousands of shoebox clones, low-rider customs, resto, and street and rat rods at annual auto shows. Recent automobile customizing is leaning toward "conservative personalizing," with smartly coordinated accessories, special top treatments, tuck-and-roll upholstery, and ground effects that often are combined with meticulous detailing and show-car paint finishes. "New Age" customizing is bent on blending the genuine custom icon appearance with contemporary (and powerful) drive trains and suspensions to create a combination custom/street rod.

The following photographs illustrate various custom styling trends—some of which are even kooky—that began in the 1940s and onward. Many of these design elements influenced auto makers in the past and even now are mimicked for the "neo retro" look.

This $18,000 1963 Rivera was restyled by George Barris for the movie, "For Those Who Think Young," starring James Darren and Nancy Sinatra.

Cars That Never Grow Old

Popular open-wheel street rods such as the Deuce, "T," and custom-bodied machines shown here are often referred to as classic or traditional hot rods—all with ancestry that arguably traces back to track roadsters of old. Although some retro rods and T-buckets may appear outwardly similar, paradoxically they all will be different—a tradition of hot rodding.

Rather than build another cloned kit car, Alex Augustin handcrafted this handsome pink scalloped street "T" over a two-year period. High road clearance and side curtains make the Chevy-powered car usable year round in rural New York State.

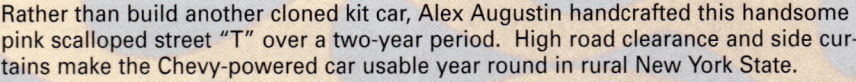

"Lead Sled" Mercurys

The venerable 1949–1951 shoebox Mercurys were among the most memorable customs ever built. Often, these cars were called "lead sleds" because of the liberal use of molten lead for filling seams and creating new body lines. A favorite among nostalgic customizers even today, the gently rounded bathtub bodies of lead sleds lend themselves to a variety of restyling treatments such as mild dechroming, nosing the hood, frenching the headlights, swapping grilles, adding spots, special tail lights, and fat cruiser or bubble skirts. Retaining or adding Buick side spears helped highlight two-tone paint schemes. The entire bodies rarely were sectioned; however, top chops of as much as four inches were popular. Sometimes the hood also was sectioned two inches and the headlights extended and dropped. These photos show variations of "old school" restyling themes.

Front Facelifts

Good-Looking Rears

Realizing that a car should look good both going and coming is why most enthusiasts gave top priority to rear-end restyling. In addition to shapely fender skirts, a nuance of much desired individuality was accomplished simply by swapping tail lights from another model car, by installing special lens, or by "frenching in" custom tail-light assemblies. Shaving the rear deck by removing extra chrome trim, emblems, and the trunk handle, plus repositioning the license plate, was a "must" (with the push-button electric or manual opener hidden).

Similar to today, no self-respecting rodder of old would be seen or heard without rumbling dual exhaust pipes. Customizing enthusiasts, who would not settle for only fancy chrome pencil-tip extensions, mounted lakes pipes outside the vehicle. For drag racing, functional types were hooked to exhaust headers with removable plugs. The more radical approach was to push the exhaust plumbing through portholes in the body or through each end of the bumpers. Either way was exhausting!

At the peak of the tail-fin craze in the late 1950s, some customizers added even sharper shapes by using either the "Caddy-style" bolt-ons or by fiberglassing on higher, longer, and wilder custom fins in what seemed to be a contest of bad taste and "one-upmanship." Bolder, more timeless customizing treatments included resculpturing of the rear fender and trunk area to complement the basic styling of the vehicle.

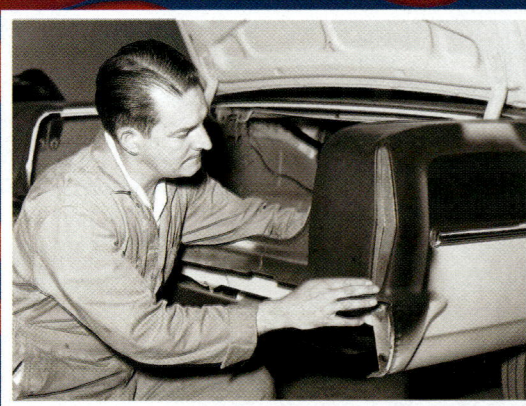

Street Rod Evolution

Bygone memories of 1950s and 1960s fenderless hot rods, candy-coated customs, and throaty motored muscle cars have sparked the baby-boomer nostalgia that drives even contemporary trends in street rodding. Since the 1970s, a renaissance of "modernized" street rods and customs has occurred, with the appearance of hybrids that combine modern, environmentally friendly, late-model powertrains with old-style Deuces, 1937 roadsters, 1950s shoeboxes, and even 1960s and later smoothies. The high cost of most of these cars (often more than $100,000) threatens to change the face of street rodding. When auto appraiser Tony Monopoli (bottom right) modernized this neat 1934 Ford Coupe, he valued it at more than $25,000.

A promising and often less costly way to meet emission requirements is to graft either a custom body or retro parts onto a late-model chassis and powertrain. A good example is a T-Bird fitted with the front and rear sheet metal of a 1949 Ford, upper right.

The low-slung "Oldie" (lower right) exhibits timeless sports car styling despite its extra jump seat.

Trick Dream Trucks

Customization of pickup trucks, vans, and SUV's began in the 1980s and is now part of America's rapidly growing truck culture. Wild paint jobs, fender flares, spoilers, air dams, and special wheels abound, but the covered pickup boxes carry little more than air.

Making dream cars for movies often has mixed rewards — such as the $200,000 Dean Jeffries was paid for the futuristic van he built for the 1977 film, "Damnation Alley." Fans hated the movie but loved Dean's van that could practically climb straight up a mountain.

By the mid-1950s, even before Dean became a movie stunt driver and automobile customizer, he had created a gallery of colorful pinstripes, scallops, flames, and exotic paint jobs with a unique brand of originality that brought him national recognition.

DEAN JEFFRIES
More Than a Pinstriper

The 1933 scalloped red-and-white chopped Ford is still Dean Jeffries' everyday car. Dean built the futuristic space vehicle (shown in rear) for the movie "Space Balls."

In the 1960s, when dune buggies were the craze, Dean (standing) manufactured more than 500 of his handsomely styled Coyote Dune Buggies that were powered by a hopped-up VW engine and componentry.

"Hanging around racing people as a young guy changed my life," Dean says. "After I painted 23 out of the 33 starting cars for the 1961 Indy race, I became a close friend of A.J. Foyt and served on his pit crew."

Ultimately, the handsome and charismatic Dean broke into the TV and movie industry, not as an actor but as a stunt driver. Dubbing for Frankie Avalon, Dean drove a Jeffries-designed, sleek bubble-top creation in the film "Bikini Beach Party." After injuring his back, Dean moved on to build movie car creations, such as a moon buggy for James Bond in "Diamonds are Forever," the Landmaster for the science-fiction flick called "Damnation Alley," and a trolley car from an old RTD bus for "Who Framed Roger Rabbit?," the sexy marriage of colorful animation and live drama.

Dean, the former twelve-year-old who had hot-wired and smashed his father's car, dis-

Dean Jeffries poses with his show-winning bubble-top Mantaray. The walls of Dean's office are papered with cars he has designed and with mementos of his days as a Hollywood stuntman.

plays in his office the Mantaray, the first car he ever built. Thirty-five years ago, he had entered that sleek automobile in a car-building contest and received the coveted Oakland Tournament of Fame trophy for the best hot rod. From his first-place win, Dean received a free trip to Europe and the cover spot on a 1964 *Hot Rod Magazine*. That same futuristic car was exhibited in the 50th edition of the Grand National Roadster Show. Whoever said "Old hot rodders never die" surely was right.

East Versus West: Drag Racing's Surprising Comeuppance

"The East has not only caught up with the West in hot rod development—but it's gone ahead," exclaimed an early 1960s article in *Popular Hot Rodding* magazine. At the start, acceleration contests, which began in Southern California, were light-years ahead of the rest of the country in organization, workmanship, and performance. In addition to their head start in many technical developments such as supercharging, twin engines, nitro fuels, and even "slingshot" dragsters, the West Coast boys had year-round strip operation and better weather on their side. However, the national trend to muscle cars and late-model sedans that replaced interest in the traditional open hot rod (i.e., 1932 roadster) soon carried over into drag strip meets, where stock car classes became the core activity and eventually the major attraction.

With the quiet evolution that resulted in astounding performances of "non-California" cars, the face of drag racing soon changed forever through catch-up that began in the mid-to-late 1950s, with virtual unknowns trouncing top Western draggers, often on their own strips. Future greats such as Don Garlits, Art Malone, Melvin Heath, Setto Postoian, and George Montgomery were early examples of Eastern superiority that had received little publicity. Instead, the naturally biased California-based car magazines buttressed and dramatized nearby events and played up their "favorite sons."

After the "gasser wars," which included the long reign of George Montgomery, another surprise came for top dragster stars. They had to move over for popular super stocks, F/X, and funny cars, which came from almost every state.

In less than ten years, drag racing had become a national sport, with fiercely contested events that spread from coast to coast. The beginning of world recognition for drag racing came in 1965 when the NHRA became a member club in ACCUS, the American Competition Committee for the Federation Internationale de l'Automobile (FIA). This initiated a World Championship Series of events across the country, and drag racing was on its way as a world motorsport.

Art Arfons (left) of Ohio is shown here with his GE J-47-powered Cyclops. Art pioneered aircraft engines in drag machines. Despite its heavy weight, the streamlined Green Monster dragster (on the truck) did the quarter-mile in 6 seconds at 270 mph.

The Nifty Fifties

NHRA NATIONAL CHAMPIONSHIP RESULTS

NAME			E.T.	SPEED
Top Eliminator				
AA/Dragster		Rodney Singer, Houston, Tex.	9.76	152.00
Little Eliminator				
A	269	George Montgomery, Dayton, O.	12.17	120.00
Gas Coupes and Sedans				
A	269	George Montgomery, Dayton, O.	12.17	120.00
B	226	James Whitaker, Indianapolis, Ind.	14.19	92.23
C/gas	299	Doug Cook, Compton, Cal.	12.92	106.64
D	261	Billie Norris Hiett, Nashville, Tenn.	15.05	95.23
E	21	John Loper, Phoenix, Ariz.	14.94	93.45
Sports Cars				
AA	366	Bill Jones, Oak Hill, Mich.	Lone run, no E.T. or Time	
A	28	Jack Horsley Jr., Miami, Fla.	12.11	122.78
B	100	Andrew Zanca, New Orleans, La.	12.96	110.02
C	509	Gilbert Dunne, Rockaway, N.J.	16.99	81.00
D	None	None		
E	30	Ernest W. Grimm Jr., Tulsa, Okla.	18.04	75.56
Dragster				
AA	504	Rodney Singer, Houston, Tex.	9.70	151.00
A	161	Jack Chrisman, Long Beach, Calif.	9.58	153.06
B	183	Lewis Garden, Birmingham, Ala.	10.47	133.13
C	329	Walter Markert, Amherst, Mass.	11.42	128.53
D	507	William Schlott, Lincoln, Penna.	11.77	108.95
Modified Roadster				
A	452	John Klein, Northboro, Mass.	11.76	116.27
B	217	Fred Dobney, Roll, Ariz.	12.41	115.23
Competition Coupe and Sedan				
A	146	Jiggs Shamblin, Akron, O.	10.56	135.15
B	12	Robert Daniels, Marysville, O.	12.74	113.49
Roadster				
Middle Eliminator				
A	47	Otis Smith, Akron, O.	10.61	148.27
B	116	Dave Whalen, Cleveland, O.	12.05	117.03
Altered Coupe and Sedan				
A	317	Gabby Bleeker, Chicago, Ill.	11.02	127.84
B	138	Don Breithaupt, Grand Prairie, Tex.	11.91	118.42
C	93	Billie Rasmusson, Fort Worth, Tex.	12.80	107.14
Street Roadsters				
A	120	G. A. Jones, Lamesa, Tex.	11.61	125.87
B	82	Peter Mattei, New Orleans, La.	13.05	107.91
Four Barrel				
A	176	Ray Huckabee, Houston, Tex.	12.78	105.63
Dragster				
AA	111	Art Arfons, Akron, O.		170.45
A	192	Alan R. Thompson, Aurora, Ill.		159.85
B	191	Fred J. Smith, Elgin, Ill.		136.98
C	329	Walter C. Markert, Amherst, Mass.		129.49
Four Barrel				
X	176	Raymond Huckabee, Houston, Tex.		106.25
Modified Roadster				
A	361	Dudley D. Proctor, Wayne, Mich.		136.98
B	217	Fred Dobney, Roll, Ariz.		118.42
Roadster				
A	47	Otis Smith, Akron, O.		145.39
		The following two entries tied.		
B	116	David Whalen, Cleveland, O.		115.83
B	66	Willis Ragsdale, Pasadena, Tex.		115.83
Street Roadster				
A	120	G. A. Jones, Lamesa, Tex.		124.30
B	82	Peter Mattei, New Orleans, La.		109.89
Altered Coupe and Sedan				
A	128	Walter J. Knoch, Dearborn, Mich.		138.67
B	138	Don Breithaupt, Grand Prairie, Tex.		111.80
C	325	Herman Mozer, Detroit, Mich.		109.75
Competition Coupe and Sedan				
A	146	Jiggs Shamblin, Akron, O.		140.62
B	12	Bob Daniels, Marysville, O.		112.78
Sports Cars				
A	373	Jim White, Toledo, O.		114.33
B	374	Ronald G. Frost, Center Line, Mich.		109.89
C	509	Gilbert Dunne, Rockaway, N.J.		83.33
D	None	None		
E	30	Ernest W. Grimm Jr., Tulsa, Okla.		76.07

DECEMBER, 1959

By the late 1950s, the West Coast had lost its dominance in drag racing. Don Garlits had whipped all taunting challengers at the big Texas showdown in 1958. There was only one winner from California in all major classes in the 1959 AHRA championships, and only two in the mighty NHRA Nationals. However, that year, it was a shutout for hot California entries at the Bonneville Speed Trials.

Before members of the Michigan Ramchargers club won the NHRA Stock Eliminator title in 1963, they had beaten 90% of the competition. Their Dodge gets the jump on Dyno Don Nickelson's Chevy (super stock winner in 1961 and 1962) in this match race.

Pete Robinson of Georgia, a clever specialist in lightweight fuel dragsters, was unknown until he won the 1961 NHRA Nationals at Indianapolis with the first 8-second run. Later, Pete was killed in a front-engine slingshot fueler.

ART MALONE
The Eastern Star

Art Malone was a multi-dimensional driver turned successful drag racing promoter.

Before mid-century, few forms of racing had any safety standards or uniform rules.

"As a kid in 1946, I remember seeing Bill France race a stock car on a muddy Florida horse track that was so rough that the doors, hood, and trunk lid fell off," recalled Art Malone.

Seven years later in 1953, when Art began driving a modified stock car, NASCAR-sanctioned tracks were smooth and well-maintained by Bill France, who by then had organized NASCAR to protect drivers and cars.

In 1959, Art turned pro and began driving for his old friend, Don Garlits, who had been badly burned in his top fuel dragster. Soon, the team of Garlits and Malone was winning almost every drag race entered, including several Top Fuel records with the "Swamp Rat" dragster. After going solo, Art held world records seven different times in Top Fuel, clinching the U.S. Fuel Championship of 1963 with his supercharged Chrysler.

In 1961, when NASCAR offered $10,000 to the first driver who could run 180 mph on the new Daytona Speedway, only the toughest drivers responded. After struggling through weeks of trial runs, Art drove Bob Osieki's Indy roadster at more than 181 mph to win the jackpot and the world's oval track record, which held for seven years.

Between drag racing events, Art drove the NASCAR stock circuit for Lee Petty and Jack Smith with moderate success. In the mid-1960s, Art drove the famous Novi for STP's Andy Granatelli, but the powerful Novi race cars were plagued with mechanical problems.

When Art took over the closed Sunshine Drag Strip in St. Petersburg and ran the Desoto Memorial Drag Strip, he sponsored championship events, including NHRA and Winston. He initiated racing events such as "King of the Street" and "Drag Nationals," hoping to encourage a return to the sporting exhilaration of yesteryear.

In 1997, Art was elected into the International Drag Racing Hall of Fame.

The Top Fuel champ in 1963, Easterner Art Malone's Chrysler-powered dragster won with 8.14 e.t. and 191 mph.

The Nifty Fifties

Art won $10,000 and a Daytona track record driving Bob Osieki's Kurtis-built 2300-pound Indy roadster. The car was powered by a 1961 Chrysler 1000-hp engine. The side-mounted airfoils prevented the car from being airborne.

Always a star at drag racing events, Art added excitement with his jet dragster powered with a J-46 Westinghouse engine that reached 260 mph in the quarter-mile.

In 1962, Art kicked up the dust at this Phoenix one-mile dirt track event. In other open-wheel attempts during the next four years, Art drove at Indy and finished eleventh driving a Novi in 1964.

"TV" TOMMY IVO
Mr. Showman

This Ivo-built fuel car had a long 123-inch wheelbase to prevent "wheelies." The blown Chrysler developed almost 1000 hp on fuel. It ran in the seven-second brackets, with speeds near 200 mph. How Tommy handled that car really attracted the ladies. (Circa 1963)

In one lifetime, Tommy Ivo has done more than most people will ever do. He even drove more "show-boating" quarters than anyone has driven. He has been a child movie actor, a television star, a drag racing driver, a builder, and a touring pro. Through showmanship, he magnificently hyped the drag racing circuit.

Likable Tommy was born in 1936 in Denver, Colorado, but moved to Burbank, California, where he excelled in local theater and TV shows. As television grew in the 1940s and 1950s, so did Tommy's lucrative career in more than 200 TV shows, which earned him a sizable nest egg. During Tommy's role in the TV show "My Little Margie," the studio tried to nix his racing passion, considering it far too dangerous for their illustrious protégé. When the series ended, Tommy left the world of "make believe" for even faster-paced action.

From the beginning, Tommy was a rebel, using the then-unpopular Buick engine to power his first slingshot dragster. In 1958, Kent Fuller supplied the chassis and aluminum body for only $500.

Tommy's next project was a twin-engine dragster with one Buick engine that ran backwards using a reverse-billet cam. By meshing the flywheel ring gears of both engines, one gear would drive the other using one clutch assembly off one of the flywheels. Each Hilborn fuel-injected engine delivered 450 hp, eventually turning 184 mph.

By 1960, Tommy barnstormed the country as the first, full-time professional drag racer. Besides making record times, he was the consummate showman and tried various tricks, such as tiresmoking fiery burnouts, which pleased the crowds both here and abroad.

"I thought if the crowd liked two motors, they would really love a four-engine car. After touring for 15 years, the only things that wore out were the tires and my welcome at the studio," said Tommy about his behemoth.

Tommy also raced Top Fuel with Chrysler Hemi-engines built by Dave Zeushel, and he became one of the first to run under six seconds in 1972. When interest in Top Fuel began to wane in the mid-1970s, Tommy switched to funny car racing. After a stint at

Moneybagged Tommy had the prophetic guts to try out new drag doozies such as rear multiple engines, stretched chassis, wings, and spoilers.

The Nifty Fifties

driving a jet-powered car in 1980, Tommy brought back his four-engine Showboat for his last tour in 1982. By then, he believed he had more quarter-mile runs than anyone alive. In the process, he had logged more than two million miles and owned 18 different race cars.

Tommy has been runner-up at several NHRA nationals, has won many world records, and has been honored with numerous awards such as the International Drag Racing Hall of Fame. However, he will be best remembered for his tiresmoking four-engine crowd-pleaser.

Although Tommy Ivo can look back on hundreds of movie and television roles, he prides himself on his 30 years as a drag racing pro. To him, that's a blast!

Airfoil wing and wheel fairings improved the aerodynamics of this rear-engine, Hemi-powered dragster. (Circa 1967)

Tommy's Plymouth funny car makes a smokey start at a Midwestern strip in the late 1970s.

"TV" Tommy Ivo on the set of "My Little Margie" with his magnificent four-engine Showboat.

Tommy's unblown twin-Buick engine dragster was the first gasser to make the magic 8 seconds. The top speed was 180 mph in 1960.

The Ivo entourage, which included this glass-sided display trailer, toured approximately 70,000 miles a year nationally.

Hot Rod Pioneers

Years–Ahead Styling with a New Fiberglass Body

The front cover of *Mechanix Illustrated* magazine in 1957 promised that readers could mount a fiberglass body on an old chassis and have a new sports car for $600! Not long afterward, national publicity in dozens of publications made Almquist Engineering the country's largest manufacturer of aerodynamically styled fiberglass body shells, which were designed by Harry Heim and Ed Almquist to bolt-on to practically any car chassis—from a Ford to a Fiat. Lighter and stronger than aluminum, the smoothly molded fiberglass bodies were wind-tunnel tested and met FIA competition regulations. Introduced in the mid-1950s, the six sporty models featured "years-ahead" styling at unbelievably low prices of $295 to $495.

It may be hard to believe, but this sporty speedster once was a 1940 Ford Sedan. The Almquist fiberglass body cost only $495 in 1957 and fit Fords, Chevys, and other cars, shortened to 90- to 106-inch wheelbases. Harold Cron is the driver.

The Nifty Fifties

These molded fiberglass bodies came ready to mount either on existing stripped car chassis platforms or to shortened frame members to which outriggers were welded for body attachment. The Ford required both frame and shaft to be shortened eight inches. The floorboard, firewall, and inside panels were fabricated from sheet metal.

The racy Almquist Sabre two-seater body fit Fiat, Crosley, and cars with 76- to 87-inch wheelbases. The cost was only $295 in 1956.

Hot Rod Pioneers

The common hot rod lowering method of "Z-ing" the front and rear frame members was done on both the Fiat and Crosley shown here. The Fiat was further modified for competition.

The streamlined Thunderbolt body was years ahead of its time in styling and fit any chassis from 96 to 116 inches. The price was only $425. A folding top and windshield frame cost an additional $29.95.

The Nifty Fifties

After this Corvette-Eater hardtop was modified to fit a shortened 1958 Cadillac chassis, its owner, Harry Heim, won 22 trophies in Eastern auto shows.

Beginning in 1956, Almquist fiberglass bodies were manufactured for more than a decade.

"OHIO GEORGE" MONTGOMERY
Traitor or Hero of the Wild Gasser Wars?

"Competitors screamed foul and even said I would kill the gas class when I switched from my little Willys to a Mustang cammer," said George Montgomery after he won the 1967 Springnationals. Until then, the gassers had mostly been oddball pre-war cars with big V8 power plants. Although George's fiberglass-bodied Mustang was completely legal, other hard-core gassers regarded it as a funny car rather than an AA/GS car.

The reason for the tumult was twofold. Throughout most of the previous decade, the ancient-bodied gas coupes and sedans had been the big guns in the colorful "East vs. West" gasser wars of the era. For many spectators, the smaller pre-1952 cars offered timeless appeal, especially the famed 1933 Willys that was George's long-time trademark machine. However, the new generation of fans probably could care less.

In the early years, George Montgomery was an often-feared underdog. However, time and again, he trounced the best of the gassers in the most competitive of all drag racing categories. Each year, the West Coast contingent, inclusive of "Big John" Mazmanian, Doug Cook, Junior Thompson, and other hotshots who sought the coveted Supercharged Gas trophy, would be soundly whipped by George's docile-looking Willys. The soft-spoken man from Dayton, Ohio, who started amateur racing in the early 1950s with a 1934 Ford street/drag machine, began nabbing local trophies even before he had swapped the flathead for a blown Cadillac mill with a handmade manifold.

A week after building his Cadillac-powered 1933 Willys, George was the big surprise winner at the 1959 NHRA Nationals in Detroit, where he took both "A" Gas and Little Eliminator class trophies, as well as $800 in merchandise. Consecutive class honors followed in 1960 and 1961, when the Nationals moved to its permanent home at Indianapolis Raceway Park. After acquiring his moniker, "Ohio George" replaced his tired Caddy with a blown small-block Chevy

George Montgomery blasts off in the "World's Wildest Willys"—the early car that brought him big-time notoriety. With a GMC supercharger and fuel injection, the jacked-up 1959 Caddy engine produced more than 600 hp. Photo shows the starter giving the Fiat driver a nod in this exhibition run in Covington, Georgia, in 1960.

In the 1959 A/G runoff win against Jack Kulp's Olds/Willys, George's Willys did 129.87 mph in 11.72 seconds. "Running with M&H nine-inch tread racing slicks was like driving on ice, and my tires burned rubber all the way through the quarter, so I added concrete inside the spare tire to improve traction," said George. "The NHRA soon outlawed the neat speed trick." Note the "concrete spare tire hump" on the rear of the car.

The Nifty Fifties

The Sanair Dragway in St. Pie, Quebec, was well attended on the 1976 summer day when George Montgomery raced for top honors with his 1974 Pinto. The little 2000-cc (around 120 inches) four-cylinder turbocharged engine put out more than 450 hp.

power plant that promptly won Middle Eliminator at the 1963 Nationals. Soon after another class win in 1964, George toured Europe as a member of the U.S. Drag Racing Team. Following a close loss in 1965 to K.S. Pittman, George changed to Ford's cammer engine and won his class in both the 1966 Springnationals and Nationals.

When the popularity of pro-stockers and funny cars began taking over, it marked the end of the "old gasser" era. With the help of funding from Ford, George switched to a fiberglass-bodied SOHC 427-equipped Mustang, and the new controversial configuration carried him to Super Eliminator titles at the Springnationals (1969), the U.S. National, and two Gatornationals (1973 and 1974).

Probably no one in racing has perfected as many unconventional engine combinations, including the fire-breathing Boss 429 twin-turbocharged 1500-hp A/GS Mustang, whose performance was so phenomenal that the NHRA slapped on a half-second handicap after the 1974 record. After George won his seventh NHRA Nationals title in 1974, he briefly campaigned an alcohol funny car (also with two turbos), which received such a stiff penalty from NHRA that the mighty car was rendered noncompetitive. This literally forced the great Ohio George Montgomery out of the very sport he helped build.

Long known for his mechanical wizardry and meticulous workmanship, George then expanded his racing engine business to include preparing power plants for Winston Cup and tractor pulls. For the past 14 years, George and his son Gregg have been the sole supplier for Buick's specification V6 racing engines that were built for the Indy Lites racing series. Ohio George was inducted into the International Drag Racing Hall of Fame in 1992.

George Montgomery (top) and son Gregg now build 4.2-liter V6 Buick engines for the Indy Lites racing, which have established a new benchmark for reliability. The 425-hp engines are calibrated to perform identically and are then sealed to prevent tampering. An earlier photo (above right) shows George in his speed shop installing a blower belt on his 1967 Mustang.

By the end of the 1960s, the face of the supercharged gas classes changed. Here Ohio George's SOHC Mustang beats the popular Stone, Woods & Cook Hemi-powered Mustang at Thompson Drag Raceway in Ohio. George's later, more powerful Mustangs struck such fear in NHRA officials that they penalized the cars out of competition. (Circa 1969)

Rubber Battles of Racing

Even in the Tin Lizzie days, the crease on great-granddad's pants often lasted longer than the tread on his tires, and that same lack of durability applied to old-style racing tires. Coupled with rapid wear, which was another early drag racing problem, especially in the upper car classes, was lack of sufficient traction to fully harness the output of the powerful engines. Racers would try everything, including chemically softening the rubber and even mounting four rear tires instead of two.

The first drag slicks sold in the early 1950s were simply wide treadless recaps made by tire retreaders such as Inglewood and Bruce. Drawing from their past oval track experience, they simply capped casings with a softer rubber camelback.

One of the biggest breakthroughs in drag racing came in 1958 with the introduction of the M&H tire, developed by Harry and Marvin Rifchin of Watertown, Massachusetts. Theirs was the first purpose-built drag racing tire, which involved much trial-and-error experimentation before developing super-wide slicks molded from a special soft compound. Soon the hotdog engine builders were concocting ways to take advantage of the new drag slick's extra "bite." Late in 1963, Goodyear introduced its new drag racing tire after Connie Kalitta had secretly torture-tested the slicks under fully "loaded" track conditions. Later, Firestone, Hoosier, Mickey Thompson, and others joined the

Called the "Giant Killer," Marvin Rifchin, with his little storefront factory, made the first true drag racing tire in 1958, and then took on and often beat the big tire companies.

Firestone was a leader in racing tires in the 1930s. Then, the Indy tire had a narrow but stiff, smooth tread surface to permit drift and skid on the turns. Early Bonneville LSR cars used similarly constructed tires but with shorter sidewall profiles to reduce distortion at high speeds. Here, an AAA official observes a Bendix balancing procedure.

Around 1960, Mickey Thompson sped to many new records at Bonneville with Goodyear LSR tires. However, he had so much difficulty finding tires capable of 400-mph-plus speeds that he vowed to someday build his own racing tires. Eventually, Goodyear's Gene McMannis (who was granted the first patents on bias-belted and low-profile tires) joined M/T in producing the famous line of Mickey Thompson High Performance Tires for strip and street. Their "wrinkle-wall" became the first soft "street-legal" tire.

M&H Often Walloped the Tire Giants

frey, each adding to drag tire technology, with Mickey Thompson starting the "sticky tire" trend.

By the mid-1960s, NHRA tire rules remained lax, but rear cheater slick tread width was limited to seven inches in stock classes. To put more rubber down lengthwise on the strips, racers began letting so much air out of the rear slicks of the car that the tires had to be cemented and screwed onto the wheel rim to keep the wheels from spinning in the tire. After tiremakers saw improved elapsed times, they began offering special "flex-wall" casings.

Later, wider racing slicks were permitted, and M&H introduced the first nine-inch slicks. This was a "turning point that brought instant success." By the mid-1960s, cars were going almost 20% faster because of new tread compounds and lower tolerable air pressure. Today, top fuel cars are going about 3 to 4 mph faster each year, and tire-makers struggle to keep up with cooler running tires of increased load capacity.

As in the past, tire-tuning often is an art of compromise for the tire and car to work in unison. Ideally, 'rollout' circumference of both rear tires should not exceed 0.25 to 0.50 inch from one another. Depending on use, tire pressures must be higher on heavy cars and lower on light cars; however, some manufacturers now recommend no less than 12 psi. Because improper burnouts dramatically shorten the life of a tire due to excessive heat buildup, experts now opt for spinning the tires an equal amount of time before each run, just until the rubber surface begins to 'haze.'

At 83, Marvin remains the "hands-on" leader of M&H Company, which now produces 100 tires a day for major racing venues. Marvin Rifchin told *Trackside Magazine* that shortly after his induction into the Drag Racing Hall of Fame, he "became more interested in drag racing than in circle track because the challenges are greater. People don't realize it, but straight-line acceleration, for a tire, is a much more demanding thing than going around in a circle." Drag racing as Marvin remembers it is also a "much more gentlemanly sport."

The quiet tire wars began heating up after Goodyear and others entered the drag racing field. This 1965 ad for M&H Racemaster tires promised a better bite. Soon afterward, the tires blew off the competition and became the choice of top racers such as Don Garlits, who made the first NHRA-sanctioned 200-mph run on M&H tires.

Tire durability remains a costly problem across the racing spectrum. Tires for top Winston Cup stock cars (such as those shown here) may last only an hour, whereas up to 50 or more passes are possible from a good drag racing tire.

GEORGE HURST
From a Backyard Garage to a $10 Million Business

George Hurst, 1927–1986. In his time, George was a monumental enigma to the hot rod enthusiast. He was a daring and brilliant innovator who changed the face of the automotive world. George possessed an ego that grew faster than his many enterprises.

What car enthusiast has not heard of the legendary Hurst shifter? Although George Hurst and his associates changed the face of the automotive performance industry, the Hurst story explodes with exciting innovation, fame, fortune, self-inflicted abuse, disappointment, and finally death. All in all, Hurst's most memorable creation was simply George Hurst himself—the paramour and the paradox. Although some people described him as a womanizer, an alcoholic, and an egomaniac, George clearly was a mechanical genius, a promotional wizard, and an optimist.

In the early years of the Hurst empire, my partner and I became co-founders of Anco Industries, later renamed Hurst Performance Products. During the pivotal time of 1959, the young enterprise operated as Hurst-Campbell, with Bill Campbell as an equal partner.

George was the grand dreamer, and Bill was the consummate engineer who turned dreams into parts. As a third party, I was either in the midst of total mayhem or in a perfectly sound "idea factory."

Our first challenge came when a California company knocked off the Hurst-Campbell motor mounts and began peddling inferior quality at cut-rate prices. While trying to recoup the lost motor mount business, we created the illusion of a totally new motor mount. After a clever adjustability feature was added, "Adjusta-Torque" was born, and George bragged about drumming up new customers for the motor mount business in a cross-country trip. To minimize expenditures, George slept in his automobile and ate candy bars for sustenance. In the 11,000-mile sales trip, orders for $100,000 worth of automobile engine mounts were secured, and the fear of bankruptcy finally disappeared.

At business meetings alternating between Milford and Abington, Pennsylvania, we brainstormed for new product ideas and opted for floor-mounted shifter conversions. After returning to the drawing boards, several months later in Milford, George and Bill displayed a chrome-plated prototype stick shift with a barrel-shaped knob and a flat shifter lever. The dual-pattern concept was

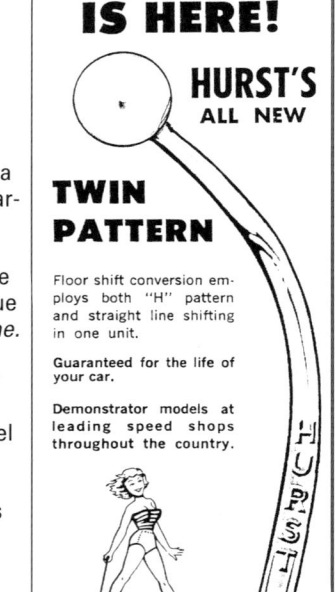

The Hurst shifter, which was actually a latecomer in the market, was formally introduced in this advertisement in the December 1960 issue of Hot Rod Magazine. The name Anco Industries was later dropped after Almquist and Anchel decided to sell out. Although the same patent attorney was used previously by Almquist, both the Hurst and Almquist shifter designs remained different as they sold successfully to non-conflicting markets.

The Nifty Fifties

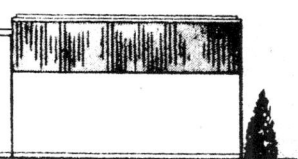

In a few years, George and Bill moved from an obscure garage in Abington to this new plant in Warminster, Pennsylvania.

George really didn't have much to smile about in the mid-1950s. His struggling backyard engine-swapping business soon would be threatened by a California firm that knocked off his line of motor mounts.

unique, but it was more costly to manufacture than the three-speed shifter design that I had succinctly envisioned.

After my partner and I refused to provide $90,000 for tooling, a friendly disagreement triggered the buy-out offer. George's blonde lady friend, Hildegarde, helped in finding an "angel" to buy us out. In late 1960, my partner and I terminated our relationship with Hurst-Campbell.

Prior to the buyout, Jack "Doc" Watson, who had connections with General Motors, suddenly appeared on the scene. An order, contingent upon prompt delivery, was secretly promised for the new Hurst floor shifter. Operating capital was provided by Lawrence Greenwald, and the Hurst organization quickly tooled up to manufacture the shifter as a non-factory branded item for the 1961 Super Duty Pontiac.

Under Bill Campbell's direction, the first shifters were produced. As George bargained for magazine advertising, he was able to squeeze free publicity from hard-nosed editors, and the skyrocket ride to success began.

Suddenly, the Hurst name began appearing everywhere in racing. Contingency awards were offered, and race cars were sponsored. Gentleman Joe Schubeck drove a winning Hurst dragster, while beautiful, scantily clad women such as Pat Flannery and Linda Vaughn posed as "Miss Golden Shifter" and made public appearances in a gold-colored Pontiac with a nine-foot shifter lever protruding from the trunk.

The Hurst organization hired talented men such as James Vaughn, Leo Kagan, and Jim Kerr for the strong national marketing teams. Jack Duffy headed the advertising and public relations department and kept the Hurst image alive with promotional gimmicks.

(Left to right) George Hurst, "Miss Golden Shifter" Pat Flannery, and Doc Watson present a new 1965 GTO to 19-year-old Alex Lampone, who won a contest sponsored by Hurst, Pontiac, and Petersen Publishing. In other promotions, George gave away cars and shifters to drag race winners.

George was no longer the T-shirted, hotshot grease monkey. Rather, he had become a briefcase-toting, sophisticated entrepreneur who promoted shifters at important racing events. Although George had not finished

Hot Rod Pioneers

grade school and had enlisted in the Navy at 17 years of age, he was articulate despite little formal education. The military service gave George his first experience with engines, but it also gave him lifelong medical problems resulting from a bomber exploding on the deck of a carrier.

In 1968, Hurst-Campbell went public, and expansion came with further product development and acquisition of other companies. Airheart anti-skid disc brake systems and Gabriel shock absorbers were added to the Hurst line, and the prestigious Schiefer Company with its famous clutch and flywheels became another button on the Hurst corporate vest. Satellite companies also were established.

In 1970, the giant Sunbeam Corporation saw an opportunity for an easy takeover bid and acquired the controlling interest in Hurst-Campbell by buying out Bill Campbell's shares for $6.2 million. It was then that many key people left the organization because of demoralizing ploys of the acquisition.

When Sunbeam discouraged the Jaws of Life rescue tool concept, George did not listen. In 1972, he demonstrated the Jaws of Life in Washington, DC, in front of U.S. Transportation Secretary John Volpe, who lauded it as a major safety breakthrough. Later, George Hurst, hoping to continue as the company's president, agonized over the winds of change and walked away from the corporation that he had founded. For George, that was the death of his soul.

Indulging himself with alcohol, George lived in a state of self-inflicted abuse. He spent the final years of his life chasing various other automotive-related dreams, including a failed franchise chain of transmission/performance shops that bore his trademark name. When new competition, the influx of muscle cars from Detroit, and the industry's market share for speed equipment slowed, the overbuilt Hurst organization began to bleed painfully.

In November 1981, Allegheny Interaction bought out Sunbeam. A year later, Cars and Concepts purchased Hurst Performance from Allegheny Interaction. Richard Chrysler, who had started as a floor sweeper at the Hurst division in Detroit, became chief executive officer before becoming a Michigan congressman.

In 1986, the Hurst organization again suffered, and all assets were then acquired by Joe Hrudka's Mr. Gasket Company.

George Hurst died on May 15, 1986. He was blatantly bold, and bold men often succumb to violent ends.

More than 60,000 eyes watched Al Eckstrand win the Stock Eliminator title at the 1963 NHRA Winternationals in his push-button 1963 Dodge. He was one of many prominent racers that George Hurst wisely helped cosponsor in exchange for product endorsement.

From the beginning, dragsters showcased the rodder's imagination and were handmade, often haphazardly, with little or no protection for the driver. Eventually, many of the amateurs' construction snafus were eliminated when Gordon "Scotty" Fenn and others began to manufacture complete, prebuilt, ready-to-race dragsters.

By 1958, Scotty had built a half dozen chassis kits, and his firm, Chassis Research Company, was rolling. Emphasizing safety, Scotty was responsible for many new innovations in front-engine dragster design. In his heyday, he produced a dozen or more frames and chassis a week, marketing them for as little as $300. Always a nonconformist, Scotty bragged, "I can supply any part on a dragster—from a bolt to a complete car—even a trained driver."

This rigid tubular frame, built by Woody Gilmore for Art Malone's funny car, had a relatively short wheelbase of 120 inches—less than half the length of later dragster chassis. The hood of the fiberglass flip-up Mustang body was stretched to accommodate the engine that was mounted 25% rearward (then NHRA legal).

Dragster Chassis Design — Ready-to-Race Rails

Later, prominent "production line" chassis builders such as Kent Fuller, Nye Frank, Don Garlits, Connie Swingle, Woody Gilmore, Jim Nelson, Dode Martin, and top fuel driver Lefty Mudersback entered the fray. Hoping for the ultimate state-of-the-art design, fabricators constantly were experimenting with different (and sometimes wild) combinations of hardware, suspension, cockpit location, engine placement, and frame widths and shapes (either rigid or flexible) with varied overall lengths.

Unfortunately, the demise of volume dragster production arrived when a judgment was awarded to a badly burned top fuel driver against a chassis maker, although the frame was not responsible for the accident. Rather than risk liability, chassis makers simply stopped construction, and subsequently that creative era ended.

Chassis having long wheelbases became a lasting trend when top fuelers discovered that the stretched slingshot functioned as a lever, thus preventing wheelstands—especially with more powerful mid-engines. Don Garlits' Swamp Rat dragsters grew from 108 inches to 300 inches, which was the maximum allowed.

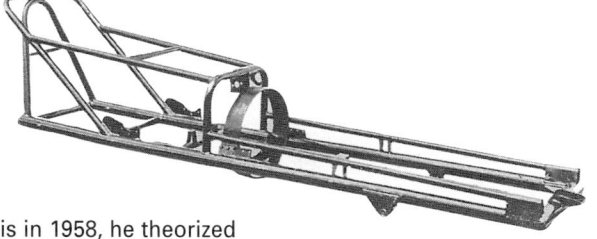

When Scotty Fenn designed the first "mass-produced" dragster chassis in 1958, he theorized that a short chassis would allow more weight to be transferred to the rear slicks for improved traction.

The longer 210-inch wheelbase chrome moly chassis of Carl Casper's Young American dragster helped win Top Fuel at the 1971 AHRA Winternationals. It was a masterpiece of ingenuity and beauty.

BILL CAMPBELL
The Pragmatic Brains Behind the Hurst Empire

Bill Campbell is the second half of the Hurst-Campbell company that produced the famous Hurst shifter. Campbell was instrumental, along with designer Jim Hobbins, in developing the Jaws of Life, the internationally known rescue tool that has spared thousands of people from the cold grip of death. Although Bill's role in the George Hurst empire has been downplayed over the years, Bill now shyly concedes his rightful place in the history of the Hurst empire as a vital arm in the production of Hurst products.

Bill never thought of himself as a hard-core hot rodder, but he was always a brilliant innovator. This humble mechanical wizard picked up the other half of the scissors in the original Hurst enterprise in a fateful way. In 1956, Bill had responded to an ad that George Hurst had placed in a local publication for a part-time production engineer for Volkswagen bumper guards and motor mounts for popular engine swaps. After working for two years, Bill and George formed Hurst-Campbell Incorporated.

"George had a backyard garage, and I had my expertise and my time," explained Bill. "We put in what we had, since we had little or no money."

As soon as the partners saw a glimmer of profits, they both pocketed a paltry $25 a week. But fleeting success nose-dived after a competitor copied and undersold their engine conversions.

When near-bankruptcy forced Bill and George to take in the business partners of Ed Almquist and John Anchel, the newly formed Anco Industries became the marketing arm for Hurst-Campbell. Then the enterprise began a climatic upswing. In 1960, Almquist and Anchel were bought out, and Hurst Performance Incorporated was established.

After Bill and George were joined by a moneyed associate, a bevy of new products was introduced and acquisitions were made. No matter what Bill and George did, they suddenly had that Midas touch. Bill, who was as tough as nails, always appeared collegiate, wearing his spectacles and having his shirt pockets filled with pencils and a slide rule. Forever humble, Bill let George, the public relations man, receive all the glory.

"George and I were different in a lot of ways, but we both agreed to make only products of

Bill Campbell made automotive performance history with George Hurst. Here, Bill checks the controls on an Indy race car. This car was unlike the radical twin-bodied Offy built for Hurst by Smokey Yunick, which failed to place in the 1964 Indianapolis 500.

uncompromising quality, and we guaranteed them 100%, no matter what," said Bill about the enterprise, which eventually employed hundreds of production workers under his supervision.

"George's showmanship, the high quality of our shifters, the products, our presence at racing events, the traveling machine shop, and the "golden girls" all helped to promote the hot rod industry," said Bill.

According to Jack "Doc" Watson, the major transformation in the Hurst-Campbell operation occurred when the company's stock was distributed to co-workers. What began as an altruistic gesture became a life-changing event for the company and its employees.

"When the company had sales of $10 million per year, I left," said Bill.

From 1969 until recently, Bill, who had received his Bachelor of Science degree in Mechanical Engineering from Lehigh University, became part owner of Provost Car, Incorporated, a Canadian bus manufacturer.

"When I joined the company in 1969, it was doing about 1.5 million Canadian dollars a year in sales. When we sold it to Volvo in 1995, we sold it for $140 million," said Bill, an avid Harley-Davidson motorcyclist who has ventured into the RV business as well as the fiberglass fabrication world.

With only a few tools, a vehicle, and a notion that a dream can come true, Bill Campbell saw his empty pockets fill with an abundance of money, excitement, and personal fulfillment. As a remarkable catalyst in automotive performance, Bill provided the hot rod and drag racing world with new designs and lasting products.

Pro—*The Allied Autosport*

In the words of Roger Huntington, which he wrote in the July 1957 issue of *Speed Mechanics* magazine, "Don't kid yourself! A lot of fellows are bending that quarter-mile into a neat oval and knocking themselves out to get around it faster than the next guy. Modified stock car racing has grown in the last five years (since 1952), especially in the East and Midwest where drag strips have also been slow developing. It's now an important new phase of the 'autosport' that must be considered right alongside drag racing, lakes straightway competition, customizings, hopping-up engines, etc. In fact, there probably are as many or more fellows in stock track racing than are competing regularly at the drag strips around the country." Huntington hyped both types of racing as a great way to blow off steam under relatively safe conditions.

Con—*What Kind of Racing Is This?*

This unflattering caption appeared in a 1950 article in *Hot Rod Magazine*. Other publications also ridiculed early stock car racing as mayhem on wheels and as a serious threat to the more dignified open-wheel forms of racing. Would you believe after a few years, *Hot Rod Magazine* acknowledged a "kissing cousin" kinship with rodders by regularly featuring stock car racing as a *bona fide* sport? Then, in 1960, the gap closed even more when NHRA joined NASCAR in joint promotion of the week-long 1960 Winternational Drag Races. Wow, how do things change!

Wild action and spectacular flips such as this helped popularize stock car racing in the early 1950s. (Photograph from the Ed Otto Collection)

JACK "DOC" WATSON

Hurst's 'Shifty Doctor'

Beginning in 1963, "Shifty Doctor" Jack Watson was present at championship drag events with the Hurst-equipped repair shop on wheels.

Jack "Doc" Watson worked his way through medical school by doing engine swapping, but he preferred fixing hot rods rather than people. While dabbling in designs for engine conversion kits in 1959, Doc met George Hurst and Bill Campbell, who were then in a slump because their Volkswagen "Nerfing Bar" bumper guard business had bombed.

During the time of the AMA racing ban, when Almquist's Anco Industries had rescued the Hurst-Campbell organization, Doc had belonged to the elite but somewhat secret General Motors' "Super Duty Group." Publicly, the automakers agreed with the AMA racing ban. However, privately they maintained clandestine racing teams, often cheating under the cloak of "an aftermarket entity."

Although Watson was hardly more than a "go-for" at the time, he boldly suggested keeping General Motors discreetly in high performance by working through an outside company such as Hurst. At the time, the shabby Corvette shifter needed a good OEM replacement, and Watson saw a golden opportunity for Hurst as an outside vendor. The rest is history.

Doc, with his bold chutzpah, soon became a valued asset of the Hurst organization in product development and marketing. Doc and his "performance practitioner" traveled to all the championship drag meets as the "Shifty Doctor." A trailer served as a mobile machine shop where racers could repair their equipment free of charge.

In 1965, the Hurst Performance Research Center was established near Detroit under the direction of Doc Watson, who began transforming stock vehicles into drag racing sensations.

Doc's promotional masterpiece was the tricked-out "Hemi Under Glass" 1965 Barracuda with a Chrysler 426 Hemi engine installed in the rear under the long rear window of the fastback. The next was the smoking "Hairy Olds," with two 1050-hp engines.

Another marketing coup was the famous Hurst/Olds 4-4-2, which boosted a 400 c.i. V8 and a Hurst-shifted four-speed gearbox. A T-top (originally patented by Gordon Buehrig) was reinvented and added distinctiveness to the handsomely striped body. With media hype and Doc's proficient marketing skills, the sporty performance car was a resounding success, with more than 500 sold in 1968 alone.

Jack Watson attributes the success of the Hurst enterprise to the unifying "spirit" of the company. When the Hurst operation went public, that spirit began dying.

"Some terrible things happened to that company when it came into big money. The destruction started with Sunbeam's buyout bid and eventually its restructuring," said Watson, who participated in the success of the Jaws of Life and other products.

A mover and shaker even today, Jack "Doc" Watson continues the Hurst saga with Hurst Heritage Vehicles.

Smiling Doc Watson is proud of the 1968 Hurst/Olds 4-4-2, the first and most famous of the limited production cars that was silver with striping as curvy as racing's sex symbol, Linda Vaughn. One of the rally-striped models served as an Indy 500 pace car.

LINDA VAUGHN
Queen of Motorsports

Because of her knockout beauty, sparkling personality, and marketing savvy, Linda Vaughn became the top sex symbol for racing and its greatest spokesperson.

Born in Dalton, Georgia, Linda left her job as a dental hygienist for a three-year reign as Miss Firebird for the Pure Oil Company. In 1966, Linda won the title "Miss Golden Shifter," a contest sponsored by the Hurst Company, wherein she became its public relations ambassador.

Endowed with Southern charm, the blond bombshell rose to celebrity status through guest appearances at all major U.S. drag races. After years at the tracks, Linda, a natural speaker for all types of drag racing and performance-improving products, was an A+ promoter for the sport of drag racing because she could hype attendance, pep up the fans, and spur on the racers.

After a third of a century specializing in motorsports' promotions and publicity, Linda amazingly exudes enthusiasm despite the grueling grind of covering 100 national auto events annually.

When she was inducted into the SEMA Hall of Fame as spokesperson for the performance industry and sport, Linda said, "I'll lobby and fight for racing because it draws more attendance than any sport in the world."

Linda has received many honors over the years. When she was the first recipient of the annual NHRA Public Relations Award in 1970, Wally Parks said, "Linda has been an energetic supporter of drag racing for many years. It is her sincere concern for the welfare of all contestants, regardless of fame, and her indomitable spirit and her natural enthusiasm for racing that have won her acceptance in all fields of motorsports."

After more than 30 years, Linda Vaughn continues as the leading lady of racing. She is still smiling, hugging, and kissing.

Who received more attention at a recent auto show—Linda Vaughn, or Joe Hrudka's gorgeous "Shazoom" custom Chevy?

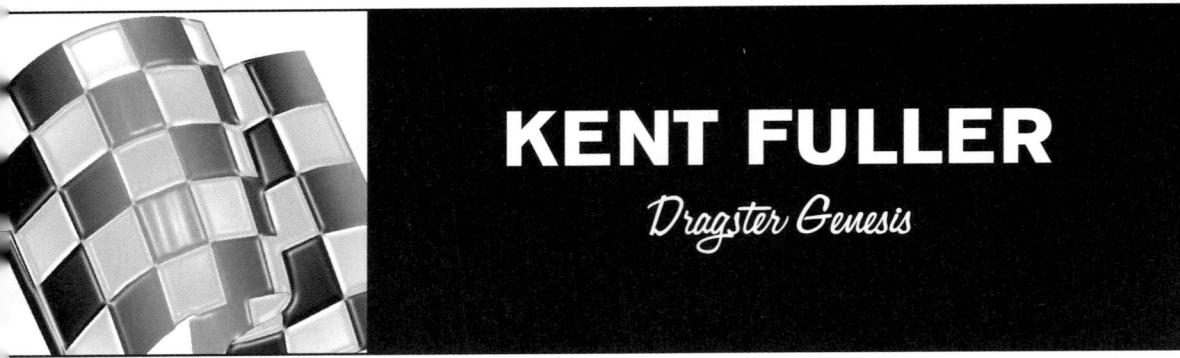

KENT FULLER
Dragster Genesis

At one time, a Fuller dragster chassis was considered unbeatable. As one of the most prominent chassis designers of the early 1960s, Kent Fuller pioneered several construction techniques that were adopted by other builders. Some say Kent's ideas were the reason early dragsters appeared as they did.

Despite being dyslexic, Kent demonstrated exceptional talent for combining beauty of form with winning function. *American Rodder* wrote, "Some rodders work on the cutting edge of the car-building art. Kent Fuller designed the edge, made it, and moved on."

His slingshot designs featured a graceful chrome-moly tubing chassis with a long wheelbase, considerable engine setback, a rugged roll cage, and trick suspension. Kent was the first volume chassis maker to use light torsion-suspended front ends, replacing heavy transverse spring mounting. This allowed quick adjustment so the front end could be tuned to different strips.

Kent may have built fewer cars than his competitors, but a larger percentage of his cars achieved championship status. Among them were cars for Tommy Ivo, Ed Donovan, Tony Nancy, Chris Karamesines, and Don Prudhomme.

A Kent Fuller chassis similar to this one made heroes of many drag racers of the 1950s and 1960s. This prototype "rail" evolved into Kent Fuller's ready-to-race tubular chassis with leading link torsion bar and fraction shocks built into the cross member. Later models had a longer wheelbase.

After achieving recognition for his stylish, super-fast dragsters, Kent became involved in other projects such as a prototype for the "Volksrod," a Volkswagen-powered dune-buggy-type kit car resembling a shrunken T-bucket roadster. Another was the unforgettable four-engined dragster and also the beautiful Magwinder, a sidewinder built in 1959 for Chuck Jones.

Kent Fuller (left) buttons up an early 1960s dragster chassis as famed engine builder Dave Zeushel (right) watches. Dave's Chrysler engines were terrors that powered many top drag racers.

Kent was inducted into the International Drag Racing Hall of Fame in 1996, a long-overdue tribute to a talented hot rodder.

Kent Fuller today works on fun projects such as restoring a rare 1947 Bentley Cabriolet, which, when completed, will have a value of more than a million dollars.

Don Prudhomme's (seated) first big win was in a Fuller-made dragster with a blown Chrysler built by Dave Zeushel (standing, right). Their car won Top Fuel Eliminator in Bakersfield in 1962.

Art Chrisman

The big Chrisman & Cannon Hustler, driven by dragster pioneer Art Chrisman, was the first Top Fuel at the Bakersfield 1959 meet and one of the first quarter-milers to exceed 180 mph. The supercharger atop the Chrysler mill pops through its sleek streamlined body.

JIM TRAVIS
Bringing the Past to Life

For almost half a century, Jim Travis' career has been the envy of many gearheads. Jim has always been in the fast lane as racer, mechanic, and builder of hot rods and racing cars. Originally from Bradford, Pennsylvania, Jim obtained his first car at the age of 16. After joining two different car clubs in the early 1950s, Jim, similar to his friends, spent most of his time keeping his jalopy running while learning from his mistakes. In those carefree days, cruising around town to all the drive-in restaurants was the "in thing."

"Cars were first, and girls were second, I hate to admit," recalls Jim, whose motor kept revving on high as he swapped engines, built cars, and raced in the East.

Still in military khakis, Jim helped run the first drag race in Montgomery, New York, for NASCAR in 1958. Don Garlits was paid only $350 for a public appearance there and for making three rip-roaring passes.

After receiving an associate degree in automotive science, Jim worked for Ford as a full-time warranty mechanic for 19 years. Jim, a folksy and likable speed merchant, ran a speed shop in Pico Rivera, California, from 1963 until 1966.

"I found myself doing too much bench racing and giving away too much free tech-advice, so I welcomed a new challenge as an automotive machinist," said Jim.

In addition to crewing on stock, sports, midget, sprint, and Indy cars, Jim says, "I've been behind the wheel of almost everything that runs for NHRA competition, but I prefer the oldies."

In 1956, Jim built his first competition roadster, which was a 1929 Model A powered by a flathead Ford V8 engine. Jim raced at all the major California strips and broke eight speed records in the half mile and flying mile in 1969 through 1973 at El Mirage and Bonneville. The car received feverish public-

After acquiring the 1934 Ford flathead SoCal coupe, Jim Travis set the record in 1973 at 142 and 146 mph at Bonneville. After installing a blown 300 c.i. small-block Chevy, he ran it at 230 mph.

ity in many magazines, car shows, movies, and TV spots.

As a change, Jim directed the NHRA staging lanes for 17 years. Even at retirement age, Jim boasts that hot rodding pumps his blood and keeps him percolating. Preserving the history of hot rod and circle track racing keeps Jim forever young.

This flathead powered 1929 Ford set eight records in 1969 through 1973 at El Mirage, Riverside, and Bonneville. Flexible hoses outside the radiator fed cold air to the carburetors.

The Nifty Fifties

As a director of Petersen's Museum, Jim Travis enjoys restoring historical vehicles in his well-equipped shop in Whittier, California.

This is a replica of *Hot Rod Magazine's* Suddenly. The original Plymouth was built by Jim Travis and did a record 183 mph at Bonneville in 1960.

Mickey Thompson's dragsters, Assault I and Assault II, were meticulously restored for museum showings by Jim Travis.

Jim Travis' 1940 Ford gets gas for an old-timers' reliability run at an old-style filling station in 1956.

Rodney Singer

Top Eliminator of the 1959 NHRA National Drags in Detroit was Texan Rodney Singer, shown here taking off in his class AA/Dragster. The 1958 supercharged 480 c.i. Lincoln engine shot him through the quarter-mile at 9.12/172.

The Big "Little Go's"
Beating the Clock in the "Boonies"

Some of the best regional championship drag meets had already spread eastward through the heartlands to the Atlantic Coast when New England's first big event in 1955 in Orange, Massachusetts, was rated by NHRA godfather Wally Parks as "championship quality equal to the best anywhere." Another hotbed of Yankee hot rodding centered around a small independent drag strip in Sanford, Maine, while a walloping regional "Go" held at the Charleston Dragway in Rhode Island produced many class champions after drawing top contestants from Maine to the Gulf Coast.

Even before the five-state Mid-Atlantic Timing Association was established in 1956, some top touring pros were pulling in appearance money, while newcomers were showing up many of the old pros from the West in the gas classes. However, it was not until the storming fuelers were running in the rest of the country that the East finally joined the big "digger shootouts." Rising stars such as Florida's Don Garlits made up for lost time by winning most of his races—including a historic victory over the famous Cook-Bedwell fuel dragster in 1957.

When NHRA banned all fuel except gasoline in 1957–1963, the ban threatened to split the drag racing world in half. However, the decree actually helped the grassroots gasser boys who rarely could afford a high cost per run of more than $5 per gallon for other fuel. As stockers dominated, track efficiency improved. One small club in Lancaster, Pennsylvania, was sanctioned by the Eastern Hot Rod Association and ran weekly dual-lane events. This club electronically timed as many as three hundred cars in fourteen classes in three hours.

In the early 1950s when abandoned airport runways often served as drag strips, a flag-wielding starter and timekeeper with a stopwatch often ran the show. Here, a lone dragster waits its turn at an airstrip near Allentown, Pennsylvania. At this meet, the participants outnumbered the spectators.

Nick Nicoli in his flathead V8-powered drag roadster is neck to neck with Ned Gibbons at the start of this 1955 match race in New England. (Photo by Jack Pennell)

Roger Walling, driver of the Strokers Club dragster, poses with some of the hundreds of trophies won through 1963 in competitions from Maine to Rhode Island. The car recently restored by Tom Shea and Roger Walling represented the "state-of-the-art in Northeast dragster technology in 1957."

Similar to other hot rod clubs popping up throughout the country in the early 1950s, the TY-RODS of Concord, Massachusetts, were typical. Members represented varied interests: some in all-out performance trials, others strictly high tech, and others a mix of rods and customs. All were pledged to safe driving and improving the new sport.

A big national breakthrough occurred for the sport when the first NHRA Records Regional was held in May 1960 at the Inyokern Drag Strip near Death Valley, California. This successful test eventually became a foolproof bonus points method, whereby all rodders would compete for coveted National records on a local level without costly travel to the Nationals. As expected, many strips throughout the nation began clamoring for record meets. Amazingly, the new Amarillo drag strip's Grand Opening in the Texas panhandle boasted nineteen new National records for its NHRA bonus regional.

Another long-time rodding hotbed since the early 1950s was located in York, Pennsylvania. Its popular U.S. 30 Drag-O-Way took the 1963 "biggie" Northeast U.S. Divisional Championship, where 19 records were rewritten by more than 600 competitors from 29 states and Canada. Then, dragsters and gassers were the strongest entries on the points trail.

In its short history, drag racing grew from localized drag meets to regional shootouts to huge national events, where the quickest of the quick would compete for the most coveted titles in the United States.

The Strokers Club, formed in 1952, built its first dragster in 1956, followed by another in 1958. After the Scotty Fenn chassis kit was assembled and welded by members, it ran with a Hilborn-injected Hemi in the A-Dragster class. With a 671 blower added, its speed increased to 165 mph in the low 9's.

Hot Rod Pioneers

Little Giant, a 1951 Crosley with a 283-injected Chevy run by George and Joe Brazee, was a regional and B/Alt trophy winner and a regular at many Eastern drag strips.

Waiting to blast off the line is the fast AF/D A&B Speed Shop slingshot, which was a seven-second threat at New England strips in the 1960s.

This "Killer" 1933 Willys powered with a small-block Chevy blew the doors off big-block gassers in the mid-1960s. It was owned by Jack Merkel of Ridgewood, New York, who later ran his famous B/GS 1939 Willys to many class wins.

Vinnie Tarrantola's C/GS 1940 Willys with a blown small-block Chevy was an NHRA e.t. record holder in 1963. Vinnie was known for his Vitar power shift hydro transmissions.

Team partners Pete Stager and Bill Novotny were hard to beat on East Coast tracks in the gasser era. Their B/Gas 1937 Willys with a 283 Chevy engine was Hilborn injected. An earlier A/GS Olds-powered 1934 Ford narrowly lost to "Ohio George" Montgomery in the 1960 Nationals. (Circa 1963)

JIM DEIST
The First Drag Chute Maker

As dragster speeds increased in the mid-1950s, so did the problem of stopping those screaming monsters to ensure that they would not run off the strip. After hearing of a racer's near-fatal crash during a record run, Jim Deist became obsessed with the need to develop a reliable slowing device. From his experience at the Irving Air Chute Company, Jim believed that a parachute activated by a driver at the end of each run could serve as a fail-safe emergency brake that would decelerate the vehicle with less dependence on heavy braking mechanisms. Parabrakes, the ribbon-type parachutes found in military aircraft as deceleration aids, would have to be redesigned to function as a drag chute for straight-line stopping.

"Jim worked diligently without pay for a long time and suffered through a lot of disappointments," said Marian, Jim Deist's wife. "Over 75 runs were made on Abe Carson's dragster, but even then, some critics doubted that a parachute could be used on a race car successfully."

In addition to conducting tedious research, Jim participated in a lot of bold experimentation and actual testing where the slightest error of a malfunctioning chute could easily wipe out a car or, even worse, a human life. By the late 1950s, Jim was manufacturing drag chutes, and top racers were using them—even calling them "the world's fastest brakes."

According to Jim, who now has a complete safety testing lab, a drag chute must be "custom engineered for each application—especially the car's speed. If a solid chute's diameter is too large on a light, fast car, a driver could receive broken ribs or damaged vertebrae from the sudden opening shock. Drag coefficient, porosity, deployment, inflation time, operating stability, and attachment location on the vehicle determine drag chute function."

"As the industry grows, so must we," explained Jim. "As cars go faster, we expect to develop the required safety equipment. Originally, just a lap belt and parachute was enough. Now, we provide five-point harness, Nomex clothing, fire systems, and time will tell what else will be needed."

Today, the sport owes a debt of gratitude to dedicated pioneers such as Jim Deist, who in 1984 was inducted into the SEMA Hall of Fame for helping to make drag strip competition safer.

Jim Deist points to his wraparound lap belt/shoulder harness that has a quick-release arm restraint.

Stopping a four-engine dragster after a fast quarter-mile run required a custom-made drag chute strong enough to decelerate from 175 mph to 30 mph within 1500 feet.

JACK MENDENHALL
Belly-Up to Gas Pumps

Being a dry lakes racing whiz and curator of what has to be America's most unique auto-oriented museum keeps the fires blazing in Jack Mendenhall's active life. One of the nation's largest historical collections of gas pumps, globes, signs, and other memorabilia is housed at Jack's Petroleum Museum, which is located in Buellton, California (home of the Dry Lakes Racing Hall of Fame).

Every year, old-time racing greats and aficionados trek to the annual induction ceremonies sponsored by the Gold Coast Roadster and Racing Club. Bench racing with wise-cracking attendees such as Ed Iskenderian and Paul Vanderley adds colorful lore to the all-day "Remember when...?" contests.

In the late 1940s, Jack's love for the sport began with a modified Ford sedan. Throughout the next two decades, he won 84 first-place trophies in various classes of drag racing. After competing successfully in circle track racing with flathead Ford and Chevy V8 machines, Jack began off-road racing and finished fourth in the Mexican 1000-mile Baja in 1973, co-driving an Olds Cutlass.

Jack's preference for the 1980s was racing at the Bonneville salt flats. One of his "highboy" roadsters even won first place for its class every time it was entered. In 1991, Jack became a life member of the prestigious 200-MPH Club after setting the D/Gas Roadster speed record with a 207-mph average.

Now retired, Jack says, "I plan to race at Bonneville every year as long as I live. That and gas pumps keep me young and alive, and that's the way I like it."

In 1992, Jack took this small-block Chevy-powered roadster to the salt flats and did 205 mph on gas for a nostalgic run.

At age 19, Jack raced this fenderless roadster in 1949 on one of the world's first drag strips in Goleta, California. With a stock flathead V8, he clocked only 71 mph.

Driver E.E. Claybaugh (left) shared driving with Jack Mendenhall (right) in their 1969 Olds for the 1971 Baja 500.

The Nifty Fifties

The world's largest museum of gasoline pumps and other petroleum memorabilia is Jack Mendenhall's brainchild.

Jack and his son Mark co-drove this 1958 pickup in a Mexican 1000-mile road race. (Circa 1971)

Ed Cortopassi

The first international (FIA) recognition of an American hot rodder occurred in 1958 when Ed Cortopassi drove the Glass Slipper Chevy-engined dragster through the kilo with a top speed of 168 mph and a two-way average of 116.43. The canopied-cockpit streamliner was featured on the cover of *Speed Mechanics* magazine.

Jack, in the No. 24 Chevy, passes to win a trophy dash. (Circa 1967) (Courtesy of Dick McDonald)

ERMIE IMMERSO
Always a Rodder

Ermie Immerso's 354 Chrysler-powered Valley Auto Special set a new D Class Lakester record at 213 mph, which in 1956 was the fastest recorded at Bonneville for an open-wheel car. The following year, Ermie raised his own record to 217 mph.

By 1956, the Bonneville National Speed Trials had grown to national prominence when Ermie Immerso, a new kid from Phoenix, Arizona, set a new Class D record with his little lakester. Much to the surprise of the California hotshots, he became a distinguished member of the 200-MPH Club.

Born in New York in 1925 to a dairy farmer, Ermie's thirst for speed progressed from a motor scooter to a Ford V8 souped-up Model A roadster that scorched local tracks. After serving as a machinist's mate in the Navy, Ermie drove track roadsters while "mechanicking" in the Arizona midget racing circuit. His rodding roots took Ermie and his buddies to the 1954 Indy 500, where he crewed with his friend, Jimmy Bryan, whose Dean Van Lines Special finished second.

Although he earned many trophies, Ermie took a job as engine shop foreman for Carroll Shelby. By 1965, Ermie was working with Holman-Moody-Stroppe and building blown Ford 427s for off-shore, powerboat racing winners.

Two years later, in Long Beach, California, Ermie started Thunderbird Products, which mainly fabricated high-efficiency exhaust systems for RVs, trucks, cycles, and street rods.

Ermie was always true to the slogan, "Once a rodder, always a rodder." After building a 1932 street rod, Ermie's flame-painted Deuce won Oakland's "World's Most Beautiful Roadster" title in 1988 and 1989. In subsequent years, that little beauty reaped more awards. Since then, Ermie has covered more than 250,000 miles trailering his gorgeous show cars that won numerous trophies. The trophies are now displayed in his home overlooking the Pacific Ocean.

Today, Ermie is bedecked with a frosty white beard and denims. He continues to do the show car circuit simply for the fun of it.

"I'm most proud of breaking into the 200-MPH Club in 1956 and then, 12 years later, seeing my son Marvin make it with a 230-mph class record," said Ermie.

At the 1962 Bonneville Speed Trials, Ermie with his sleek streamliner that had four 427 c.i. Ford engines did only 283 mph on gas. The following year, he coaxed the liner's speed to 302 mph as it burned alcohol.

The Nifty Fifties

Ermie Immerso (right) and Danny Ames (left) check the four Ford big-block engines on the 1962 streamliner.

Getting ready for the next run at Bonneville, Bill Stroppe (with his back to the camera) and crew members (including Fran Hernandez) listen to chief mechanic Ermie Immerso explain why the Lincoln-powered DVL Special was running too rich.

Chris Karamesines

As a touring pro, veteran Chris "The Golden Greek" Karamesines terrorized the country with several top fuel dragsters such as this beauty (below) with its Kent Fuller chassis. (Note the comical key in the rear.) Chris was among the first to break the 200-mph barrier, and he won many Top Fuel Eliminator titles, including the AHRA Nationals in 1959, 1966, and 1968. Always an experimenter, Chris reportedly had some success with exotic fuels such as hydrazine. Hydrazine was more propulsive than nitro; however, it was unpredictably dangerous and ultimately was banned. Because of his daredevil showmanship and skill as a dragster and funny car driver, Chris remained a favorite among fans for almost three decades. Chris' speed shop (right) in Chicago was signed as the "Home of the World's Fastest Engine Specialist."

Supercharging Magic
Spurring the Horsepower Race

Although the idea of forcing more fuel-air mixture into an engine than it can inhale with atmospheric pressure alone was first used on a race car by Mercedes in 1921, supercharging didn't catch on in the hot rod field until well into the 1950s.

Although a few dry lakers such as Tom Beatty, Ernie Hashim, and the Spalding brothers successfully experimented with forced induction, drag racing really kicked off supercharging as a quick and easy way to boost top speed.

At first, "blowing an engine" seemed to be a "bucket of worms," even for experienced mechanics. A strong conversion began with "beefing up the engine," plus coordinating changes in carburetion, compression, and cam timing. Too much boost or even detonation could scrap an engine in seconds. The old-style centrifugal units with a three-pound boost were troublesome and useless at low speeds, and most vane-type and axial-flow blowers also had wear problems. However, the sturdy GMC Roots-type positive displacement blowers, originally designed for diesel engines, proved both dependable and effective but required special end plates and a manifold for mounting either on top of the engine (driven by a notched belt or chain) or on the end of the engine driven by the crankshaft.

The popular 6-71 GMC blower with a theoretical displacement of 420 inches could pump up to a 15- or 20-psi boost on the biggest engines; the smaller 4-71 "Jimmy" pumped 280 c.i. per revolution. Excellent cloned Roots systems that include the larger 8-71 size continue to be made by several firms.

Improvement in design and materials of the newer Roots-type and centrifugal supercharger and turbocharger kits, which are now available for stock engines, promise power gains of 25% to 40% with good driveability.

When installed in combination with other modifications, the competition model Roots-type superchargers will more than double the horsepower over normally aspirated engines. That's a lot of extra ponies!

Note: Information on the AiResearch turbocharger can be found elsewhere in this book in the chapter about Robert De Bisschop.

Supercharging remained a black art in the mid-1920s when this intercooled Roots-type supercharger and large Stromberg downdraft carburetor fed this old Duray-powered race car. Note the two spark plugs per cylinder and huge exhaust headers. (Photograph by F.M. Kirkpatrick)

The successful adaptation of the big GMC 6-71 blower to the Chrysler engine helped the Hemi become a winner in the high drag classes. Here, Kent Fuller and helper are installing a Jimmy-blown Hemi in a Fuller-chassied dragster in the early 1960s.

The Nifty Fifties

Although ineffective at low speeds, an early centrifugal supercharger helped this Winfield-designed OHV racing engine deliver more than 550 hp. By locating the three carburetors on the intake side of the blower, the fuel's latent heat of vaporization plus the intercooler surrounding the intake manifold substantially reduced the temperature of the incoming fuel charge (Circa 1949)

At full throttle and full boost, a competition engine such as this requires 50% more air than if it is unblown; therefore, it needs more carburetion to produce full power. Simply switching drive-pulleys changes the blower pressure. Optimum boost level is usually 5 to 8 psi for compression ratios of 8 to 8.5:1.

"Mr. Flathead" John Bradley grabbed many Top Eliminator wins (even against big OHV Chryslers) through the 1950s with his old fuel-burning Mercury engine dragster. Note the box atop the homemade "manifold" that houses four Stromberg 97's. A 4-71 GMC blower (not on the car when photographed) pushes air to the carb box via the large flex hose.

JOE SCHUBECK
Gentleman Joe

As a popular drag racer in the 1960s, Joe Schubeck's classy attire and impeccable manners earned him the nickname "Gentleman Joe."

Clutch/flywheel explosions in the early 1960s became a racer's nightmare. As bigger engines revved at higher rpm, typical cast aluminum bellhousings would disintegrate and catapult shrapnel-like debris into the driver and even innocent bystanders. After a drag racer's foot was ripped apart by a devastating clutch explosion, Joe Schubeck vowed he would build a stronger bellhousing to couple the engine to the drive line. After stumbling on the new aerospace process of shaping metal, known as hydroforming, Joe, with the proverbial "fire in his belly" began producing "Lakewood Bellhousings" that promised lifesaving strength.

Joe's heavy-duty J-Bolt traction bars and various suspension and safety-improving products followed. However, by the early 1970s, the skyrocketing growth of the high-performance industry had attracted big-time conglomerates. One even successively purchased Joe's Lakewood Industries enterprise, only to spin it off in 1975 to Mr. Gasket Company.

Joe's next venture was a dual spark plug, aluminum Hemi cylinder head that is now a virtual standard in top fuel competition.

An innovator since his high school years in Lakewood, Ohio, Joe built his own flathead Ford-powered slingshot drag racer in the 1950s. The car burned nitro at $1.50 a gallon. After building three more racers with mechanic partner Butch Scarpelli, Joe became co-holder of the World's Gas Dragster title at the International Timing Association meet in 1959. Joe campaigned dragsters successfully in the 1960s and piloted the famed "Hurst Hairy Olds" exhibition vehicle in a handsomely tailored gold driving suit.

At an age when most men think of retiring, Joe put muscle into his 30-year-old idea with a desktop CAD and designed a quad V8 engine with an engineering envelope akin to that of the fabled Chrysler Hemi. This big-league project went from concept to production drawings in only nine months—an astoundingly short time compared to Detroit standards, where lead time is measured in years. Perhaps it will play a part in history!

Driver Joe Schubeck with his happy crew had just won a drag race in Akron, Ohio, with this blown Chevy small-block slingshot dragster. Joe's 170 mph run with Hemi-power won the International Timing Association Gas Championship in 1959.

The Nifty Fifties

The Eagle engine is lighter and stronger than a Hemi, with four valves per cylinder and two cams on each head. The 588 c.i.d. version for marine use produces more than 750 hp. The block is expandable to 800 c.i.d. In full drag-racing form (with a Roots supercharger and burning nitro), the Eagle tested almost 3000 hp.

Joe blasts off on a winning run at Carlsbad Raceway. Earlier, his supercharged Hemi A/Fuel dragster was the recipient of the Best Engineering Car Award at the 1965 U.S. Fuel and Gas Championship.

Bob Daniels

On a winning streak, Ohio native Bob Daniels drove his Chevy-powered "Fire-Fly" II Fiat to A/A class wins in 1959. In collecting almost 100 trophies, Bob set class records at the U.S. Nationals and won the C-Altered at the fifth Nationals in Detroit. (Bob's wife Eileen was a consistent Powder Puff class winner and member of the all-woman Piper Puffs Club, organized in 1956.) In his 40-year career, Bob helped build the then-infant NHRA into the leading drag racing organization. As general manager of Indianapolis Raceway Park, he helped turn the failing facility into a world-class motorsports center. (Circa 1959)

Hot Rod Pioneers

Chevy's Hot One

Although sometimes called the "Mouse Motor," the original small-block Chevy V8 engine changed the face of hot rodding forever. Almost overnight, the 1955 Chevrolet lifted itself from its stodgy stovebolt-six image to becoming a young man's pride and joy. An early ad warned, "Don't wear heavy shoes. For most driving, you can forget the bottom half of the accelerator; that's for emergencies. But when you need it, mister, you've got it."

The secret of the "little mouse" was its versatility that grew from 265 to 400 cubes, without any changes in the external dimensions. Its horsepower rating rocketed from 162 hp in 1955 to 375 hp in 1964 and even more horses with aftermarket racing hardware added.

Vince Piggins, who was credited with the creation of the legendary 1967 Z28 Camaro, passed away in 1983. However, he and Duntov will be remembered as the men who kept the fires of Chevy performance burning during the dark days.

The Z28 Camaro was Chevy's super-quick response to the Mustang. Its potent 302 small block went from 0 to 60 mph in only 6 seconds. Of the 602 Z28 Camaros made in 1967, many were raced. The following year, Mark Donohue road-raced one to Trans Am glory.

The Nifty Fifties

The 350 small block debuted in 1967 and continued as a second-generation LT1 in 1992, delivering 300 hp. This cutaway shows smooth flow passages and room for modification magic. An arsenal of aftermarket speed equipment made the Chevy engine the "winningest" of all.

The history of the small-block V8 resides with the talented engineers who worked with visionary Ed Cole to create a masterpiece in industrial engineering. One of the heroes was Vince Piggins, who labored largely in secrecy to put Duntov's performance ideas onto the track. At the behest of Chevrolet, Vince established SEDCO (Southern Engineering Development Co.) in 1956 to create race-ready stock cars. The garage was a remote outpost—similar to Smokey Yunick's place—where the preparations for Chevy's *de facto* racing team's cars were conducted in private.

By the time the Automobile Manufacturers Association pulled the plug on auto racing, Piggins' group had birthed numerous NASCAR winners. In four decades, 65 million small-block Chevys were produced. The "hot ones" have flattened hills, conquered Pike's Peak, set records at Daytona Beach time trials, and won National drag race and NASCAR series.

The 1955 small block 265 V8 featured cutting-edge technology for its time. It was compact, lightweight, efficient, and reliable. Its high-revving valve train utilized stamped steel rocker arms that were unlike competitors' heavy, complex rocker arm shafts. With fuel injection, a 283-cube version produced 1 hp per inch. However, speed tuners quickly learned how to double that horsepower.

RAMPAGING RAMCHARGERS
Dynasty of Fuelers, Funnies, and Stockers

"From the beginning, most outsiders suspected that the Ramchargers was not a typical car club," recalls Bill Shoppe, who was then a Chrysler engineer. "Rumors even spread that the club was a sneaky way for Chrysler to violate its "no racing" agreement with GM and Ford. Boy, were they wrong! We minimized our connection with Chrysler simply because we were scared that we might be fired."

The Ramchargers club originated in 1958 with a small group of car enthusiasts who worked for Chrysler. Members were angered by Chevy's domination of almost every NHRA class record (except top dragster), and they selected the C/Altered competition class for their first club project car. However, at the time, all they could afford was a used 1949 Plymouth business coupe that cost $50 but required extensive modification if they ever hoped to embarrass all those Chevy owners. A Chrysler Hemi engine had to be installed while still getting enough tractive weight on the rear wheels without violating the engine setback rule. Therefore, the only way to make the old car competitive was by raising the center of gravity (CG) to help the transfer of weight to the rear wheels during acceleration. The idea worked, and the comical-looking coupe turned 109.75 mph for the best C/Altered speed at the 1959 Nationals in Detroit.

The following year, after Dick Maxwell and Doug Patterson with Herman Mozer drove the car to several records, a new rule change put the Altered out of business. The club finagled a factory-fresh 1961 Dart from Dodge and began competing in the super stock class which was only beginning to heat up. Their first national event victory came when Al Eckstrand won the Top Stock Eliminator title at the 1963 Winternationals and later was a perennial match-race winner with his Lawman's Dodge.

Without the advantage of large budgets enjoyed by other companies, dedicated club members made the most of the racing work on their own time.

"We would be at the Woodside Garage by 6:30 on weeknights and work until midnight, and then race all weekend," recalls Mike Buckel, who like other engineers doubled as driver and pit crew.

Many of the Ramchargers' best developments came from the original group of factory engineers, whose innovations included the first tunnel ram manifold and their mighty funny

Jim Thornton, who at age 26 was one of the most famous of the Ramcharger drivers, led the team to many regional and national records—beginning with big wins in 1962. Big Jim was the first Ramcharger to crack the eleven-second barrier at 128.98 in 1964.

Mike Buckel, driving a Dodge Super Stocker, captured Top Eliminator (11.02/1231) at the AHRA Nationals in 1965. Mike quit driving two years later when a funny car transmission exploded and tore up his foot.

The Nifty Fifties

The teams jacked up a 1949 Plymouth, dubbed the High and Mighty, and set an NHRA national record in 1959. However, insiders say that the NHRA promptly "legislated" the car out of future competition with a new 24-inch crankshaft centerline rule. The 354-Hemi had eight big ear-splitting megaphones, and the home-brewed intake (two Carter AFB carburetors mounted on a sheet-metal plenum box connected with rubber radiator hose runners) was the first high-rise tunnel ram intake manifold.

cars. Race engine guru Tom Hoover helped design the new Hemi (that dominates drag racing even today), while Jim Thornton designed the club's first F/X altered wheelbase funny car (built by the Alexander brothers) in 1964.

"Our cars were like test mules, and we viewed every race as a chance to try something new in camshafts, tires, springs, fuels, what-have-you," said Buckel, who like engine expert Dan Mancini functioned as a jack-of-all-trades. "Some radical ideas worked,"

Youthful bravado, engineering skills, and eventually corporate blessings gave the Ramchargers the edge over competition. Bob Cahill later admitted, "It's true that teams like the Ramchargers got the new parts first, because they were the ones that developed them. Later parts received numbers and were made available to all racers."

Despite their many successes in 1967, the overly taxed original group disbanded and sold out to fellow dragster campaigners, including Dan Knapp and Dave Rockwell, who continued racing under the Ramchargers' name and colors.

Over a seven-year period, the Ramchargers' professional drag racing organization, campaigning top fuel dragsters, funny cars, altereds, and super and F/X stocks, won 90% of all races entered and four national championships. Team president and main driver Jim Thornton, who also was responsible for vehicle design, said, "I feel our biggest accomplishment was waking up Chrysler to the importance of performance! This led to changing Plymouth's and Dodge's image from 'an old man's car to a young man's car.'"

Dr. David Rockwell has written a book on "The Original Ramchargers" and details major triumphs of more than thirty team members.

Don Westerdale's hot Dodge AA Fuel dragster smokes off the line to set the low e.t. record at the Eleventh Annual NHRA Nationals. The hot rail did 7.507/210.66. Don drove winners in 1964 to 1966.

Hot Rod Pioneers

This ram-tubed 1965 Dodge Funny Car had a radically altered wheelbase that increased weight transfer to the rear tires under acceleration. The wheels were moved forward as much as 10 inches and rearward 15 inches.

Club members drove a variety of Hemi-powered cars, mostly in stock classes, and set performance marks all over the eastern United States. However, a few drivers, led by team member Dan Knapp, concentrated on top fuel dragsters powered by the new design Hemi. This soon led to triumphant eight-second clockings with speeds near 200 mph.

Ramcharger club members are: (standing, left to right) Tom Coddington, president Jim Thornton, Dan Knapp, and (squatting) Dan Mancini. They are in their "secret" garage, premiering their new dragster with a Gilmore chassis, Kruse body, and a Dodge Hemi engine. Features include adjustable traction and dual rear wheel brakes.

Bobby Darrin

"Bobby Darrin's $150,000 Dream Car" was designed by Andy DiDia and fabricated in the late 1950s by Bob Kaiser and Ron Clark. No donor car was used. Instead, the radically finned aluminum body and tubular chassis were handmade.

The Nifty Fifties

Hot Rods, Not Shot Rods

"Hot rodding is a menace and must be stopped."

"Hot rodder involved in fatal accident."

"Terrorizing teens who drive old cars are hot rodders."

Those screaming headlines of ignorance are clips from mainstream media of the past. A Syracuse, New York, judge added to this blatant prejudice when he suggested a statute to outlaw souped-up cars in 1953! This was similar to outlawing automotive progress. Imagine if that had been the case, performance and safety contributions made by the hot rod industry since then might never have come to fruition. No wonder hot rodders received a bad name!

In the early days, members of hot rod clubs were desperately trying to think of a new word that would eradicate the black-balled swearword image of hot rodding. They opted to educate the press and the public on the true and forthright intentions of hot rodding. Eventually, respect came to hot rodding in the 1950s as the sport was being organized. Unfortunately, even today, some law enforcement officials and the public remain misinformed.

"My association with hot rodding, until a few weeks ago, was extremely limited," wrote actor Walter Brennan, who starred on "The Real McCoys" television series, in 1959. "If I thought about it at all, I envisioned reckless, high-strung youngsters in black jackets, driving hell-bent down the highway and endangering others…"

"But the young rodders and the officials of the National Hot Rod Association convinced me otherwise," continued Walter,

Walter Brennan's Endorsement

who worked with members of the NHRA for the ABC network television show. "…Their motto is 'Dedicated to Safety,' and, I might add, to courtesy."

"Since my 'education,' I now believe that hot rodders are the safest group of young drivers in the country," stated Walter. "I could only wish that all youngsters shared that interest in automobiles. The result would be greater road courtesy, less abuse of cars, and fewer accidents."

Today, hot rod clubs strive for a "good guy" image and continue to encourage responsible and safe driving as violators are disciplined, thereby making club members among the safest drivers on the highway.

This is what America's fastest-growing hobby deserves!

In the early days, many law enforcement agencies wisely endorsed hot rod clubs and organized drag races because they encouraged both safe highway driving and safe cars. Here, a street in Glen Cove, New York, is closed for races on a summer Sunday in 1954.

Acts of courtesy to stranded motorists earned praise for early rodders, whose club motto was "Dedicated to Safety." The slogan was printed on some cards similar to the one shown here.

YOU HAVE BEEN ASSISTED BY A MEMBER OF "ROAD KNIGHTS"

A "HOT ROD" organization formed by a group of responsible auto enthusiasts, dedicated to promote interest in the sport, wherever it may be found, and who some day hope to unveil to the public the true meaning of the word ------ "HOT ROD."

GENE MOONEYHAM
Supercharger

When personable Gene Mooneyham, a name synonymous with early drag racing, moved to New Orleans in 1968 to help manage a plush speed emporium called Car Shop, Inc., he surprisingly found the rest of the country was catching up to and even surpassing California in drag racing technology and competition.

"Californians were no longer top dogs or major winners at national events. Being competitive then was tougher because a racer had to be more of a do-it-yourselfer. Even though the money was here, the parts were often 2,000 miles away," recalled Gene, who prepared a Chrysler-powered AA/FD car driven by Dave Chenevert to win the first Gatornationals in 1970.

Gene's first serious competition car was a 1934 Ford sedan that he coaxed to 120 mph at El Mirage dry lakes in 1951. After bolting on a 1934 Ford coupe body, he ran 127.29 mph at Bonneville in 1953. Later, after installing a blown Chrysler Hemi, the car turned winning strip times at Russetta meets.

Gene's driving ended when his car's brakes failed and he raced off a San Diego drag strip through a barbed wire fence that tore off most of the top of the coupe. After Gene's bruises healed, "Jungle Larry" Faust piloted most of Gene's cars.

While working with engine builder Keith Black, Gene saw that most top competition cars were GMC blown. Therefore, he decided to go into the business of modifying GMC superchargers with drive kits that supplied 8 to 15 pounds pressure. Today, Gene manufactures his own Roots-type supercharger, which is favored by many top drag racing names.

Gene (center) is pictured here with his son Fred (left) and Li'l Gene (right). The immaculate A/F dragster went 209.30 mph in 7.22 seconds at the Long Beach drag strip in 1967. The 1956 Chrysler mill had a 6.71 GMC blower, Enderle direct port fuel injection, and a Clay Smith roller cam. The "Blue Car" represented the talents of Gene, Jerry Jackson, Wayne Ferguson, and driver Larry Faust.

The famous #554 Mooneyham and Sharp Hemi-powered 1934 Ford/Fuel coupe turned 8 sec./170 plus times in the early 1960s. (Lions' Drag Strip photo)

The Nifty Fifties

Pioneer speed equipment maker Al Sharp (right) collaborated with Gene (left) on many successful projects. Al built the fuel injector that fed the chain-driven blower of the Hemi. (Curt Hamilton photo)

Gene (left) is interviewed as honoree at a 1996 NHRA reunion by famed auto racing announcer Dave McClelland (right).

The Mooneyham and Sharp A/F slingshot dragster was another hot performer in the 1960s. Its best quarter-mile times were 6.98 e.t. and 205.94 mph.

ART AND WALT ARFONS
Creators of the Green Monster Jets

Aerodynamic engineers scoffed at the unorthodox jet-powered car built by Art Arfons. However, by the end of 1964, the skeptics stopped laughing when Art won the world's land speed record of 536.71 mph at the Bonneville salt flats.

In the early 1950s, Art Arfons and his brother Walt began their racing careers by making a dragster fashioned from junk airplane parts and a war-surplus, Allison-aircraft engine at a total cost of only $500. Lacking another color, they painted the car a John Deere green and took it to the races. The track announcer thought the car was so ugly that he labeled it the Green Monster, the first of a long line of Green Monsters.

From Akron, Ohio, the Arfons brothers went on to build several more dragsters with aircraft engines. Although powerful, the wild cars often lacked traction and were too heavy to get off the starting line quickly. Nevertheless, Art and Walt regularly rode off with top speed honors. Art laid claim to becoming the first man to reach 150 mph in the standing quarter-mile in 1956. Several years later, he was among the first to break drag racing's 250-mph barrier.

After making more than 800 quarter-mile runs at speeds up to and exceeding 200 mph, Walt's career as a dragster driver ended in 1962 because of health reasons. When the NHRA ban on jets was lifted, Walt built his own car propelled by a Westinghouse J-46 jet engine with an afterburner. The Wing Foot Express, driven by Tom Green, briefly held the land speed record of 413 mph in 1964.

Here, Art Arfons is working on his Anteater. Its huge Allison supercharged liquid-cooled V12 aircraft engine was rated at 1500 hp on gasoline, which was then double or triple the power of conventional dragsters. The brothers always had enough brute horsepower, but their problem frequently was available traction.

The Nifty Fifties

That same year, Art returned to the salt flats with his own uniquely designed jet racer. After the record changed hands several times, it was finally his—first at 434.02 mph, and then at 536.71 mph. In 1965, Art regained the record with a 576.55-mph average, which held for less than a week.

By the 1970s, the heavy flaming jets could no longer match the elapsed times of fuel dragsters, and Art switched to the new sport of tractor pulling with another unusual creation—a turbine-powered tractor with a homemade rear end and wheelie bars. Art became a national tractor-pulling champion several times and a roaring crowd-pleaser with his blazing-signature fire show.

In the 1970s, Art's son Tim performed in stunt shows as a jet car driver who catapulted over ten cars, ramp to ramp. In 1984, Art's daughter Dusty began competing in tractor pulling.

"It's great when you can take your hobby and make it your living," said Art, an inductee into the Motorsports Hall of Fame and the International Drag Racing Hall of Fame.

The 21-foot Green Monster was the fastest thing on wheels in 1964 at 536.76 mph. The record holder was built using hand tools and scrap iron. Its surplus but new J-79 turbo jet engine, which developed 17,000 pounds of thrust, cost only $750. The huge frontal area admits air to feed the hungry jet mill that gulped 67 gallons of kerosene per minute.

This bathing beauty (Walter's daughter Patricia) stands on Arfon's experimental car called Baloney Slicer. The car was pushed by a propeller, driven by a Ranger 12-cylinder 700-hp engine.

A blast from the past, Cyclops once held the open wheel–open cockpit land speed record at 366 mph in 1961. After stints of mainly crowd-pleasing exhibition runs, the car crashed.

Super Cyclops went 294 mph in 5.5 seconds. An extra strong parachute was made to slow the behemoth dragster after its quarter-mile run.

Road Warriors More than Parts Peddlers

"No bull...nothing happens until a sale is made" might be an old cliché, but it rings true for the marketing people who helped propagate the new motorsport, right from the start. Many early speed equipment makers initially relied on mail-order ads for promoting business on a national scale because a full-time traveling sales staff would be too expensive.

An alternative route was to team with manufacturers' reps who "lived or died" by the lucidity of their tongues. As sweet-talking hustlers, they earned a chunk of the pie, often receiving a 10% commission for simply writing factory-direct orders.

In the early 1950s when most aftermarket sales reps thought of rodding as a passing fad, George "Hot Rod" Ellis, a middle-aged, gravel-voiced New Yorker, saw gold in mom-and-pop speed shops that were springing up across the United States. He became one of the first reps to earn his living solely by peddling for and to the growing hot rod industry. Hobnobbing with hard-drinking, quick-cussing, speed merchants and representing several small manufacturers (including Weiand) was his forte.

As a stereotypical, lead-footed traveling salesman, Ellis would drive all day to coax up an order or to sell inventory that was stashed in the trunk of his Hudson Hornet. Because he simply refused to take "No" for an answer, he often would maneuver a "swap" transaction for a disgruntled buyer, thereby avoiding a painful cash refund.

Literally putting fledgling companies on the map were other tireless reps, such as Harvey Goldstein, who hit the road with cunning marketing programs. Today's reps definitely enjoy an advantage with fax machines, cell phones, and high-tech audio-visual communication systems.

More than pesky order takers, modern independent reps such as Jim Kerr, Dennis Holding, and Scott Cowern function as esoteric sales consultants while wholesaling factory products to speed shops, auto parts stores, and performance centers.

Now tweaking buyers instead of engines, "Smiling Jim" Kerr, pictured here as a sharp young racing mechanic, was inducted into the SEMA Hall of Fame in 1996.

At early auto shows, factory reps often helped speed shop exhibitors peddle parts. Here, Frankie Del Roy and helpers "tech-talk" to listeners at a New York Armory in the early 1950s.

RAY STILLWELL
Lone Rodder Humbles Factory Teams

Ray in his 1932 Ford Almquist Special is about to set a 1958 record of the experimental car one-mile standing-start acceleration meet sanctioned by NASCAR.

"Working in a small-town garage in Slate Hill, New York, for $2 an hour in the 1950s didn't leave me much time or money for my hot rod hobby" said Ray Stillwell, a razor-sharp young mechanic who often was too busy for idle noncar conversation. Ray preferred talking with his right foot, as he did in 1958 while winning the Daytona Beach standing-start acceleration one-mile experimental class competition with a 1932 Ford sedan.

The previous year, Ray tied for fourth place with a 1932 Ford roadster. However, what made both contests almost humorous was that newcomer Ray dared to pit his lowly home-built daily drivers against the *crème de la crème* of California drivers and their hot factory "racing specials," some costing 20 times more than Ray's car.

Originally purchased for $75, the flathead in the 1932 sedan was replaced with a more potent 1954 Chrysler Hemi of stock bore and stroke. Compression was raised to 10:1 with pop-up pistons. The cam was a special track grind, and four #97 Stromberg carburetors fed the engine. The drive-train combination was right on the money with the 1939 Ford transmission and a 1940 Ford rear reworked with Zephyr gears.

Self-taught Ray elevated speed mechanics to the highest level of competency. Practicing his trade for almost half a century, Ray possessed exceptional innovative skills. "If there had been lawn mower races when we were

After driving to Daytona Beach, Ray Stillwell removed the headlights and switched Hemi engines the hard way.

A happy Ray Stillwel now trailers his little "World Beater" to shows and meets.

The Nifty Fifties

kids," said a rodding friend Ed Ardler, "Ray Stillwell would have been unbeatable. I remember when he once hand-ground a single-cylinder engine's camshaft. Gosh, you had to run behind that souped-up mower."

Although Ray insists he is retired now, he is restoring his eight-car collection of 1932 Ford coupes, roadsters, and sedans. (The going price for show-quality coupes and Hi-Boys today ranges from $25,000 to $50,000 and higher.) While the Deuce is the timeless favorite of the hot rodder, Ray's Daytona winner remains priceless to him.

Art Chrisman, in the new Mercury Mermaid factory experimental, gets the starter's flag for the second fastest average (93.4 mph) at the NASCAR 1957 one-mile acceleration meet.

Cool Coupes

The Coupe classes have always been favorites, especially in cold climates. For racing, numbers and letters were taped on the side and the cycle-type fenders were removed on George Miller's chopped-top 1933 Ford street/strip coupe. (Jack Pennell photo, circa 1950)

BRUCE LARSON
First with a Fiberglass Funny Car

Bruce Larson was one pioneer of drag racing's Funny Car class, which began in 1965 and quickly caught on with fans who identified with the stock-like appearance of the car. Bruce was from Dauphin, Pennsylvania, and in 1954 competed at the first officially sanctioned drag race on the East Coast in Linden, New Jersey. However, it took Bruce another 35 years to become the 1989 NHRA Top Fuel Funny Car champion.

Bruce's career gained momentum in 1965 when he introduced the first fiberglass-bodied funny car, the USA-1 Chevrolet, and when he forged a funny car trail across the country before winning the 1989 Winston Funny Car title. The lightweight fiberglass-laminated body gave incredible performance to the 1000+ hp, one-ton vehicle, which was actually a high-powered dragster disguised as a standard automobile. Bruce's revolutionary

Bruce Larson's USA-1 funny car with an injected 454 Chevelle was painted patriotic red, white, and blue.

Bruce Larson won the 1989 NHRA World Funny Car championship finals with a stunning 5.28/283.28 win. Bruce (left) shared dual victory honors with many-time Top Fuel champion Joe Amato (right) at the big drag racing rumble in Columbus, Ohio. During the mid-1950s, Joe and his brothers began the huge Keystone enterprise where his father left off with A&A Speed Shop in Moosic, Pennsylvania.

injected fueler had a state-of-the-art-chassis, but the engine in front and the shorter wheelbase often made the car difficult to keep in the lane while accelerating to high strip speeds.

Adept at chauffeuring both dragsters and stock cars, Bruce won 12 national event titles in his run to the Funny Car championship. While driving the Garlits Fuel Dragster, Bruce's best e.t. for the quarter-mile was 4.88 seconds, with a top speed of 299.30 mph.

Bruce was an Eastern Top Fueler and donated his fully restored championship Oldsmobile Cutlass funny car to the Smithsonian Institution in Washington, DC. However, over the years, he did not receive the publicity that he deserved, partly because of his inability to lure big-buck sponsors.

Easterner Bill Winterbottom collected more than a hundred trophies in the East with this supercharged Buick B-Class dragster. Bill's crowning achievement came in 1960 when he shattered all existing 1/8-mile acceleration records at Langhorne, Pennsylvania, with an officially timed run of 7.23 seconds at 127.80 mph.

Bill Winterbottom

Early dragsters often made up in performance what they lacked in appearance. Bill Winterbottom's Ardun OHV converted Ford flathead consistently did more than 125 mph in the early 1950s. The wheel covers, intake manifold, and exhaust were homemade, as was the body, which sat on Model T rails. This photo was taken at an airstrip in York, Pennsylvania.

The big complaint Bill Winterbottom had when he began drag racing in the mid-Atlantic states was weak competition. "Many times, I'd go to a drag strip, and nobody could go over 90 mile per hour, even in the mid-1950s," said Bill. "My five-carb GMC six dragster [shown] would go over 120 miles per hour."

FRED CARRILLO
His Rods Make a Better Connection

Back in 1940 when "money was tight and junk cars were cheap," Los Angeles-born Fred Carrillo fell in love with hot rods, as did most youngsters.

"Trying to get [the cars] to run fast was the challenge," said Fred, "though once running, their life expectancy was anybody's guess."

During his stint in the U.S. Air Force, Fred teamed with his future brother-in-law Bob Betz and started collecting parts for their '27 T-Roadster. The car was later stolen when Fred took his bride to Venice Beach, only to have the body and frame recovered a year later. Bob and Fred began rebuilding the flathead into a rear-engine roadster utilizing a belly tank section for the nose. At the first 1949 Bonneville time trials, the roadster turned 135 mph as a B Class Modified Roadster and took second place.

For the 1950 Bonneville time trials, Fred and Bob further modified the roadster and set A, B, and C Class records in the first few days.

"Then the protests came, and the were records taken away," said Fred. "On the fifth night, we taped cardboard to return [the car] to the original dimensions and recovered the B Class record at 172 mph. Afterward, Wally Parks awarded me the Sportsmanship Trophy.

"At the 1951 Bonneville meet," continued Fred, "Earl Evans and I recorded the fastest ever C Modified Roadster speed of 178 mph.

The first meet at El Mirage Dry Lake in 1947 included this channeled 1932 roadster run by Fred Carrillo and Vic Peterman.

Thank God, Don Waite was in the service, so we did not have him and the Edelbrock gang to contend with."

When it was rumored that a world-class time trial would occur in 1953, Fred and Bob, with the help of Allan McDougal, fashioned a streamliner by copying the aerodynamics from a P-51 Fighter and by narrowing the rear tread to 28 inches. (This narrow tread is standard for dragsters today.) In the first attempt at trying for 300 mph, the left front wheel exploded, and the car flipped and rolled for 1500 feet. After skin and bone grafts, Fred eventually lost part of his left leg.

Fred Carrillo at the dry lakes in 1947, standing beside Earl Evans' tanker with its cockpit located ahead of the three-carb flathead V8 engine.

The Nifty Fifties

With the help of Dan Gurney and Art Sparks, in 1963 Fred developed a superior-strength, lightweight steel connecting rod called the Carrillo Rod, which is now used worldwide in every venue of racing. Fred's name lives on in Carrillo Industries, Inc.

In 1969, Fred teamed with Doug Champlin running Formula I but finished second for the 1971 Formula 5000. They took fifth at the Indy 500 in 1972 with this Cosworth. Shown here are (left to right) Sam Posey (driver), Jack McCormack, Ken Norris, Doug Champlin, and Fred Carrillo.

Fred Carrillo ran a car in the USAC series and qualified for the Indy 500 each year from 1972 to 1978. Here, #73 is running at the Pocono 500.

Hot Rod Pioneers

Showtime A Microcosm of a Hobby

Dreamy illusions, carried over from boyhood fantasies, probably have added to the aura and fun for many diehards who attend auto shows. For one person, it might be the fastest machine that lures them; for another, it might be a customized or retro show car. However, for most enthusiasts, viewing the creative examples of automotive individuality is reward in itself.

Nostalgia associated with auto memorabilia, original coach work, and antique vehicles has always been a spectator draw. Promoters quickly recognized the moneymaking potential of showing off attractive street rods, racing machines, and customized vehicles to "car-hungry" audiences.

Although professionally sponsored "Autoramas" and other hot rod shows were beginning to culminate around the country in the early 1950s, smaller previous shows such as the Detroit Motor Racing Expo (1947) and the first SCTA Show (1948) set the stage for the famed perennial National Roadster Show in Oakland, California (1949) and the Motorama LA (1950), which were among the first well-staged show spectaculars.

When the first large hot rod show occurred in the East in 1951 in Linden, New Jersey, it attracted thousands at an admission fee of $1.00 per person. By the following year, Herb Shriner, the famous radio and television comedian, consulted Ed Almquist for planning the "even-bigger" New York City extravaganza. Today, more than a thousand special-interest shows occur every year in North America, and well-planned themes capture the true personae of the cars on display.

Modern-day promoter Terry Cook, who plays up clever words, has created a fantastic concoction of annual Eastern custom car shows that he calls Lead East, the biggest "Fifties party" in the world. (Lead, formerly used to fill seams, is a synonym for customized cars, i.e., lead sleds.) Each event "turns back the clock," with some of the best customized cars, street rods, and famed auto personalities in the country.

Program books for early rod and custom shows cost 25 to 50 cents each. The poster for the first Oakland Roadster Show in 1949 pictured an eight-foot trophy for the most beautiful roadster in America.

Cam grinder Harry Weber (right) is shown here with Ed Almquist (left) at the first major hot rod show in the East in 1951. At that time, Almquist was the only large manufacturer of racing equipment east of the Rocky Mountains. (Linden, New Jersey, Armory)

The Nifty Fifties

Ed Reavie watches as two-time Indy winner Gordon Johncock speaks at the St. Ignace Show awards banquet.

Gary Meadors' love of hot rods began on the farm in Dinuba, California, in the mid-1950s, with this 1947 Plymouth coupe that he lowered, scalloped, and customized.

Robert Larivee, Sr., poses with a comical T-bucket that appeared in one of his ICAS shows. Bob turned his hobby into a profession, saying, "Anybody who can turn his greatest interest into a life's work is fortunate."

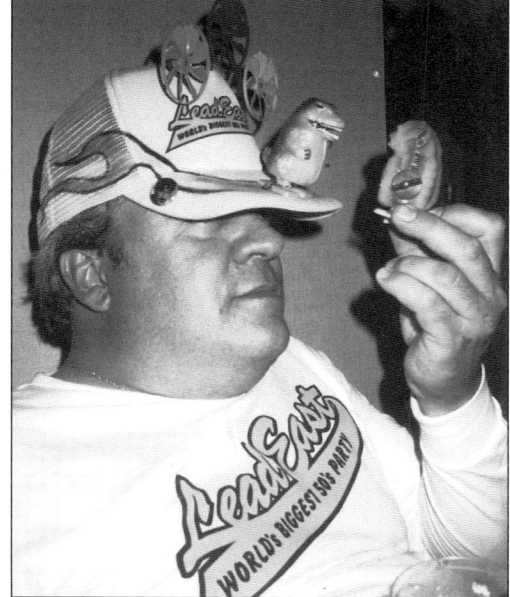

Terry Cook is seen in regalia and an outrageous hat at a "Lost-in-the-'50s" auto show event. The talented Terry, who was the first starter at Island Dragway, New Jersey, before becoming a tell-it-like-it-is writer and editor of *Hot Rod and Car Craft*, is now a successful entrepreneur who has been staging show spectaculars similar to Lead East since 1983.

For a quarter of a century, promoter Ed Reavie has been running successful nostalgic auto shows every summer at St. Ignace, a sleepy little village in upper Michigan, which becomes flooded with more than 100,000 visitors. Three miles of main street into "Memory Lane" become packed with 3,500 vehicles and vendor exhibits, while an endless rolling parade of early antiques, classics, muscle cars, street rods, and mild-to-wild customized cars and trucks "make the past come alive."

In the late 1950s, when early car shows had lax judging and few ground rules, Bob Larivee, who revolutionized the auto show business by introducing a standardized format for fairer competition, founded the International Championship Auto Show (ICAS) series. Since then, the ICAS has showcased the world's finest dream cars and trucks built by renowned customizers such as Barris, Starbird, Winfield, Jeffries, Bradley, Roth, Casper, and the Alexander brothers. Bob Larivee continues to present the nostalgic "Hot Rod Heritage" display at the yearly SEMA show.

Today, gung-ho icon Gary Meadors promotes nostalgia street rod and custom car events called "Goodguys."

"But the real fun is driving hot rods—not just looking at 'em," says Gary. "When muscle cars became the fad, street rodding really never died. In the late Sixties, old rod revivals motivated me and my buddies to form the Nor-Cal Early Iron Club, which helped kick off the first West Coast Street Rod Mini Nationals in 1973." Gary continues to produce as many as 30 events nationwide, attracting hundreds of thousands of folks who grew up in the 1950s, 1960s, and 1970s. All of them happily relive those carefree days.

JOHN FITCH
Honored for Saving Lives

A tall, cultured man of multiple talents, John Fitch epitomizes the hot rodder's penchant for innovation. After serving as a decorated American fighter pilot in World War II, John turned his love of speed into another venue. First, he raced sports cars in England before driving in U.S. competition. Then, in 1949, John built by hand a Type B racing roadster powered with a Ford V8-60 engine (called the "Fitch Bitch"), which trumped second place at Watkins Glen in 1950 and won in its class at Bridgehampton the following year.

Called "Fearless Fitch" by competitors, John was heralded as one of the top American drivers by *Road & Track* magazine in 1958. Added to John's triumphs were the World's Sports Car Championship race at Sebring in 1953 in a Cunningham, the GT class in the Mille Miglia in a Mercedes Gull Wing, and the 1955 Tourist Trophy in Ireland while co-driving with Sterling Moss for Mercedes.

Among John's numerous inventions is the familiar Fitch Inertial Barrier, consisting of

This hot sports rod, built by Fitch Enterprises, outclassed its competition in 1951. It had a lightweight, all-aluminum competition body on a Jaguar XX120 chassis.

yellow energy-absorbing crash cushions located in front of today's highway hazards in all 50 states and overseas. To convince doubting government transportation officials of its effectiveness, John often served as a live driver in dangerous crash tests.

Today, at the golden age of 80, John Fitch continues to live in the future and is working on a bright idea for a modern Model T, a basic car for the Third World to go anywhere at any time.

Forever a racer and entrepreneur, John Fitch poses with his 1962 Corvair Sprint, which was then priced less than $3,000. The GT conversion included fast ratio steering, beefed-up suspension, and a four-carburetor high-performance rear engine.

The Nifty Fifties

The Fitch Phoenix was America's first rear-engine sports car. Designed by John Fitch in 1966, the handsome two-seater GT featured a roll bar, padded dash, and contoured air-cooled seats (another Fitch invention). The hot Corvair engine with four carburetors and 9-1/4:1 compression delivered 170 bhp at 5200 rpm. Priced at $8,760, the car cruised at more than 100 mph.

The Fitch energy-absorbing crash barriers commonly seen at highway intersections or in front of bridge abutments will probably save many more thousands of lives—perhaps even yours. The patented, yellow foam-core plastic barrels, usually positioned in a "V" formation, are filled with sand in varying amounts. When an errant car hits the lighter barrels that are holding less sand, the vehicle progressively slows down as it plows into the larger, fully-filled barrels of sand.

Ted Cyr

After goofing at the starting line (and losing), Ted Cyr (right) drove another backup (but unblown) dragster to win the 1958 NHRA Nationals Top Eliminator title plus a new Chevy Apache pickup. The new blown Chrysler dragster (shown with partner Bill Hopper on the left) has a Scotty Fenn frame, to which was added a three-leaf transverse front spring that replaced the solid suspension. The front axle was filled with lead to keep the front wheels on the ground.

BOB AND DON SPAR

B&M Picked Up Where GM Left Off

"Real rods and sports cars don't have automatic transmissions" was once a tired but justified cliché of the auto world, but Bob and Don Spar and Mort Schuman changed that bias when they invented the "smart" Hydro-Stick and put critics in their place.

In early drag racing, engine power for the rear wheels was relayed through a standard transmission or a "high-gear-only" clutch setup. However, in 1959, a new era for automatics began with B&M's patented Hydro-Stick, which was redesigned to enable the driver to shift manually for racing or automatically for street driving.

By the mid-1960s when automatics were catching up to and even outperforming sticks, B&M and others began building shifters, high-stall torque converters, and high-performance transmissions for competition and street use. B&M's Torkmaster (a self-contained fluid coupling with a companion small dog-clutch in/out box), developed for top fuelers, was on Frank Cannon's early-record dragster.

"By the end of the decade," recalls Bob Spar, "every fuel funny car in the country was running B&M transmissions."

By the next decade, the B&M enterprise diversified. However, in 1988 when the trade was erroneously informed that B&M was no longer in racing transmissions, sales bottomed out. Afterward, the Spar brothers returned with a marketing vengeance.

"It has taken several years," said Bob Spar proudly, "but now B&M Automotive is, once again, the leader in automatic racing transmissions and torque converters."

By the mid 1960s, automatic transmissions were no longer weak-performing "slushboxes." They were being super-modified, with virtually all important components redesigned to withstand the strain of much higher horsepower drag racing machines. Since then, B&M and other major transmission shops also began offering more trick hop-up products such as shift improver kits, dash-controlled vacuum modulators, windage trays, special adapter plates, and improved floor shifters for street or strip.

In 1958, when this picture was taken, Mort Schuman (left), Bob Spar (center), and Don Spar (right) did not realize that their smart Hydro-Stick would start the unprecedented switch to automatic transmissions for drag racing. (Circa 1959)

JOCKO JOHNSON
Still Ahead of His Time?

With a passion for aerodynamics, Jocko Johnson became an inventor of things dealing with better airflow. His skill with an ordinary grinder made Jocko a legend from the mid-1950s onward, simply because he made engines breathe better. In fact, Jocko's Porting Service helped put more than 200 racers in the record books.

Robert Johnson received his nickname of "Jocko" after relieving his "persistent, personal itch" while working at the famed Barris Kustom Shop during the heyday of radical customizing. As a teenager in 1953, Jocko became so fascinated with aesthetic streamlining that he began studying aerodynamics. After Jocko teamed with Scotty Fenn, where he gained initial experience in the popular art of engine porting, Jocko taught himself how to work fiberglass and weld steel.

"After 18 months of hard work, I had built the most streamlined dragster in the world," said Jocko, who was one of the first to recognize the importance of aerodynamics, which is expressed in today's wind tunnel generated cars and aero accessory packages (e.g., front air dams and rear spoilers).

"When 'Jazzy' Nelson, then known as 'Mr. Drag Race,' saw my new untried car, he was so impressed that I let him drive it—and he broke the world record for elapsed time at Riverside," said Jocko. "The next time out, the fiberglass body broke apart, but I was convinced that aero was the right path, so I started to hand-form an aluminum body with the same shape."

Five years later, Jocko finished the second car with an Allison V12 aircraft engine that Emery Cook drove to a world record speed of 193 mph on gasoline in 1965.

"Streamlining a dragster, back then, was an oxymoron, but who knew?" exclaimed Jocko. "My cars made a statement about air working for you in racing. Too bad I disguised them so well, because people who saw them only saw a streamlined shape, not the downforce they produced. Nowadays, if you don't pay attention to aerodynamics, you lose big time."

Today, Jocko is somewhat of a recluse and thrives, tucked away in his desert home near Twentynine Palms, California. There the former racing guru spends his days working on a new 30-foot, wasp-like streamliner he hopes will break the present land speed record with a revolutionary three-cycle radial engine called the "PoweRRing," which has 18 cylinders arranged around a large 12-lobe camshaft (instead of a crankshaft). For every 360 degrees of rotation, 216 ignition firings occur, with 6 cylinders firing simultaneously by every 10 degrees of rotation. If successful, this unique engine could revolutionize the automotive industry.

The International Drag Racing Hall of Fame inducted Jocko in 1997. However, when he was notified, he asked to be removed, calling it the "Hall of Shame" and refusing to have his name placed on a stone statue. The selectors promptly removed Jocko's name from the list. "Too bad," said a director of the International Drag Racing Hall of Fame, "because Jocko has made a lot of contributions to our sport and definitely deserves the honor."

Jocko's beautiful but short wheelbase streamliner was the first dragster to hit the strip with a full-envelope body. With Jazzy Nelson driving, the Hemi-powered machine broke the world record for fastest elapsed time at 8.35 seconds at 178 mph in 1959.

BOB LARIVEE
The Force Behind Championship Auto Shows

However, in 1959, when Bob and his brother Marvin kicked off the famous International Championship Auto Show (ICAS) series, they found the going rough. At that time, the public knew little about hot rods and custom vehicles during the subculture years of hot rodding.

Over a span of almost 40 years, Bob saw his idea flourish from a disappointing start-up to the longest-running, sanctioning auto show

When the Michigan Hot Rod Association, with its 22 clubs, wanted to build a concrete-surfaced drag strip in 1952, a series of car show benefits called Detroit Autorama was held with promotional help from Bob Larivee, who was then a gutsy weekend racer.

After Bob foresaw the need for standardized rules to discourage unfair competition, he conceptualized a plan that would later revolutionize the auto show business.

This flashy Barracuda won the ICAS grand prize in 1972. It was a unique Harry Bradley/Church Miller creation.

This 1974 ICAS program cover displays a show truck built for TV star Redd Foxx.

In recognition of more than four decades of leadership in automotive photojournalism, Bob Larivee (right) presents the ICAS Auto Show Award to Eric Rickman (left). (Circa 1990)

organization in the world. With millions of spectators attending the ICAS network of professional championship shows in major cities in the United States and Canada, thousands of the world's finest automobiles and dream machines have been showcased—many by the most talented artists in the nation. After four decades, Bob continues to help clubs by promoting successful hot rod shows that enable the sport to continue growing.

The Nifty Fifties

The Red Baron, one of the wildest show cars ever created, was designed by Tom Daniel for Monogram Models before a full-size replica was built by Chuck Miller.

The Outlaw, built by Ed Roth, was a featured ICAS show roadster in the early 1960s.

Connie Swingle

The swashbuckling Connie Swingle was hardly more than a beginning drag racer from Oklahoma when he walked into Don Garlits' shop in Tampa, Florida, in the late 1950s. Connie was looking for either a job or a ride, whichever came first. With Don Garlits, he received both and became one of the best fabricators in Garlits' dragster-making business.

In 1961, Connie drove Don Garlits' Swamp Rat III, terrorizing the drag circuit while winning against the best. Similar to Don Garlits, Connie was a true showman who could electrify the fans.

Junk is all that remained of the top fuel dragster after a fast strip run in Emporia, Virginia. There, Connie's chute failed to open and the left rear wheel of the dragster hit a pine tree. Both the dragster and Connie were torn up. What a close call!

DARRYL STARBIRD
Created Cars of the Future

Once dubbed the "Bubble Top King" for his futuristic cars, Darryl Starbird has created more than 500 metal vehicles in the 45-plus years he has reigned as one of America's top customizers who continues to perpetuate the rare art of automobile restyling.

At first, following his dream was not easy for Darryl because he had no commercial body-shop experience. However, his metal shaping expertise quickly caught up with his blossoming imaginative talent, thus justifying the opening of his Star Kustom Shop in 1954 in Wichita, Kansas.

When Darryl's first show car, a 1955 Plymouth, was featured on a magazine cover, Darryl decided to concentrate on more futuristic styling that would make Detroit car makers take notice—and they did! The car that eventually brought him national prominence was a 1957 Thunderbird that copped top honors at the NHRA custom car show in 1959.

Darryl's first bubble topped vehicle, the Predicta, was featured on the cover of four magazines and took home dozens of awards. After that, there was no stopping Darryl, whose creative imagination appeared limitless. He then built 15 more futuristic bubble top vehicles plus hundreds of fantastic custom cars.

In 1963, Darryl became a design consultant for Monogram Models, where 15 of his original creations were done in 1/24th and 1/8th scale. More than one million kits were sold.

Recognizing the need to showcase the products of his and other customizers' talents, Darryl began promoting road and custom shows in 15 major cities. This effort led him to co-produce the famed Oakland Roadster Show for a while. As a leader in the sport, Darryl also was involved in founding of the National Rod and Custom Association, and he served a two-year stint managing the International Specialty Car Association.

A visionary builder even today, Darryl now operates the National Rod and Custom Hall of Fame museum in Eighteen, Oklahoma, where exotic creations of Starbird and other customizers are displayed.

Starbird's famous Lil Coffin later was a Monogram model.

Darryl Starbird's beautiful Star Trek coupe is an entirely handmade futuristic dream machine with a "floating" air scoop feeding its 350 Chevy engine.

The Nifty Fifties

These are two beauties. This red show stopper is one of Starbird's more radical street rods.

Predicta was Starbird's first bubble top custom car and now has a historical value of more than a half-million dollars.

The aerodynamic styling skirts and gullwing doors tend to hide the identity of this Datsun 280Z.

The Illusion is a 1976 AMC Pacer with an all hand-built metal body and Plexiglas® bubble.

Hot Rod Pioneers

Sports Rod Specials

"Make it light, low, and fast like a hot rod—with tight suspension for cornering—and if you have any money left over, make it sharp looking."

That was good advice. In 1948, famed Detroit designer Alex de Saknoffsky prophesied that the prototype for a protocol U.S. mass-produced sports car would come from hot rodding. He was correct, because the Corvette was on the scene by 1953 and was hot rodded later by Duntov. Previous to that, Frank Kurtis had marketed a competition kit car before he designed the prototype for the sporty Muntz Jet.

When the era of American sports car specials began around the mid-1950s, hot-rod rooted competition techniques were again utilized, with thousands of the hybrids built for road, strip, and track. Many had fiberglass car bodies produced by Devin and Almquist. These photographs show the evolution from the stark functionalism of early roadsters to later and more imaginatively styled road racers, street rods, and fiberglass-bodied sports rods.

Will the "sports rods" or similar hot-rod-type hybrid ever return? A look at the twenty-first century sports model concept cars suggests at least a return to some original back-to-basics design styles with modern interpretation.

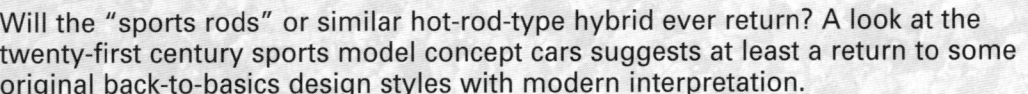

The Nifty Fifties

263

THE SEXY SIXTIES

The booming Sixties brought more mobility, the Beatles, and the mighty muscle car. The new Interstate Highway System made the legendary Route 66 obsolete and hyped a coast-to-coast drive without stop signs. Sadly, the last Burma-Shave sign disappeared from the roadside, too.

Drive-in movies, often called "passion pits," disappeared and were replaced by shopping malls and fast-food joints. Traffic jams and speed traps became a motorist's nightmare. Rock 'n roll music became louder, and cars were glorified in records, such as the "Little Deuce Coupe" sung by the Beach Boys.

Mickey Thompson broke fourteen international speed records in one day in 1961, which was only the beginning of many more accomplishments before he was murdered. By 1964, there were more than 6,000 drag meets with three-quarters of a million entries and more than ten million spectators.

The ban on rocket fuel, or nitromethane, was lifted by the NHRA in 1964, allowing the direction of drag racing to accelerate. The "East vs. West" drag war was raging, and "Big Daddy" Don Garlits' fueler was the first to smash the 200-mph barrier that year. A year later, Funny Cars, camouflaged as stock-bodied dragsters, received top billing.

Enthusiasts started replacing their sloppy column shifters with floor-mounted conversions for quicker shifting. Through the 1960s, Ansen, Hurst, and Almquist could barely keep up with the growing demand. Today, manual transmissions are nearing extinction, falling to 10% of the market, down from the 28.6% in 1960.

After the 1964 Pontiac GTO kicked off the "muscle car" wars with pumped-up horsepower to grab the ever-expanding youth market, other auto makers followed by stuffing big engines into small cars and giving "Pony Cars" macho names, such as Mustang, Cougar, Cobra, and Barracuda. In 1964, the sporty Mustang sold a record one-half million at a starting price of only $2,368.

NASCAR, with its "roundy-round" racing growing faster than anyone's charge account balance, started its short-lived Drag Race Division in 1966.

"Beatlemania" arrived with the Volkswagen "Bug," which was the "in" car for students in the mid-1960s. A replacement fender cost only $12, and cheap hop-up equipment made the little car a snarling drag racer or off-road dune buggy RV.

By 1968, one in six Americans reportedly made, sold, maintained, or drove a motor vehicle for a living. Sixty percent of marriage proposals took place in cars, according to a survey that failed to reveal statistics on other "intimate" things that sometimes occurred in cars.

Continuing to ignore the racing ban, factory hot rodding often became clandestine, with car makers farming out their race efforts. The reintroduced 426 Hemi-head engine was outlawed by NASCAR in 1965, but not until Hemi-powered Dodges and Plymouths won 111 of 131 major stock car races.

Chrysler drag racing teams also "cleaned up." The practically race-ready, Hemi-powered Barracuda and Dodge Dart (priced at $5,150 and higher) easily ran 11.25/120 in the quarter-mile.

Funny Car racing, similar to Super Stocks and F/Xers, had reached the "big time," with moneyed corporate sponsors. The idea was "Race on Sunday—Sell on Monday."

By 1968, all cars had seat belts for all passengers, front-seat headrests, and side-marker lights. However, the onslaught of environmental, economic, and safety directives, plus inept moves from Detroit, began to make the end of the decade miserable for motorists.

DEAN MOON
A Rebel with a Cause

Auto buffs will remember the famous "Moon Eyes" as one of the most recognizable logos in the performance industry in the 1960s. Dean Moon made the comical "eyeballs" his company's trademark, not realizing that eventually the decals and stickers would become better known than his original Moon products.

From the start, Moon's life was colorful. At a tender age in Minnesota (from his dad's Phillips 66 gas station), he would bootleg moonshine to even the elite folk of the community. One day, his Pa was caught red-handed with the outlaw high-octane juice.

Later, in the 1930s in Santa Fe Springs, California, Dean was a busboy and short-order cook in his father's "Moon Cafe." There, Lana Turner, Spencer Tracy, and James Cagney would stop for lunch while filming at the nearby oil fields.

In a fix-it shop across from the cafe, Dean concocted parts for hopping-up his Model A coupe and his buddies' street racers. At that time, when youngsters challenged each other to illegal street racing, Dean got busted and had to spend a few days in jail.

Dean's first products were fuel blocks, spun aluminum fuel tanks, and foot-shaped accel-

After an engine swap, Dean Moon tuned the 1951 Studebaker OHV mill, which was squeezed in his immaculate 1934 Ford Coupe. He copped an easy class win at 117 mph at El Mirage in 1953.

erator pedals. The shiny "Moon Disc" wheel covers that were intended to reduce wind resistance on Bonneville racers were sold by the thousands for street use.

During his varied career, which included a stint as a photographer for auto magazines, Dean was always in the thick of things as an activist. Seeing the need for a trade organization for the growing speed equipment industry, he became a forceful advocate in establishing SEMA. In 1964, Dean was elected as the second president of SEMA.

Throughout the 1950s and into the 1980s, Dean's cars ran at Bonneville. In 1964, Dean made his first appearance with an organized U.S. Drag Racing Team in England. "The Moonbeam," a modified sports car, was driven by Dante Duce.

In 1987, Dean was posthumously inducted into the SEMA Hall of Fame. The Moon speed equipment business now falls under the name "MoonEyes USA" and is Japanese-owned.

Dean Moon built his "Moonbeam" sports car in 1959. The yellow Devin fiberglass body is glued and bolted on a rugged handmade tubular frame. A GMC-blown Chevy mill goosed the car to more than 200 mph on the salt.

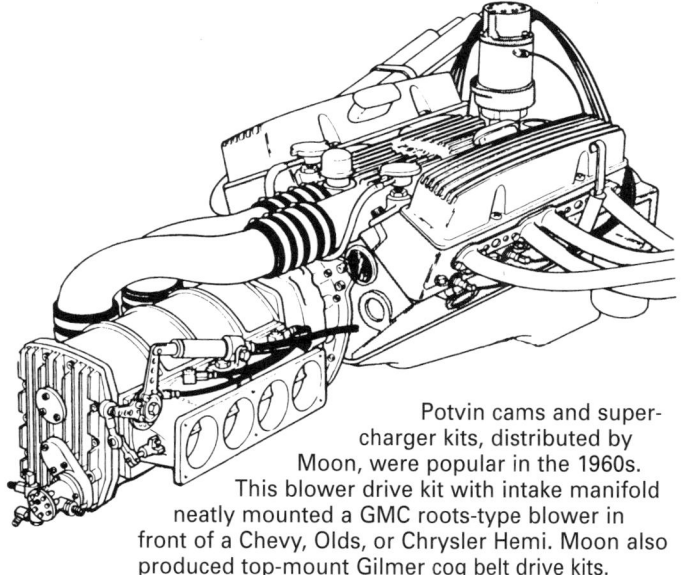
Potvin cams and supercharger kits, distributed by Moon, were popular in the 1960s. This blower drive kit with intake manifold neatly mounted a GMC roots-type blower in front of a Chevy, Olds, or Chrysler Hemi. Moon also produced top-mount Gilmer cog belt drive kits.

Land Speed Record Cars

Clever Streamlining Boosted Speed

As the fight for "fastest man on earth" continued in the 1960s, the U.S. Auto Club and its world affiliate FIA divided the land speed contests into two categories: wheel-and-piston-driven and unlimited, jet-powered, thrust vehicles. After the remarkable piston-driven Bonneville record speeds of more than 409 mph were achieved by the Summers brothers' and Al Teague's streamliners, the speed envelope soon was jet-pushed to top records by Breedlove, Arfons, Gabelich, and Noble. In 1979, Stan Barrett's Rocket even did 638 mph but was not recognized by sanctioning bodies. On October 15, 1997, Andy Green exceeded the speed of sound with an official 763 mph for the world's land speed record.

Aerodynamic streamlining to reduce wind resistance was first discovered and practiced by the racing fraternity. Early attempts at streamlining were based on the "droplet of water" idea, wherein the best body shape was thought to resemble a drop of water falling freely in the air, with a rounded front that tapered to the tail. Variations of the teardrop configuration were seen in some early racers such as Frank Lockhart's Stutz Special.

These photographs show the evolution of major land speed record cars—many that had full-bodied wind-tunnel-tested streamlining that reduced wind resistance. Old streamliners progressed from exposed wheels to partial or full wheel and frontal fairings.

Frank Lockhart's Stutz Special is an excellent example of early attempts at aerodynamic streamlining of vehicles in a teardrop configuration.

Sir Malcolm Campbell (left) raised his Bluebird's world record from 272 to 301 mph in 1935 by employing a fully enhanced body instead of the earlier semi-opened wheel version shown here. Immediately after this photo was taken, Sir Campbell easily beat the airplane in a publicity stunt race at Daytona Beach, Florida, in 1932.

The canopy on Marvin Lee's B Streamliner reduced wind resistance. Here, the car is shown with its Wayne-equipped, Chevy six-engine exposed. The car set a Class B record of 153.5 mph at El Mirage in 1949.

The Sexy Sixties

The plexiglass canopy cuts wind drag on this teardrop-shaped Class A Lakester that won numerous SCTA records from 1951 onward for the Alex Xydias racing team.

When Lee Chapel's "Tornado" streamliner, driven by Sonny Rogers, streaked to a new Class C record at 224.144 mph in 1952, it was a multiple triumph for Lee, who again proved that his own Tornado overhead-valve cylinder heads really could make a Ford or Mercury flathead go. An innovative pioneer Los Angeles speed shop owner, Lee Chapel later moved to Oakland, California, where he peddled speed equipment for many decades.

This early 1940s streamliner also had a fully enclosed aluminum body with open wheels. Owned by Ralph Schenck, it had a four-cylinder Chevy with an Olds three-port cylinder head. The car ran in the range of 120 mph. By 1950, this type of open-wheeler would be reclassified as a Lakester.

Mal Hooper (second from left) is pictured here with his happy crew and the Shadoff Special that took the Class C Streamliner honors at the 1953 Bonneville meet at 231.66 mph. Later, a new International C Streamliner record was set for the measured mile at 236.36 mph. The same car, with a larger engine, set an FIA B Class record at 272 mph, with Bob Bowen as driver.

The formula for land speed record chasers has always been massive horsepower in the most slick body envelope. On the salt flats in 1960, talented Mickey Thompson clocked more than 400 mph one way in his four-wheel-drive Challenger I, which was powered by four Pontiac engines.

Hot Rod Pioneers

In 1965, Bob and Bill Summers' "Goldenrod" streamliner (driven by Bob) broke the world land speed record with a blazing two-way average of 409.277 mph—a record that stood for 25 years. Numerous wind tunnel tests were made before perfecting the cigar-shaped, aluminum-skinned, 30-ft.-long streamliner that weighed 3 tons. Its front and rear wheels received power from four in-line, unblown, 426-Hemi engines that produced 2400 hp at 7000 rpm.

After drag racer Gary Gabelich (left) drove the "Blue Flame Special," a thrust-driven streamliner, 622 mph for the official one-mile FIA land speed record in 1970, the record stood until 1983, when Britisher Richard Noble copped the title at 634 mph. A daredevil who also drove top fuelers and jets, Gabelich was innovative. Note the large stabilizing fin and outside rear wheels (right). The car ran on liquid natural gas.

In addition to occasional dry lakes racing, Sandy Belond ran this handsome streamliner with an enclosed cockpit to advertise his Belond exhaust systems. It finished fourteenth in a 1955 IMS blast.

By the 1970s, most dragsters were utilizing some form of aerodynamics for stability and downward force. Large rear wings that acted as levers were elevated as much as seven feet off the ground to exert maximum downward force. The front wheel "pants" and frontal fairing on this blown-fueler reduced turbulence around the axle and front wheels; however, the NHRA later outlawed them. James Warren (who teamed with Roger Coburn) drove "Rain for Rent" and other top fuelers to many NHRA victories, including Bakersfield in 1975, 1976, and 1977.

HOLMAN AND MOODY
The Odd Couple

Did divine providence link Holman with Moody to be builders of champions? Former race driver Ralph Moody knew how to build winning cars. Former salvage yard hustler John Holman, who was faster than a rum runner while "kiting checks" between Texas and California, was a fantastic promoter and a wheeler-dealer proficient in squeezing real dollars from wooden nickels. In 1957, when the AMA banned factories from racing participation, the then-unemployed partners acquired Pete De Paolo's Ford stock car racing team.

"The first thing we did was to stop racing all the existing cars. They weren't winning, so why race them?" said former Midwestern native John, who later witnessed to the semi-knockout blow from a windshield that hit his partner Ralph at the NASCAR track in Arkansas. Surprisingly, Ralph, a typical Yankee from Massachusetts, took the win.

"They were in high cotton and got dirt-cheap parts to build a ready-to-race stock car for

John Holman (right) holds the trophy that smiling "Fireball" Roberts had received for driving a Holman and Moody stock car to an easy win at Langhorne, Pennsylvania. Roberts was a NASCAR hot shot in the 1950s.

about $10,000," said all-time-great Smokey Yunick about the skilled team that became Ford's semi-independent racing shop when the AMA racing ban ended.

Moody was the race team coach who molded NASCAR champion Dan Gurney and "Fearless Fred" Lorenzen—who was, in 1967 when he retired, the biggest money-winner in the history of NASCAR.

By 1965, when both stock car and drag racing again were becoming full-fledged factory wars, Holman and Moody's sprawling speed shop was busy building stockers, drag racers, Can-Am road racers, and the big-block Ford GT40, which included the 1966 winner of LeMans.

Moody, the former New England midget and sprint car winner and Midwestern USAC stock car winner, sold out in 1972. Three years later, John Holman—the wild, golden-nuggetted carefree man who could laugh while outrunning fire-belching trains with his huge truckloads of salvage—had a fatal heart attack at the wheel of his beloved big rig. That ended the enterprises of Holman and Moody, the men who "kicked butt" with their racing cars, their high-performance tricks, and their great sense of humor.

Ralph Moody (left), holding the winner's flag, was a NASCAR star who captured five Grand National victories. His business partner was John Holman.

Some of the Holman-Moody's half-wild A/FX Mustangs, with more than double the horsepower, blasted the 1/4-mile at 128 mph in 10.92. Because they were campaigned by dealers with factory support, it was rumored that some top drivers acquired the hot pony cars for a mere dollar.

Hot Rod Pioneers

Exhibition Cars

Wheelstanders such as this Dodge added to the excitement at drag racing events in the 1960s. Casters were added to prevent tearing up the rear of the car.

The "Little Red Wagon," driven by drag racer Bill "Maverick" Golden, was a big crowd pleaser at drag strips in the United States and Canada. The pickup could remain upright for a full quarter-mile—often while exceeding 140 mph!

Jet-powered thrust exhibition vehicles were a hot new feature at many early 1960s tracks. Later, popular exhibition vehicles included rocket cars, jet trucks, V8-powered motorcycles, historic racers, and trick motorized oddities. Here, Doug Rose's jet dragster blows the doors off as the scorched station wagon was lifted and toppled.

ROBERT DE BISSCHOP
A Key Player in the Offy's Struggle for Survival

After almost half a century as the icon of the racing world, the mighty four-cylinder Offy engine seemed doomed when the formidable Cosworth V8 was introduced in the early 1970s. Even earlier, the new four-cam Ford V8 racing engine had threatened to put the Offy out of business. It was a "do-or-die" situation in 1965 when Dale Drake, Louie Meyer, and Leo Goosen decided to try a new type of supercharging. Help came in the nick of time from a team of hot rodders headed by Bob De Bisschop, an AiResearch turbocharger engineer.

Bob combined his visionary skill with master mechanic Herb Porter and Stuart Hilborn who had developed a fuel injector for superchargers. The breakthrough was finalized after extensive beefing-up by Goosen so that the Offenhauser engine would withstand the turbo's boost of 15 psi or more required to deliver 625 hp at 8000 rpm. The AiResearch turbocharger, which was fed with Hilborn injectors and priced as low as $3,500 in 1965, helped the almost obsolete Offy remain competitive for another decade. Shortly after the 1979 Indy races, the Drake-Offy shop was closed permanently.

In 1968, Bobby Unser was the first to win an Indy race with a turbocharged Offy engine. Turbos ran at Indy from 1966 to 1996, when the rules changed to normally aspirated engines.

Bob's racing career began in 1947 when he built a flathead Ford V8-powered track roadster that future Indy champ Bob Sweikert raced seven nights a week. Later, Bob built one of the first Ford flathead V8-engine dragsters.

After working his way through engineering college, Bob worked for Frank Kurtis, producer of the classic Kurtis Kraft race cars. There he hooked up with A.J. Watson, the famous wrench wizard who developed technically advanced midget, sprint, and Indy cars. In 1955 and 1956, Bob was co-mechanic for the John Zink cars driven by Sweikert and Flaherty, both of whom won the Indy 500. After that, Bob and Dennie Moore built two 1957 Pontiacs—one set a 141-mph straightway record on the sands of Daytona.

While working for Garrett's AiResearch Division, Bob helped develop the first turbocharged production car for Oldsmobile in 1962. Since then, the small but mighty turbo has become the wave of the future as a prized, hop-up option and as standard equipment for passenger cars.

Future Indy champ Bob Sweikert seated in the track roadster that Bob De Bisschop built in 1947.

Bob De Bisschop today.

Bob De Bisschop looks at the turbocharger he helped develop in 1965. The small turbo and Hilborn fuel injection helped the Offy engine remain in the winner's circle for another decade.

DAN GURNEY
Mr. Versatility

Dan Gurney and the winning Eagle Formula One car in 1967.

Legendary Dan Gurney began racing hot rods in 1950, clocking a modest 130.43 mph in a 1932 roadster at the Bonneville Salt Flats on the Utah/Nevada line. Then he graduated to big-time sports car, stock car, Formula One, Indy, and Grand Prix racing and became one of the most versatile race drivers in the world. Dan Gurney's 1967 Belgian Grand Prix victory with his Eagle was the first Grand Prix victory by an American driver in an American car in 46 years.

Dan Gurney is said to be similar to his trophies—understated elegance.

"Racing's been a labor of love for me and probably will continue to be. The fact that we frequently win makes it even better," notes Dan Gurney, who has been a successful driver, designer, builder, and team owner.

Daniel Sexton Gurney was born in 1931 on Long Island to Roma Sexton and New York Metropolitan Opera star John Gurney. He graduated from Manhasset High School in New York and Menlo Junior College in California before serving two years in the military—mostly in Korea.

Like most hot rodders who were fascinated with racing, Dan learned the ropes through personal involvement. He joined the Ferrari F1 factory team after a sports car race win in a Porsche in 1956 and after successfully piloting a Corvette and an Arciero-Ferrari.

During the next 15 years, Dan raced in four major types of competition: Formula One, sports cars, Indy cars, and NASCAR races.

A triumphant Dan Gurney after winning the Belgian Grand Prix (SPA) in 1967.

The Sexy Sixties

After competing in 25 different makes of cars, Dan Gurney retired with a record of 37 racing wins in 10 different countries. His legacy includes seven Formula One Grand Prix races (four World Championship events) and numerous stock car, Can-Am, and sports car races, with wins at Daytona and Sebring.

Twice, Dan placed second in the Indianapolis 500. Teamed with A.J. Foyt, Dan won the famous 24 hours of Le Mans. When Dan was handed the customary bottle of champagne upon the victory podium, he accidentally sprayed all the surrounding dignitaries, including a startled but delighted Henry Ford II. This incident caught the eye of the press and rocked the sports world, thus clinching the winner's ceremonial act for posterity.

In 1965, with the initial funding from Goodyear, Dan began capitalizing on his mechanical talents with his production of championship-caliber racing cars. Thereby, All-American Racers (AAR) was formed, with Carroll Shelby as partner. In 1968, Dan took over as sole proprietor.

From 1966 and through the 1970s, AAR Eagles became the equivalent of today's

Toyota Eagles, such as this MK111 driven by Juan Fangio II and P.J. Jones (son of Parnelli Jones), won more than 20 GTP races.

Lolas, the consummate customer car. The AAR cars have won more than 24 major races in the United States and abroad, including wins at the Indianapolis 500 with Bobby Unser and USAC National, GTO, and GTP championships.

The massive AAR complex contains a state-of-the-art CAD design and research and development department, two wind tunnels, and a complete composite materials department, where all the Eagle race car bodies and chassis are built. Often, AAR projects, similarly to Toyota's, become rolling laboratories in which lessons of tomorrow are learned. The Gurney-built Eagle was so dominant that the GTP (Grand Touring Prototype) class for which it was created has been assigned to history.

Dan Gurney drove this 1929 Ford channeled roadster to 130 mph in 1950.

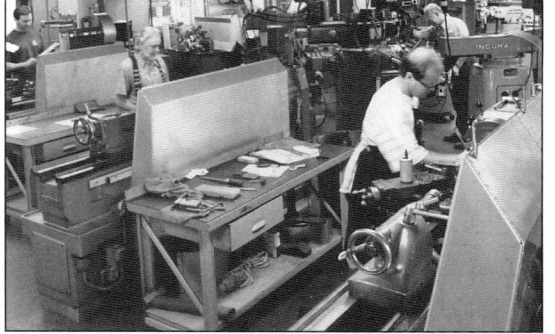
This corner of the All-American Racers (AAR) factory shows a variety of metal-working machines that can quickly fabricate parts.

Skilled workmen assemble an Eagle racer.

JOE WILHELM
Creative Customizer

One of America's lesser-known, premier automobile customizers was Joe Wilhelm, who produced scores of award-winning custom cars in the 1950s and 1960s.

"The many fender-benders I got involved in when racing modified stock cars as a kid gave me additional practice doing body work," admits Joe.

As customizing became popular, Joe began restyling with a passion. He opened his own shop in San Jose, California, where his flawless metal work earned him the reputation as a top custom car builder. Joe's "Wild Dream" custom car earned five trophies at the 1963 National Auto Show in Indianapolis, and a model kit was later made of the car.

"I put a lot of myself in it, with practically everything on it hand formed," Joe said with pride.

Because many of Joe's winning customs were designed from mental images rather than sketches, creativity became the trademark of the master craftsman who refused to follow the trendy crowd.

"The reason I got into customizing is that I wanted to create something different, with broad, public appeal," said Joe. "I also wanted to assist others with their ideas because there is great satisfaction in helping others solve their problems."

When interest in custom cars ebbed, Joe closed the shop in 1968. Customizing then became Joe's sideline as he taught auto body classes at San Jose Junior College.

Sadly, the recent resurgence in customizing and individualizing today's boring and bland look-alike cars came too late for legendary Joe Wilhelm, who died in 1996.

Joe Wilhelm's "Wild Dream" came true with a beautiful, prize-winning street roadster that took three years and $10,000 to build. The body shell was hand formed from aluminum and fitted to a completely chromed tubular frame. The grill is made of square bar stock. The body is finished in 20 coats of purple (later gold) metal flake lacquer. Black Naugahyde upholstery sets off a handmade birchwood steering wheel and grab bars. The injected 283 Chevy engine puts out more than 300 hp. A model was made and called "Wilhelm's Wonder." (Andy Southern Photo)

After attaching the sculptured body to the narrow tubular frame, Joe Wilhelm begins lengthy preparations for painting. (Andy Southern Photo)

The Sexy Sixties

This Model A pickup won sweepstakes awards at San Jose, Modesto, and Oakland Roadster shows. The flawless body sports black lacquer paint. (Andy Southern Photo)

Customized 1952 Chevy features customer Jim Greggs' ideas that were well executed by Joe. The narrowed grill is from a 1956 Chrysler; the headlights are from a Kaiser. The fender walls are rounded, and side air-scoops are functional. (Courtesy of Greg Sharp)

Mark I Mist is a handsome 1936 Ford restyled by Joe Wilhelm and lowered by sectioning 4" and channeling 6." Lincoln headlights are sculpted into 1940 Ford fenders. Rear fenders are from a 1938 Chevrolet, and taillights are from a Corvette. Horizontal bars fill the Edsel-like grill shell. (Andy Southern Photo)

TONY NANCY
The Loner

This Plymouth V8 Wedge-powered gas dragster was one of Tony's most successful vehicles. He remained Number One on *Drag News*' "Mr. Eliminator" list for two years. (Photo by C.K. Callaway)

Before the days of network television and big-dollar sponsors, most drag racers ran on shoestring budgets and drove and wrenched their own thundering machines. Tony Nancy, "the Loner," was no different and operated without even a pit crew.

In 1956, to support his racing urge, Tony opened his own automobile upholstery shop in Sherman Oaks, California, where Hollywood celebrities such as Gary Cooper, John Wayne, and Dean Martin lined up for Tony's luxurious leather interiors.

After Tom Sparks dropped a hot Sparks-Bonner nitro-burning flathead in Tony's first 1929 Ford roadster, the car began winning trophies for top time, top eliminator, and top speed, as well as West Coast national records. This car and a roadster-body dragster, which was powered by an injected and blown Buick, appeared on various *Hot Rod Magazine* covers. After setting a new A/MR Class and national record at the 1962 Winternationals, Tony decided to break with tradition and construct an awesome rear-engine dragster called Wedge I. After that car crashed, Tony built a Hemi-powered Wedge II, which became one of five cover shots for *Hot Rod Magazine*. Tony then toured overseas and became the roving ambassador of drag racing.

With a newly built top fuel dragster, in 1972 Tony blasted the top speed record for top fuel dragsters at 236.22 mph and then 238.57 mph. Later, Tony raced a rear-engine top fueler with success. However, the high costs of racing and sponsorship dollars forced Tony to retire from competition in 1979.

"Guys like Art Chrisman, Jim 'Jazzy' Nelson, Calvin Rice, and Emery Cook, who were my heroes, built different engines and chassis," said Tony. "Everything was their own idea, done totally on a fixed budget. Back then, you had to be sharper than the other guy if you really wanted to win. Now, you've got huge dollars, high-tech engines, computers, a crew chief, and a team of specialists. It's just not the same."

Teaming with his mentor, engine builder Tom Sparks, Tony used a blown flathead to power his 1929 roadster. The car had a tubular frame built by Kent Fuller. Tony's next car was a 1927 T-modified roadster that earned a new national record of 169 mph at the 1962 Winternationals. (Photo courtesy of the Tony Nancy Collection)

The Sexy Sixties

"But don't get me wrong. Drag racing is still very exciting, with the awesome ETs and fantastic top speeds. But, to me, it is still more fun when you do everything yourself," said Tony, a true artist in automobile restoration and customizing.

Award-winning automotive journalist and former racer Tom Madigan recently wrote a fascinating book, *The Loner: The Story of a Drag Racer*, based on Tony Nancy's exciting career.

Tony (in leather jacket holding a trophy) won "Best Engineered Car" and "Competition Eliminator" at the 1964 Winternationals.

Getting ready to launch, a crewman makes a last-minute adjustment on Tony's blown Hemi top fuel dragster that had a frontal wing and winglets on its sides to aid aerodynamics. Tony won the famed U.S. Fuel & Gas Championship in Bakersfield in 1970. (Photo by Leslie Lovett)

Chris Economaki

Today, the success of major sports, including drag racing, is judged on how much television exposure it can generate. The first of the best television commentators to telecast racing events was Chris Economaki, who joined ABC Network in 1961. After working on dirt track cars in the mid-1930s, Chris became a track announcer, a print journalist, and a radio broadcaster. As former editor and publisher of *National Speed Sport News*, Chris knows everybody in racing, including where the skeletons are hidden. When the Winston Cup race at Charlotte was given the red flag because of a broken sewer line, Chris ad-libbed. He told Paul Haney of TV Motorsports, "That's the first time I've seen a race stopped because a track had urinary problems."

CRAIG BREEDLOVE
The "Spirit of America"

On the morning of August 5, 1963, the thunderous roar of a jet-propelled 35-foot-long car shattered the silence of the Bonneville Salt Flats. Craig Breedlove's "Spirit of America" had eclipsed the land speed record with a whopping 407.45-mph two-way average. This surpassed the 394-mph record held by John Cobb of England and kicked off a jet-car duel with Art Arfons, who later grabbed the title doing 536 mph in 1964. The following year, Craig regained the land speed record with a 600.601-mph run and again became the "Fastest Man on Wheels."

In the early 1950s, Craig's first hot rod was a 1934 Ford that clocked 154 mph at El Mirage Dry Lakes when he was 17 years old. Next, he built a flathead Ford dragster, followed by two coupes and a belly tank that did 233 mph at Bonneville. In the next decade, he purchased a surplus J-47 jet engine for $500 and began constructing his first jet car in his home garage.

"When I started building the 'Spirit of America,' I didn't know if I was building a car or a coffin," Craig recalled. "At the speeds I was hoping to go, my trip to the salt flats could have been my last, but I was determined not to think of it too much."

Because his car had exceeded the anticipated $10,000 figure, Breedlove had to seek corporate sponsors. After Breedlove traveled throughout the nation, Shell decided to help on the financial end, and Goodyear provided the tires.

As a self-taught designer, Craig (who was the first man to exceed 400, 500, and 600 mph) tried new ideas in aerodynamics, such as outrigger wheels at the rear with an enclosed front wheel in an extremely narrow fuselage. This reduced air drag and improved handling but caused a ruckus about defining a "car."

With his resolute vision and his iron will, Craig Breedlove has earned a middle name of "Determination." The bold and daring Craig Breedlove continues to personify the spirit of America in his further attempts to break the world land speed record.

In 1968, the Breedloves were then the fastest man-woman racing team in history. Lee and Craig Breedlove, at the wheel of the AMXs, broke 106 American, national, and international speed records in Class B and C for closed cars. (American Motors photo)

The 1965 Sonic I, built and driven by Craig Breedlove, was the first car to exceed 600 mph. In an effort to exceed the speed of sound, Craig went 677 mph in 1996 in another jet-powered "Spirit of America" but failed to make a record second run.

Carl Casper

Casper's "Ghost" won America's Most Beautiful Roadster title in 1964. Carl Casper spent 2,000 hours fabricating this uniquely sculptured car with a hand-formed fiberglass body and padded cantilever top. Two GMC blowers fed the Pontiac mill. Although Carl won more than 200 trophies for his wild show rods, he found time to produce auto shows and drag races as well. His AA/FD, driven by Danny Ongais, won the 1963 AHRA Winternationals.

Yesterday's fashion often becomes today's joke, but the sculptured beauty of Carl Casper's futuristic sportster endures. The fiberglass bodied concept car won the NHRA Experimental Sweepstakes award in 1967.

Gaspar 'Gas' Ronda

The NHRA World Champion Top Stock Eliminator in 1964 was Gaspar "Gas" Ronda, a former dance instructor who gave up the light fantastic toe for a heavy foot. He won more than 85% of his match races in 1967 with a fuel-injected Mustang that sported a fiberglass body hinged to a rail-like chassis.

Ronda was one of the first to drive a stock-bodied car over 170 mph in competition, and he continued to be a consistent winner throughout the decade. However, a freak funny-car accident scorched 30% of Ronda's body with third-degree burns, ending his illustrious racing career.

Gas Ronda won most of his races in 1967 with this fuel-injected Mustang.

BILL SIMPSON
Mr. Safety

"Pour some gasoline, and light me on fire," said Bill Simpson, wearing a firesuit and challenging a "doubting Thomas" competitor at an Indianapolis Speedway press conference.

"When the fireball subsided, there were no more disbelievers," smirked Bill. "Pretty soon, everyone was wearing my protective gear."

The famous Simpson "Heat Shield Firesuit," first introduced in 1966, followed Bill's earlier aluminized cotton suit that was only minimally flame resistant. Prior to that, race drivers dipped jumpsuits in borax for an attempt at fire retardancy.

A chance meeting with astronaut Charles "Pete" Conrad convinced Bill to use a space-age material called Nomex, which had already been adopted by NASA and the U.S. Air Force. The new miracle fabric, when blended with a Beta fiber, created the most heat-resistant fabric known to mankind.

As someone who raced dragsters, Bill became concerned about slowing down those mighty machines. One day, while Bill was racing a dragster, the malfunctioning hand brake caused the car to careen through a barrier. Hoping to improve on cumbersome aircraft-type chutes, Bill began experimenting with more compact designs in the late 1950s. Starting with $20 cash and $200 borrowed from relatives, Bill began producing cross-form drag chutes. As word spread confirming the quick release of the Simpson chute, business expanded. Then, Bill introduced a high-quality racing helmet, which proved so successful that 23 out of 33 Indy qualifiers wore the RX-1 helmet in 1979.

For his contribution to the advancement of safety in motorsports, Bill received many awards and was inducted into the SEMA Hall of Fame in 1989. Both the Simpson factory and the World of Racing Museum are now located in Mooresville, North Carolina.

Thanks to the foresight of Bill Simpson and other makers of safety equipment, today's modern drag racer is safer in the cockpit of his hot machine than in the family car.

Bill Simpson's early parachute and firesuit manufacture began in this small shop.

Fire is the race driver's biggest hazard, and the Simpson firesuit shown in this flaming ad can protect a driver for approximately 50 seconds.

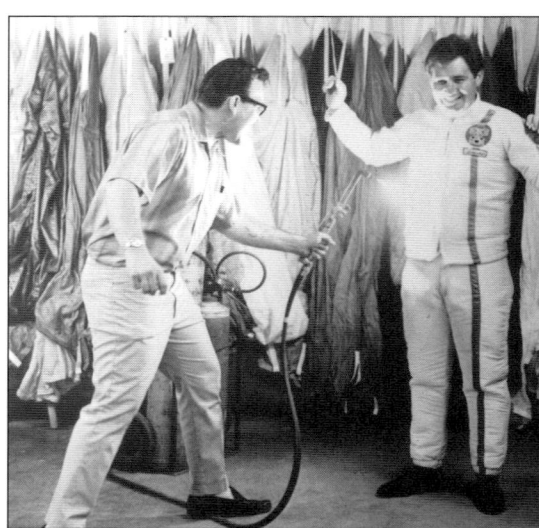

Race driver Mike Solokin is being blowtorched by Bill Simpson to prove fire resistancy of the suit.

The Sexy Sixties

Bill Simpson's fuel dragster is pictured here. Also in the 1960s, Bill successfully drove his own Indy car.

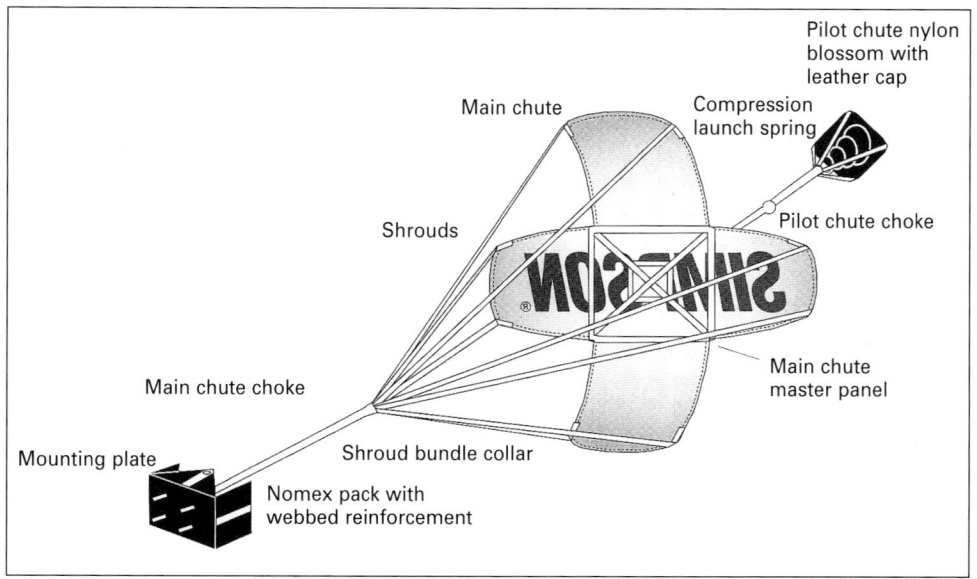

Handmade, Simpson Crossform Drag Chutes are available today in a 10-foot design for speeds less than 200 mph and a 12-foot design for speeds exceeding 300 mph.

Bill Simpson is shown here with his fuel rail in 1963.

PETE McNICHOLL
"The Slant-Six Kid"

After one traffic ticket too many cost Pete McNicholl his driver's license, he "got smart" and started racing on the drag strip. Detroit-born Pete apprenticed as a Chrysler metal model builder before becoming a 1958 Ramchargers charter member, where he did plumbing and welding on the club's original "High & Mighty" Altered.

Because he had both a family and a mortgage, Pete opted to build his own low-budget, go-fast car with a 1934 Plymouth coupe, in which he installed a 170 c.i. Valiant slant-six with milled head, reground cam, homemade headers, and a triple-carburetor ram intake. The total cost was less than $500.

"In 1961," Pete recalls, "when the hot little coupe won G/Gas class record at the NHRA Nationals at Detroit Dragway, the NHRA didn't take too kindly to my putting almost 10 mph onto the old record. Therefore, they soon changed the rules, thus eliminating the car from future tries."

Incensed by the NHRA ruling, Pete "got even" and began building even more successful slant-six motored cars, including a 1933 Dodge coupe that won eight NHRA Class E

For 1963, when a new class, C/Dragster, was created, Pete built this rail powered by a 200-inch slant-six with both 3/5-inch bore and stroke. It did about 130 mph in 10.5 seconds. This engine also featured tuned exhaust headers developed earlier by Pete. The scavenging effect of the cloverleaf collection added about 6% more horsepower over conventional log-type headers.

and Class F Modified Production records. With a new e.t. of 13.57 in 1965, Pete was the first to set an NHRA record outside the United States. From 1960 through 1966, Pete's cars set 15 NHRA national records. Although Pete did all the building and driving of his cars, he admits to being fed many helpful hints from Ramcharger members.

The "Slant-Six Kid," who had a preference for the slant-six engines, said, "I found that most of the hop-up techniques that apply to V8 engines also worked on the slant-six. In fact, smaller engines can usually be reworked to produce more power per cubic inch than larger engines. The biggest discovery I made with the slant-six is its response to exhaust system tuning."

Pete's most notable contribution toward cranking more horsepower from naturally aspirated engines was through exhaust

Called the "Galloping Gasser," Pete's G/Gas Willys (with slant-six) reset the record to 97.08/14.32 at the Fall 1961 NHRA Nationals in Indy. In 1962, when NHRA banned slant-sixes, and only pre-1960 flathead V8s and six-cylinder engines could compete in G/Gas, Pete ran in D/Gas, a class dominated by small Chevy V8s. He never lost a race—until he blew a clutch at the Indy Nationals.

system reworking, the method of exhausting that began in 1961. Most race cars of yesteryear used log-style headers, but Pete stumbled onto using a cloverleaf-style collector that ended all the exhaust stacks in a single plane. This helped to scavenge the exhaust gases. This is why everybody now uses a similar style of exhaust headers.

"I was glad to be about two years ahead of others in this collector design," says Pete.

Even before retiring from racing, Pete was an entrepreneur and a half-owner in The Hot Rod Shop in Detroit. After quitting Chrysler in 1965, Pete owned a summer resort and several recreational manufacturing companies. Now semi-retired, Pete has a welding shop called "Pro Welding." For fun, Pete collects, buys, and sells old MOPAR cars, specializing in DeSotos. Even "good old boys" still play with cars!

After winning the E/MP class at the Summer NHRA Nationals at Indy in 1964, George Hurst congratulated Pete. The car was a 1933 Plymouth four-door with a slant-six.

Leonard Harris

Right before he was killed while testing another dragster, Leonard Harris drove his competition gas coupe "Lil Red Rocket" (right) to a world record of 9.49/153. Also in 1960, Harris was the U.S. National Top Eliminator with the Albertson Olds dragster (below) (9.56/165.13).

ELS LOHN
Eelco Left a Footprint

Driver Els Lohn sits in his "Wee Eel" streamliner, awaiting his turn on the salt. He soon posted a USAC-certified, two-way record of 203.36 mph in 1964.

In the late 1950s, spunky Els Lohn dramatically attempted to show up the big "iron boys" by establishing a new speed record—the hard way—with a Lilliputian-engine car. As an eleventh-hour project, Els and his workers built a fiberglass streamliner around a 58 c.i. Morris Minor engine. The little car quickly captured the 1959 Class G Bonneville record at 105.530 mph. Later, a second and sleeker streamliner clinched a two-way record of 147.564 mph.

In 1964, the little screamer nabbed a two-way record of 203.36 mph while using a modest 91 c.i. Coventry Climax engine. By 1965, Els held two more Bonneville streamliner records in Classes G and F. The latter became a two-way average of 185.695 mph, a record that stood for more than 15 years.

Following World War II and the Korean conflict, Naval officer Els resumed hot rodding, studied at the University of Southern California, and started the Eelco Company, which specialized in "carburetor linkage that links and fittings that fit."

From humble beginnings in the family's garage, Eelco grew to a 24,000-square-foot plant with complete manufacturing facilities that produced more than 90% of its products in-house.

In the 1970s, Els' Eelco Company was acquired by a conglomerate, only to be "spun off" several years later. Because the enterprising Els hoped to regain clientele lost to corporate indifference, he subsequently purchased the Ansen brand name. Els, who was a 1985 recipient of SEMA's Hall of Fame, started again from scratch and made an astonishing comeback by building a prestigious line of performance products.

The 91-cubic-inch Coventry Climax engine runs on an H&C cam, Jahns pistons, Hilborn injectors, a Joe Hunt magneto, and an S&P blower. Headers were built by Eelco.

The Sexy Sixties

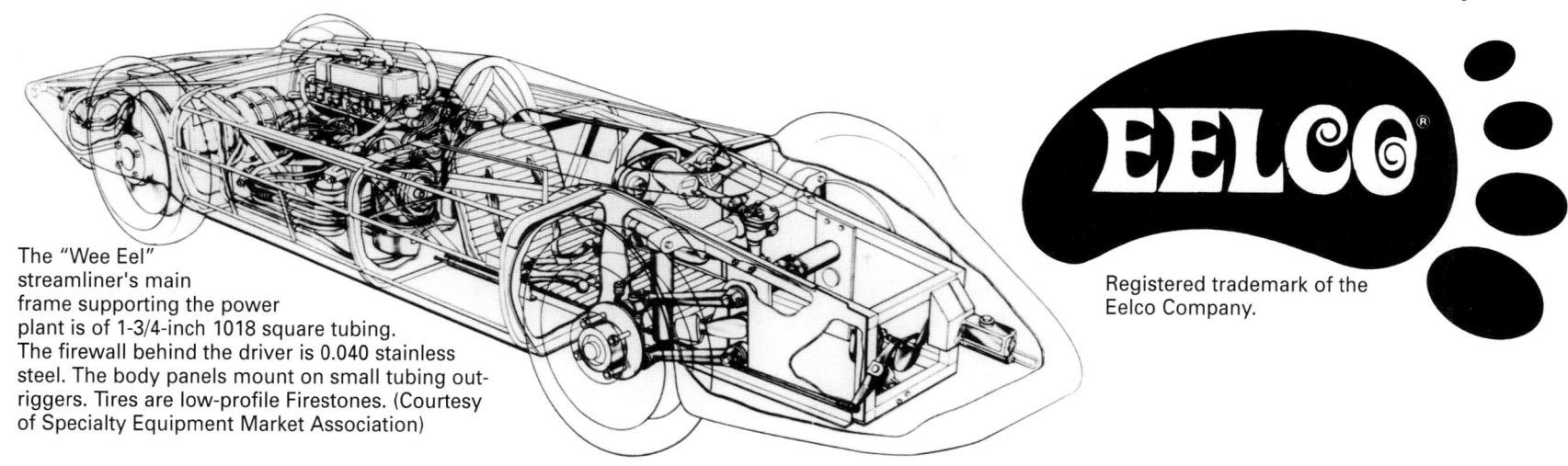

The "Wee Eel" streamliner's main frame supporting the power plant is of 1-3/4-inch 1018 square tubing. The firewall behind the driver is 0.040 stainless steel. The body panels mount on small tubing outriggers. Tires are low-profile Firestones. (Courtesy of Specialty Equipment Market Association)

Registered trademark of the Eelco Company.

Don Montgomery

Happiness is young Don Montgomery, shown installing a full-race, straight-8 Buick engine in a 1936 Cord that pushed the heavy car to 137 mph for the Russetta Class B Sedan record in 1952. The six-carburetor setup was made by mating sections of two Howard GMC manifolds. (Racer Brown photo courtesy of Don Montgomery)

After a successful career as an electronics engineer, Don Montgomery returned to his rodding roots and is now a historian/publisher of Hot Rod Books. His literary efforts developed after reading magazine articles that were historically inaccurate. His latest book, *Those Wild Fuel Altereds* (from which this fire-spewing photo was taken), is all about drag racing in the 1960s. (Photo courtesy of Roger Garten)

After dry lakes competition in the 1940s, Don Montgomery began drag racing in the 1950s. Don ran fuel coupes such as this Chrysler-powered Willys to many track triumphs until fuel was banned in 1957. He then switched to supercharged gassers until 1972 and managed the SoCal B/Gas Supercharged circuit. (Photo courtesy of Bill Fisher)

EDWARD PINK
Master Engine Builder

Ed Pink works on one of his first dragsters in 1960.

In racing, Pink is power. Ed Pink set a new standard for engine development, starting in the 1940s with circle track racing and continuing with drag racing and the Can-Am and Indy competitions of the 1960s and 1970s. Today, Ed is regarded as the "Old Master" by his peers and by top racing competitors on the international scene.

From the spawning ground of speed in Southern California, Ed "wrenched" in various speed shops and worked up to star mechanic for Lou Baney's renowned engine facility. After serving in Korea with the U.S. Army, Ed worked for the famous Eddie Meyer's speed engineering firm before opening his first shop that specialized in ignition and cylinder head work. He also did custom engine and machine work for rising "stars," such as Tommy Ivo, Tony Nancy, and John Masmanian.

When drag racing rose to fever pitch in the 1960s, Ed cornered the market for fast quarter-mile engines. In the 1970s, drag racing hit the big time, with sponsorships creating a demand for high-quality engines. Ed's customer list mushroomed, with Don Prudhomme, Tom McEwen, Shirley Muldowney, Ed McCulloch, Blue Max, Don Garlits, and Kenny Bernstein.

As a perfectionist in quality control, Ed began reworking and "blueprinting" engine components to fine tolerances. In his 6,000-square-foot shop in Van Nuys, California, he installed one of the first computerized dynamometers in the industry and began building engines for Formula 5000, endurance racing, and the Can-Am series.

"I remember when Vel Miletich and Parnelli Jones contracted for us to help on the new Cosworth engine for Indy cars," said Ed. "It was a normally aspirated, gasoline-burning engine that was to be converted to methanol fuel and turbocharged."

The "Old Master Engine Builder," Ed Pink (left) points the way to Ed McCulloch, a standout driver who switched from dragsters to funny cars in 1969.

The Sexy Sixties

Ed Pink is dyno testing the four-cylinder Ford sprint midget engine he helped develop into top racing form. (Circa 1996)

Ed Pink's first dragster was powered by a fuel-burning, blown Hemi. The car was painted pink, his trademark color. (Greg Sharp photo)

By the 1980s, Ed, an innovative dyno engineer, had become a leader in computer-generated technology. His facility housed more than a million dollars worth of sophisticated testing and R&D equipment, regularly used by the Ford, Pontiac, and Nissan Motorsports divisions.

For being a "Legend in Drag Racing," Ed received a "Lifetime Achievement Award" from the NHRA in 1995. Later, he was inducted into the International Drag Racing Hall of Fame and then into the *Superstock Magazine* Hall of Fame. In the 1970s, the readers of *Car Craft Magazine* voted Ed "Engine Builder of the Year."

"Think Pink" is more than a buzzword. Pink is perfection in the winner's circle.

Art Chrisman (left) and Lou Baney (right) cajole Ed Pink (center) as he works on a dragster at Riverside, California. (Greg Sharp photo)

HARVEY CRANE, JR.
"Professor" of Camshaft Technology

While planning to race his first hot rod, a 1932 flathead Ford V8, Harvey installed a rough idling radical regrind that prompted him to explore camshaft designs. After discovering that most of the early racing camshafts lacked consistent machining quality control, Harvey began experimenting with his avant-garde ideas for a new grind. In 1953, Harvey rented a corner of his father's machine shop and hung up a small sign that read, "Crane Engineering Company."

How did Harvey Crane, Jr., an upstart young engineer from the East, outsmart the West Coast experts to make Crane the world's largest performance camshaft maker? How did Crane then become fired from the company he built?

I clearly remember the day in 1989 when Harvey phoned me with the news.

"I've just been fired! It's unbelievable that my own board of directors just threw me out," blurted Harvey, who had unwittingly given most of his stock holdings to his employees.

By 1965, Crane had more than 40 employees and was on its way to becoming the biggest racing cam grinder in the country.

In 1961, Harvey Crane could only afford a 1-inch ad in *Hot Rod Magazine*.

"Finally, bigger ads got the ball rolling, and, by 1989, the year I was fired, we were doing over $30 million in sales," stated Harvey regarding the now solely employee-owned business that grew to an 85,000-square-foot facility in Daytona, Florida.

In the beginning, Harvey was ridiculed by jealous competitors who called his first extra high-lift racing camshaft too radical to work. However, time would prove it performed so well that its design would later be copied as the standard in the industry. Harvey later developed other revolutionary and patented cam and valve train designs that continue winning important races.

Harvey Crane at his first cam grinder in 1953. Note the large grinding wheel that tracked the master.

After purchasing a Storm-Vulcan camshaft grinder for $600 per month, Harvey added minor improvements to a Howard cam and began grinding precision cams with all lobes identical. Business was slow at first but picked up when the Korean Conflict escalated and Harvey was drafted. When many attempts were made to repossess the cam grinder, Harvey "sweet talked" the seller into allowing him to make smaller payments.

After returning from the military service, Harvey Crane purchased a Van Norman cam grinder and began producing precision camshafts.

"Developing special cam grinds for all makes of engines wasn't easy," said Harvey. "I often had to use willing customers as 'guinea pigs.'"

The Sexy Sixties

"The regrinding was a slow and painstaking trial-by-fire process that is still used today, even by General Motors. Today, the intelligent use of computers is essential, but camshaft design is still an inexact science. For every hour put in design, a hundred hours are spent doing analysis."

Harvey Crane, Jr., a talented hot rodder who gave so much to the industry, has been recognized and honored with the prestigious Fellowship Achievement Award from the Society of Automotive Engineers. He also received the SEMA Man of the Year Award in 1976 and later was elected to the SEMA Hall of Fame.

Today, Harvey Crane is an internationally known automotive consultant who operates the first and only "Cam School" in the world, near his home in Daytona. There he shares more than 40 years of camshaft, valve train, and "monkey-motion" secrets with others. Even former competitors seek Harvey's expertise.

Today, Harvey Crane (right) runs a "Camshaft College."

Ed Pink stands next to the supercharged 427 SOHC that pushed Don Prudhomme's dragster for the 1967 Winternationals Top Fuel win at 6.92 e.t. and 222.76 mph.

The "Total Performance" ad slogan launched by Ford in 1963 began an all-out racing war. By 1966, the new 427 Ford V8 SOHC engine powered winners in road racing, drag, and stock car circuits. The potent "Cammers" used a similar combustion chamber layout as the Chrysler Hemi, but with the added advantage of a single camshaft in each head. It was tough to "blueprint" to "tighter" racing tolerances; however, with fuel injection and supercharging, output exceeded 1000 hp. One builder's engines cranked out 1200 hp at 7800 rpm.

FLAK BY AK

PERFORMANCE ADVISOR

THE PERFECT COMBINATION

Of the hundreds of letters that cross my desk each month, one of the most frequent questions asked is, "How can I make my car go faster in the quarter-mile?"

With this question in mind, I wonder how many of you drag race fans realize that Ford now offers a fully equipped lightweight vehicle—designed specifically for quarter-mile competition.

This car includes lightweight fiberglass panels, aluminum bumpers, 427-cubic-inch V-8 engine, aluminum-cased 4-speed transmission and lightweight bucket seats, all put together around Ford's new fastback model.

AK MILLER
P. O. BOX 627, DEARBORN, MICHIGAN

"FLAK BY AK" hot rod ads such as this ran in 1963 by Ford and touted Ford's 427 V8 engine and lightweight body components for drag racing. Yet, other hard-core modification tips offered by Ak had this claim, "Ak Miller's advice...are his own and do not necessarily represent recommendations of Ford Motor Company."

PETE MILLAR
Hot Rod Humorist

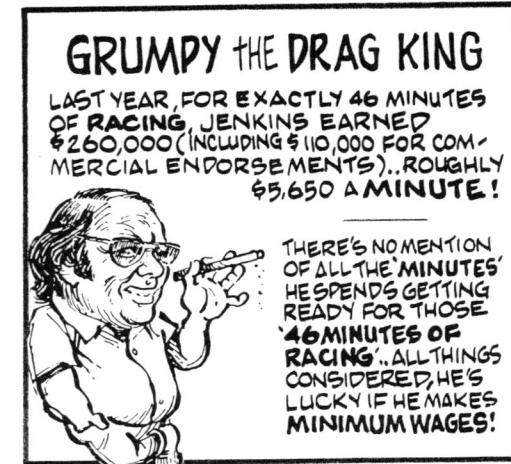

"Cartoons can do more than poke fun. They can be a quick and subtle way to sell a product, get votes, or promote a cause," said Pete Millar, whose cartoons on Isky probably sold more cams than did the Camfather's T-shirts.

From the beginning, cartoonists such as Pete Millar have enriched the fascinating culture, fun, and image of motorsports. Pete, who saw the comic side of hot rodding, masterminded his original "Drag Cartoons," which became a nostalgic reminder of the colorful lore of drag racing.

Graphic expression always came easily for Pete, a self-taught artist who was blessed with skillful hands and a sharp wit. While in the Army Air Corps, Pete created his first comic strip for the base's newspaper. The encour-

Time Magazine was out of sync with the times, especially with drag racing. Although it hyped drag racing with a little blurb, it missed the passion of the sport in its October 22, 1973 issue.

agement he received from "the boys" was the impetus to continue his artistic doodling.

In the early 1960s, Pete created "Car-toons" for Petersen Publishing Company. However, he left to start another comic series called "Drag Cartoons," which were witty and memorable caricatures of real-life racers and the gutsy folks in the industry.

As a tough drag racing competitor, Pete campaigned top-contending cars in the 1960s, including several dragsters and a "giant killer" 1940 Willys coupe. Even today, Pete loves to find humor in the rodder's daily life—in sweet victory and in crushing defeat.

A youthful Pete Millar pens a cartoon at his drawing board in 1965.

The Sexy Sixties

The first flathead Ford dragster run by Pete Millar in 1964 had four carburetors and did 135 mph at the Lions drag strip. Later, a single Scott fuel injector replaced the four Stromberg 97s.

The mighty "Gangreen" 1940 Willys driven by Pete Millar was a regular winner at the Lions drag strip in the mid-1960s.

Nye Frank

The thundering "Freight Train" was a tandem-twin Chevy-motored gas dragster run by Nye Frank and John Peters, who won both Top Gas titles at the 1962 Bakersfield and the 1963 NHRA Winternational meets. Although the two-engine dragsters delivered tremendous horsepower, the additional weight complexity and the high operating cost would prove a handicap. Eventually, the big twin-engined diggers would lose out to the improved breed of single-engine lightweights.

The 289 Ford supercharged "Chicken Coupe" dragster driven by Pete Millar was a tough competitor at the 1967 Winternationals, doing 8 e.t./200 mph.

CONRAD "CONNIE" KALITTA
The Bounty Hunter

In 1967, Connie successfully campaigned with this 173-inch-wheelbase dragster with a supercharged 427 c.i. single overhead Ford engine.

Flying fast and high have qualified Connie Kalitta as a drag-racing champion and self-made millionaire.

Growing up in the 1950s, Connie was a self-reliant, scrappy youngster who took naturally to the fast lane with his street rod. Before he was 19 years old, he had scraped together enough cash to build a Chrysler-powered rail, which he raced with only minimal success. However, with better machines that had more bite and power, Connie persevered at top drag strips such as Motor City Dragway.

In the lean years, Connie supplemented his income by working for a fellow who taught him to fly and to transport auto parts in a Cessna 310. With his racing profits and his instructor's encouragement, Connie purchased a secondhand Cessna 310 and copied his mentor. As a businessman, his success exemplified the American dream; the one-plane company he started in 1967, American International Airways, now has more than 100 commercial cargo jets.

Although he is visually and hearing impaired, Connie's determined, renegade attitude is punctuated with a Midwestern drawl loaded with guttural cuss words. Kalitta, the swashbuckling rogue, acquired fame in the 1983 movie "Heart Like a Wheel," which depicted his life. The heroine in the film portrayed the real-life, three-time NHRA champion, Shirley Muldowney. Connie worked with Muldowney for four years and was instrumental in her being crowned the first female Top Fuel Champion in 1977.

As tuner and driver, Connie continued his winning ways from 1960 to 1995 by competing in 20 NHRA championships and winning ten of them. Twice, he was the top IHRA Fuel Champ. In 1962, he was the first driver to surpass 180 mph. In 1989, he was the first driver to exceed 290 mph, after which he topped 300 mph. In 1992, he was inducted into the Motorsports Hall of Fame. At the age of 56, he won the 1994 Top Fuel at the U.S. Nationals. To this day, Connie's biggest competitor is his son Scott, who won the NHRA Winston Top Fuel crown in 1994 and 1995.

Connie Kalitta knows all too well that being a top competitor does not come easily, and it is expensive.

"Even though I am a self-taught engine builder, I still maintain an expert pit crew with a fully equipped traveling machine shop," remarked Connie. "I carry at least one spare engine, tuned and ready to run, and my crew is trained for emergencies. A while ago, they switched a motor in 25 minutes at the Nationals. Today, they might even do it quicker."

After years of courting the big prizes with only runner-up trophies, Connie made history in 1967 as the only driver to win three Top Fuel Eliminator honors at NHRA, AHRA, and NASCAR meets, catapulting him to the top of the dragster ranks.

The Sexy Sixties

Jeg Coughlin

Dubbed "Captain Quick," Jeg Coughlin stormed the competition with this injected Barracuda funny car in 1969 and held the A/FC national record for three years. To date, the Coughlin clan, which includes four sons, has accumulated 15 national event titles. Talented Jeg, Jr. is now the "biggie" in pro stock racing.

Before he started Jeg's Automotive in Columbus, Ohio, Jeg Coughlin (inset, top) raced a D/Gas 1955 Chevy, which took class honors at the 1959 Nationals. Then, Jeg campaigned numerous winners, including this Camaro. In 1972, Bob Durban drove one to a ProComp victory at the Springnationals.

The sporty "Jegster" is a fiberglass-bodied T-like roadster that is often a winner in the S/G class.

John Abbott drove this state-of-the-art Swindahl fueler to a Top Fuel win at the 1981 U.S. Nationals. Jeg Coughlin (shown doing a burnout) had driven it previously. However, after running it off the end of the track a couple times, Jeg decided to stick with stockers.

DON PRUDHOMME
The Snake

This 1968 press photo of Don Prudhomme already touts him as one of the all-time greats of drag racing. A decade later, he became the NHRA career leader with 27 wins. (Ford News Bureau)

Drag races may be won or lost in that split second when a driver hits the throttle and shoots off the starting line. According to past records, barely less than a thousandth of a second has won many a race.

Don Prudhomme is one top competitor who has been graced with a remarkable phantom-like ability to anticipate the green light. He has been able to accelerate easily off the line at the triggering of the starting light, thereby beating his opponent at the onset. His quick reflexes have rightfully earned Don the nickname, "The Snake."

Like most drag racers, Don began his competitive career as a teenager in a car that he built. At age 20, he turned pro and in 1962 won a Top Fuel victory at Bakersfield, California. After that, few records escaped the amiable young Californian as he won almost 90% of all races entered with the Greer-Black-Prudhomme dragster. Don's success continued, and in 1965 he drove another Top Fuel dragster, the "Hawaiian," to victory in both the NHRA Winternationals and the U.S. Nationals.

In the 1970s, Don parked his dragster and began driving in funny car competition. After receiving a sponsorship from the U.S. Army in 1974, Don won 13 of 16 NHRA Nationals events in one year. He then went on to win four straight NHRA/Winston championships.

In 1989, Don won the Big Bad Shootout and the U.S. Nationals (plus $110,000) in Ennis, Texas. He exited with a career best time of 5.157 seconds.

Don's illustrious driving career ended with the 1994 season. For more than three decades, he thrilled racing fans nationwide by roaring to more victories than any other driver in the Top Fuel and Funny Car fuel categories. After that, he competed only as a car owner looking to strike with his usual grace and professionalism.

"Drag racing today is a science. But there is no substitute for an intuitive driver. In top fuel dragsters, the engine in back puts more

This Zeuschel-Fuller-Prudhomme car driven by Don was the Top Fueler in 1962. Don scored more than 250 victories in this hot machine. The car is pushed for the first run at a night drag with promoter C.J. Hart (wearing a hat) in background.

weight on the rear wheels, thus providing more traction. With funny cars, the engine is in the front, and the driver's skill creates a weight transfer to the rear wheels for good traction. It's just like a teeter-totter, where the car just won't stay straight. Everything is happening so fast. You've got to be in control. It's an incredible amount of driving to do in that split second of time," said Don about driving a nitro-methane-powered 4000-hp car.

It takes an incredible person with guts and skill to drive the wild beast that is a NHRA top fuel dragster. It takes a legend such as Don "The Snake" Prudhomme.

Don drove Lou Baney's single overhead cam Ford AA/Fuel dragster to numerous championships in 1967. His top speed that year was 225 mph.

Don became a dominant funny car driver from 1974 and soon approached the 5's, matching the speeds of dragsters. In 1982, Don was the first to break the 250 mph barrier in a funny car. The U.S. Army was one of Don's many sponsors.

Jack Flynn

In his trophy-winning Corvette (AA/SP) in the early 1960s, Jack Flynn, a farm boy from Sparrowbush, New York, set many NHRA track records while dusting off top East Coast competitors. "We did race preparation in the winter and warmed our freezing hands on the sides of cows who shared the other end of the barn," recalls Jack. Jack first raced in one of the East's first drag strips in the 1950s. His previous 1957 Chevy served as a test mule for the patented Equa-Torque Traction Bars invented by Ed Almquist.

DICK LANDY
Dandy Dick, Match-Racing Icon

Dick Landy, one of the first "factory" racers, was nicknamed "Dandy Dick" for his dapper appearance. A cigar was his trademark.

Long before drag racing witnessed the hectic revolution in funny cars in the mid-1960s, forerunners of these wild machines existed almost everywhere in the country.

From the beginning, rodders frequently altered wheelbases and squeezed big engines into lightened chassis. Even later, exhibition stockers with fiberglass bodies and tubular frames could be categorized as funny cars. Therefore, after the Dodge Ramchargers won the 1963 Stock Eliminator title, Chrysler factory teams were on a roll as they continued their pursuit of quicker e.t.s with a radical modification that would inspire one of the first commercialized funny cars of that era.

In an effort to improve weight distribution of the Dodge and Plymouth drag strip cars, both front and rear axles were moved forward. This created a funny look, leading to the name "funny car." At first, the unique configuration kept the car out of NHRA competition. However, it also launched Dick Landy's long and lucrative match-race career.

In the early 1960s, Dandy Dick Landy switched from drag racing Fords to the winning Maxi-Wedge Plymouth and then to Dodges in the Super Stock class.

"That was the best move I ever made," said Dick. "It was a corporate association that lasted for more than 30 years."

With full sponsorship from Chrysler in 1965, Dick began match racing nationally with his Dodge Hemi-powered funny car. He became the NHRA champion in 1973 and 1974. After rule changes made it impossible for Hemis to compete, Dick quit racing in 1980 to be a spokesman for Chrysler Performance Clinics, the original grassroots parts program that exists today as Mopar.

Dick Landy's 1965 Hemi-powered funny car was one of the first altered wheelbase funny cars with the front axle moved forward 15 inches and the rear axle 10 inches. Drastic modifications reduced the weight to 1700 pounds to yield terrific acceleration and consistent under-eight-second passes.

The Multiple Engine Trend

The early 1960s would go down in the annals of drag racing as a time of wild experimentation that included twin-engined cars that often turned faster drag times than single-engined vehicles.

Following the lead of the Bean Bandits' dual-engined dragster in the early 1950s, a rash of twin-engined cars emerged—some mounted in tandem, similar to Peter's and Frank's 1963 Winternationals winner. However, Tommy Ivo, the young Hollywood star, did not quite make it in the 1960 NHRA Nationals with his Hilborn-injected twin Buick-powered dragster. Later, Tommy ran a monstrous, four-engined, exhibition car. Other big names ran different configurations of twin engines—many with Chrysler or Chevy engines, blown and unblown—included Bill Kenz, Jack Moss with "Too Much," Tom McEwen, Howard Johansen with "Twin Bear," and Eddie Hill. Multiple-engined machines delivered much more horsepower, but they were heavy and complex to hook up, and sometimes even unwieldy to drive primarily because of poor static weight distribution.

Although Mickey Thompson's streamliner with four Pontiac engines would be the first American car to top 400 mph, the multiple-engine configuration soon gave way to single-engined, lightweight machines.

A twin-engined car, with engines mounted in tandem.

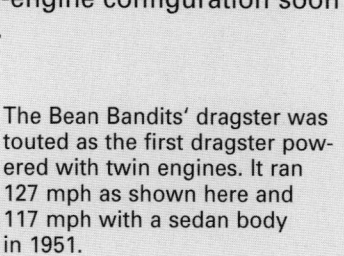

The Bean Bandits' dragster was touted as the first dragster powered with twin engines. It ran 127 mph as shown here and 117 mph with a sedan body in 1951.

Tommy Ivo ran this monstrous four-engined car.

Tommy Ivo's Hilborn-injected twin Buick-powered dragster.

Hot Rod Pioneers

Jack Clifford

In the early 1950s when the Hudson Hornet dominated NASCAR racing, Jack Clifford fell in love with the big stocker and vowed that he would, one day, race and win. By 1963, Jack, a graduate mechanical engineer, had put enough "sting" in his Hornet to enjoy consistent drag strip wins in the low stock classes. He won the L/Stock class at the Winternationals at 85.65 mph and 15.77 e.t.

Today, Jack Clifford specializes in speed equipment for inline engines. The power-boosting intake manifold and headers shown here are for 1970 and later CJ series Jeeps.

Gordon Collett

Gordon "Collecting" Collett got his nickname while earning big bucks in the 1960s as a Top Gas star who competed successfully on the West Coast and in the North Central Division. His was one of the first single engine gassers to exceed 180 mph. Hailing from Portsmouth, Ohio, Gordon's big win at the 1964 Nationals included a new Ford Mustang. After the Top Gas category ended in 1971, Gordon and his son Terry raced Pro Stock in 1972 and 1973.

Danny Ongais

Danny Ongais (left) and Connie Kalitta (center) are being interviewed by strip co-owner Jim Nelson (right) on opening day at the Carlsbad, California, raceway. Originally from Hawaii, Danny was Top Gas Eliminator at the 1964 AHRA Winternationals with a Dragmaster AA/Dragster. Later, he drove a funny car. All were early members of the 200-Miles Per Hour Club.

Jim Hall's Chaparrals

The legendary Chaparrals were the marque of downforce in the 1960s. Jim Hall's famed mid-engine Chaparrals (1966 model shown here), with movable rear wings and other breakthrough ground effects, changed the face of racing forever. Hall was a bold innovator who teamed with Chevy R&D engineers. He pioneered the use of plastic composites, automatic transmissions, and ground effects. Hall also was a derring-do driver who aced the U.S. Road Racing title in 1964. In the drag racing world, the mid-engine revolution (kicked off by top fueler Don Garlits) coincided with frenzied aerodynamic experiments with negative airfoils, dams, wings, stabilizers, and fairings—with the rear wing eventually becoming the standard.

The Sexy Sixties

Ronnie Sox and Buddy Martin

Screaming Pro Stockers were becoming major attractions when Ronnie Sox and Buddy Martin won their first major event—Factory Stock Eliminator at the 1964 NHRA Winternationals. The following year, Sox and Martin joined Plymouth to campaign the forerunner of funny cars, an altered-wheelbase Plymouth Duster, whose Hemi had twin magnetos firing two spark plugs per cylinder. For almost a decade, the Eastern team of Sox and Martin dominated—winning numerous major AHRA and NHRA championships. Their reign ended when NHRA permitted Fords and Chevys with wedge-head engines to run at a lighter weight than cars with MOPAR Hemis.

Before the 2% ruling on altering F/X wheelbases, the Sox and Martin Plymouth revealed the extreme in radical chopping, with front and rear wheels offset forward to achieve more weight transfer rearward for improved traction. Note how the rear tires distort on initial acceleration.

Bill Lawton

A popular and speedy driver in the 1960s, Bill Lawton was the New England champ of 1962 and 1963. He took first Factory Stock Eliminator honors in the NHRA Winternationals with a 1965 Mustang (10.92/128.20). Bill went on to win many important titles before going into the trucking business.

Bill Lawton was one of the quickest and most popular drivers in New England in the 1960s.

The lightweight Tasca Mustang, with a supercharged 427 mill, had a flip-top one-piece fiberglass body on a Logghe Brothers chassis.

BILL "GRUMPY" JENKINS
Underdog Pro Stocker Beats Big Team Players

Bill "Grumpy" Jenkins was the "King of Pro Stock" and its "winningest" driver in the 1970s, taking many AHRA and NHRA National Pro Stock titles plus Super Stock and other national wins.

One of the most colorful and innovative characters in drag racing history is William Tyler Jenkins, who earned the nickname "Grumpy" early in his career because of his guttural utterances and grunts when avoiding small talk during race days. Expertise in driving and building winning cars earned Jenkins respect, even among the hardcore competitors. As the lone man who outraced the favorites in Pro Stock's formative years, Jenkins quickly gained superstar status and drew huge crowds every time he pitted his small underdog Chevrolet Vega against the big factory-backed teams.

Jenkins' participation in the quarter-mile game dates back to the 1950s when he built and piloted an assortment of stockers. In the 1960s, the East Coast mechanical wizard first gained recognition in racing super stock cars.

When Jenkins took a shot at big-time driving, he won on his first outing and took home the Top Stock Eliminator title at the 1965 NHRA Winternationals in a 1965 Plymouth. After two years without major wins, Jenkins astounded the drag racing world in 1972 with his "David vs. Goliath" wins when his small Vega overtook the so-called superior technology of Ford and Chrysler factory racing teams. Jenkins hit the jackpot with $250,000 in prize, sponsorship, and appearance monies that year.

Over the years, the unflappable Jenkins helped improve the image and direction of drag racing. As a lifelong Easterner, he spurred the development of modified production cars, a product of the East. Jenkins was largely instrumental in creating the new Pro Stock class by utilizing his power as a box-office attraction to influence hardheaded NHRA rule makers. Of course, Detroit's racing people prized the new class of stockers, and it was rumored that they muscled influence.

Terry Cook, former editor of *Car Craft*, spent hours talking with and observing "Grumpy," the masterful rule bender who quipped, "Never phrase a question to a track team official in such a way that he could answer with a 'No!' "

For a while, Jenkins added Pro Stock drivers Larry Lombardo and Ken Dondero to his

"Giant Killer" Grumpy Jenkins won 12 national titles in 1972 with his Pro Stock Vega, with clockings in the high 8's at more than 150 mph. The car had a light chrome moly tube frame, with fiberglass front fenders and hood and trunk lids on its striped body. The caster-device on the rear helped prevent "wheelies."

The Sexy Sixties

racing team. Lombardo was the 1974 NHRA Summernationals champ, and Dondero won the 1975 AHRA championship. Both credible competitors brought home more strong wins.

In 1974, Jenkins opened a business called Jenkins Competition. There, Jenkins and his staff worked on preparing engines and cars for racing, as well as doing prototype work for GM. Today, Jenkins-prepared engines and cars continue to win in all forms of drag racing competition as well as open wheel, NASCAR, and Formula 5000.

"Grumpy's toy" in 1971 was this Chevy Camaro that pulled crowd-pleasing wheel-stands off the line, along with consistent nine-second blasts. Note how the rear tires are stressed out of shape.

Hot Rod Pioneers

The SEMA Saga

Still Fighting Government Regulatory Horrors

As the "Golden Age" of the rapidly growing performance parts industry developed, some forward-thinking thinkers saw the need for a trade organization. In 1963, the Speed Equipment Manufacturers Association (SEMA) was formed with 36 founding members. Many were small independent entrepreneurs, and some were hardheads with conflicting or even self-serving goals.

Initially, SEMA addressed several major concerns, including credit abuse. "We needed a good credit-checking system to screen out deadbeat customers," said Howard Douglass. Although the young industry was trying to stop prostituting itself while selling direct, most manufacturers were guilty of the same sin, which caused a collision course that harmed speed-shop customers. (Self-policing urged legitimate wholesaling with less cutthroat pricing.) SEMA began combating the NHRA's seemingly arbitrary disqualification of vehicles with certain aftermarket parts and eventually became an industry force with a Spec Program for quality, safety, and performance standards.

Since the 1960s, SEMA put up a gallant fight to limit government overregulation and anti-pollution statutes that had changed the industry forever. From its Washington office, SEMA is now a watchdog on monitoring legislation that threatens the freedom of motoring, and it has exposed regulatory horrors and unfair smog checks as well as opposing flawed "scrappage programs" that would kill the collector car hobby.

To accommodate a broader spectrum of the aftermarket, the name of SEMA was changed to Specialty Equipment Market Association, thus retaining the familiar acronym SEMA for its more than 3,500 current members.

Charles "Chuck" Blum, SEMA's capable president since 1980, predicted "continued growth in the new millennium for the innovative hot rod industry because we are one jump ahead of the automakers."

"Extravaganza of toys for big boys!" SEMA is the largest hot rod show in the world, but you cannot attend unless you are in the trade. The first SEMA show in 1965 had only 98 booths. Today's annual trade-only SEMA show has thousands of exhibitors showcasing the latest in speed and custom products. Here, the top gurus in motordom can be seen viewing the latest in hot rod technology and trends. What better place to find new ideas and inspiration?

L.R. "Pete" Robinson

Likable Pete Robinson was a 27-year-old dragster driver from Atlanta, Georgia, who fans hardly knew until he beat the best for the Top Eliminator title at the 1961 NHRA Nationals. His extremely light AA/D supercharged Chevy-engined "Tinker Toy Too" slingshot did 8.92 e.t.—the first 8-second run at the "big go." In his brilliant career, Pete won hundreds of trophies and captured the 1964 NHRA Nationals AA/GD honors and 1966 Top Fuel. That same year, Pete signed on with Ford to run its SOHC engine that he further refined to power, which was arguably the lightest slingshot in the business.

Pete was often called "Sneaky Pete" because of his maverick engineering ability to befuddle opponents with new and ingenious devices such as Jumping Jacks, which simply lifted the dragster rear wheels after staging. Then, with the engine revving, the car could be lever-actuated to drop the spinning wheels instantly for dramatically quicker starts. An even bolder innovation was an ill-fated ground effects device, which probably caused Pete's dragster crash that killed him at the young age of 37.

Pete Robinson was a young dragster driver whose life ended tragically in a fatal crash when he was only 37 years old.

Johnny Jackson

On July 12, 1969, Johnny Jackson and his bride Darlene took their wedding vows during intermission on the straightaway of the speedway in Jacksonville, Florida. Minutes later, Johnny jumped in his Chevy and finished fifth in the Sportsman feature race. Before that, Johnny had won his share of modified stock, drag, and sprint car races in the Southeast from 1955 onward.

DON ALDERSON
Co-Founder of Milodon

Don shows Mickey Thompson (left) the wild-bodied "ScrimaLiner" that debuted in 1964. The aerodynamic rear-fendered Chrysler fueler was among the fastest in its day at more than 200 mph.

When trying to beat his own B/Fuel roadster record of 210 mph during Speed Weeks 1963, Don Alderson's 1932 roadster went airborne and blew off the course. After that experience, Don turned to "motor magic."

By the 1960s, when drag racing was getting into full swing, Don's Milodon company was already associated with red-hot engines and unique but needed speed parts. Alternating between area drag strips and dry lakes, big Don (who was 6.5 feet tall and weighed 266 pounds) was a perennial trophy winner in the 1950s.

When Don's stylish, streamlined slingshot dragster served as a rolling test lab, he pulled more press than records. However, Don's reputation as a Bonneville builder peaked when he was a partner in the Herda-Knapp Milodon lakes racer. Designer Bob Herda, an aeronautical engineer, drove their Class B streamliner to many triumphs, including 326.853 mph at the 1967 meet.

Don had started Milodon Engineering in 1957 with Milo Franklin. By 1970, the business had grown from fabricating main bearing supports to making more than 100 different oil plans and oiling systems for street and strip applications.

When the aluminum-engine-block war began, Milodon's big beefed-up 426 Hemi (similar to Donovan's earlier 417) featured cast steel removable wet-sleeve cylinder liners, which allowed both "greater reliability and repairability," as Don put it. In the early 1970s, racing's drag VII Litre Hemi was "big and bold," the same as Don Alderson.

The Herda-Knapp-Milodon Class B streamliner was the fastest single-engine vehicle for more than a decade. The GMC-supercharged Hemi racer did 302 mph in 1965 and broke its own record 18 times in three years to a one-way 347 mph.

Don Alderson's 1932 "High Boy" roadster did 219 mph for the B/Fuel Roadster record at Bonneville in 1963.

"DYNO DON" NICKELSON
The Rolling Test Lab

After becoming a top star in the flip-top funny car metamorphosis, "Dyno Don" Nickelson helped propagate the new Pro Stock class that was finally established in 1969.

Don's major triumphs began with Super Stock wins with "Dyno-tuned" Chevys at the NHRA Winternationals in 1961 and 1962. He also took the unlimited Fuel Stock Eliminator title at the AHRA Winter Championships.

Later, Don became Mercury's "rolling test lab," with spectacular tricked-up Comet and Cyclone funny cars that scorched the country's quarter-mile circuits with wins in the 8- to 10-second range. Similar ready-to-race Factory Experimental (F/X) cars were offered to the public (on special order) from Lincoln-Mercury dealers.

Don, a Missouri farm boy born in 1926, learned to drive on a homemade tractor that his dad converted from an old Oakland sedan. After serving in the Navy, Don's first organized drag race was near Pasadena, California, where he and his brother Harold opened a speed shop in the early 1950s.

Before retiring, Dyno Don's spectacular tuning and driving accomplishments earned him the unofficial title of "Mr. Stock Eliminator."

One of the quickest in the Experimental Stock class in the mid-1960s was Don's orange-colored Comet compact. The SOHC 427 engine was positioned rearward to put more weight on the rear wheels for improved traction. Special shock valving in the suspension allowed the chassis to rise 5 inches under acceleration without lifting the front wheels in a "wheelie." The fiberglass body with nonfunctional doors and hood was hinged for quick access. (Courtesy of Ford Motor Company)

Crowd-pleasing Don Nickelson was one of the most feared drag racers among the stockers in the 1960s. This Mustang Cobra Jet with a 428 engine raced in the SS/E class, clocking as low as the 7's in 1968.

Drag Racing Goes International
A "Jolly Good Show" Began in Merrie Olde England

Imagine the gall of the European who once wrote, "Hot rodding should not be considered the prerogative of a small group of rich devotees in the United States. On the contrary, it is already practiced by many thousands of people, mostly young."

The drag racing sport had been alive in Canada in a big way long before America's expeditionary team of ten drag racers arrived in 1964 at the First British International Drag Racing Festival. The festival was the brainchild of sports car builder Sydney Allard and sponsored for sparking European interest in this hitherto all-American sport.

Six rounds at six different locations gave the Britishers a good idea of what makes the American "hot dog" machinery go "so bloody fast." The runs were only for exhibition because it was not fair to match big-inched American cars against small-milled English cars. The visiting U.S. contingent included Don Garlits, Mickey Thompson, Tommy Ivo, George Montgomery, Tony Nancy, Dave Strickler, Dante Duce, K.S. Pittman, Bob Keith, and Ronnie Sox.

Since then, hot rodding has spread globally, with most countries following American drag racing rules and major events sanctioned by the FIA. Today, Australia ranks second to North America in drag racing growth, with an astoundingly high per-capita attendance watching the sport. Who would have thought that hot rodding would go global?

In Sweden in 1965, this squat replicated 1932 Ford hot rod meant GO! Since then, drag racing has become a popular sport in Scandinavia, with King Gustaf being one of its ardent fans.

England's first slingshot dragster, built in 1961 by famous sports car builder Sydney Allard, received great television and press coverage that showed the 1,650-pound Chrysler-powered rail (pictured here without its body skin) and increased national interest in drag racing.

Was this an early British "hot rod"? This rugged "Butterball Special," built in the late 1940s by A.J. Butterworth for road racing and hill climbs, could do a standing quarter in 11 seconds. It had a modified Jeep suspension, four-wheel drive, a homemade five-speed progressive transmission, and a massive V8 air-cooled engine with a 14:1 compression ratio and eight motorcycle carburetors. Bill Milliken brought the hybrid to the United States for competition in the 1950s.

The Sexy Sixties

Charlie Boucher
"Bad news" probably was how competitors felt about this small-block-injected Chevy-powered Ford pickup owned by Charlie Boucher. Boucher, from Connecticut, won the C/Altered class title at the NHRA Nationals at Indy in 1964.

Bill Ireland
By 1967, Bill Ireland of Portland, Oregon, had collected more than 400 trophies, including AHRA Nationals and NASCAR Super Stock class titles. Also named the outstanding driver in the Northeast, Bill conducted clinics for youth groups. Note how the large rear tires of Bill's Fairlane stocker became distorted as the car accelerated.

Scott Wilson
The fastest fuel dragster in Canada was driven by Scott Wilson in the mid-1960s. Scott's "Time Machine" (AA/Fuel dragster pushed by Ford's 427 SOHV 1500-hp engine) reached 220 mph in 7.18 seconds, which was tops for 1967.

Tom Grove
Tom Grove has won many of the top titles in drag racing, including Super Stock in the 1964 NHRA Winternationals (1963 Plymouth), NASCAR Top Stock in 1966, and Funny Car Eliminator in the 1967 Spring Nationals (Mustang). Tom's ultralight Mustang shown here had a one-piece fiberglass "flip-top" body on a Logghe Brothers chassis. With a supercharged 427 SOHV engine, the car reached 186 mph in less than eight seconds.

Hubert Platt
Hailing from Atlanta, a southern hotbed of drag racing, Hubert Platt won the A/XS title in the 1967 Winternationals. He eventually set NHRA class records of 8.61 e.t./171 mph with a 2400-lb. exhibition Mustang. The 1968 Mustang, shown here, had a Cobra Jet 428 engine rated at 335 hp at 5400 rpm. Many speed parts from the 427 wedge engine are interchangeable.

BILL CUSHENBERY
Crafter of Cool Customs

The El Matador started life as a 1940 Ford Coupe. The top was chopped 3-1/2 inches, while a 4-1/2-inch section was removed from the body. With headlights canted, the fenders were scooped and reshaped. The car sported a candy red paint job.

As custom car building peaked during the 1950s and 1960s, business was highly competitive. When most radical customizing ideas were credited to Southern California builders, a talented young visionary from Wichita, Kansas, named Bill Cushenbery, changed that perception with some wild and new styling concepts.

After starting his own business called The Kansas Custom Shop, Bill moved to California. There he advertised his first "Cushenbery Customs" shop by transforming a 1940 Ford into the renowned El Matador. Soon other magnificent creations followed, which spanned magazine covers nationally.

In 1965 Bill was commissioned by Barris Customs to convert the Italian-built Lincoln Futura into the famous Batmobile, which had to be delivered in three weeks to the studio for the Batman series. Because Barris had no drawings, he described what should be done.

"I simply pounded the metal out and ran it through the air hammer to form the desired shape," said Bill. "Then I trimmed it and

Playboy magazine centerfold model Dee Dee Lynne poses with Bill and his 1929 Ford street rod.

Bill Cushenbery shows off a customized interior with stick floor shifter and "joystick" steering mechanism for this customized 1940 Ford.

Bill Cushenbery is "leading" the rolled rear pan of the El Matador.

The Sexy Sixties

welded it in place." Other styling projects were for Ford Motor Company and AMT, makers of models.

Today, Bill Cushenbery continues customizing and restoring works of art. His indelible mark in the field of custom cars remains alive and will continue to live on in the next century.

The Bubble-Top Dream Rod featured an 18-gauge steel outer skin that was hand-formed over tubing attached to a modified 1956 Buick chassis. The handsome air cleaner housing also served as an air scoop.

The Limelighter was a restyled 1958 Chevy Impala. Both front and rear pans were rolled, and the quad headlights were sunk deeply into the sculpted front fenders.

The beautiful Silhouette was a radical handbuilt roadster and a 15-time Oakland Show winner. Among Bill's awards was a trip to Europe.

The "World's Longest Headache" was a 40-foot stretched Cadillac made for a movie.

THE ALEXANDER BROTHERS
Pacesetters in Custom Car Design

Called the most imaginative customizers east of the Mississippi River, brothers Mike and Larry Alexander experienced their share of growing pains before creating styling trends that even car makers emulated.

After military service, both brothers utilized the G.I. Bill to learn auto body repair. Then, in 1957, they opened a small shop boldly called "Custom Car City." There, the Alexanders' superb pearlized painting and metal craftsmanship caught the eye of Harry

Mike Alexander (left) and Milt Kaltz (right) provided the basics that were the foundation for American Sunroof Company's role in reviving the convertible from its near extinction. When Mike joined ASC in 1971, their company developed specialty products and concept cars for major auto makers.

Bradley, an upcoming stylist who would later work with the brothers on show car winners.

A big break came when AMT model company hired Mike and Larry as consultants, which led to joining the Ford Custom Car Caravan.

"But as fate would have it," remembers Larry, "a 1969 highway expansion project wiped out our shop, forcing us out of business."

When the great Corvette co-designer, Larry Shinoda, who had moved to Ford, heard the Alexanders' shop was closing, he hired Mike to run Kar Kraft Design Center. After two years of making Ford show cars, Mike was pegged by Heinz Prechter, founder of American Sunroof Company, to manage the new Custom Craft division of ASC, where later as vice president, he specialized in OEM projects.

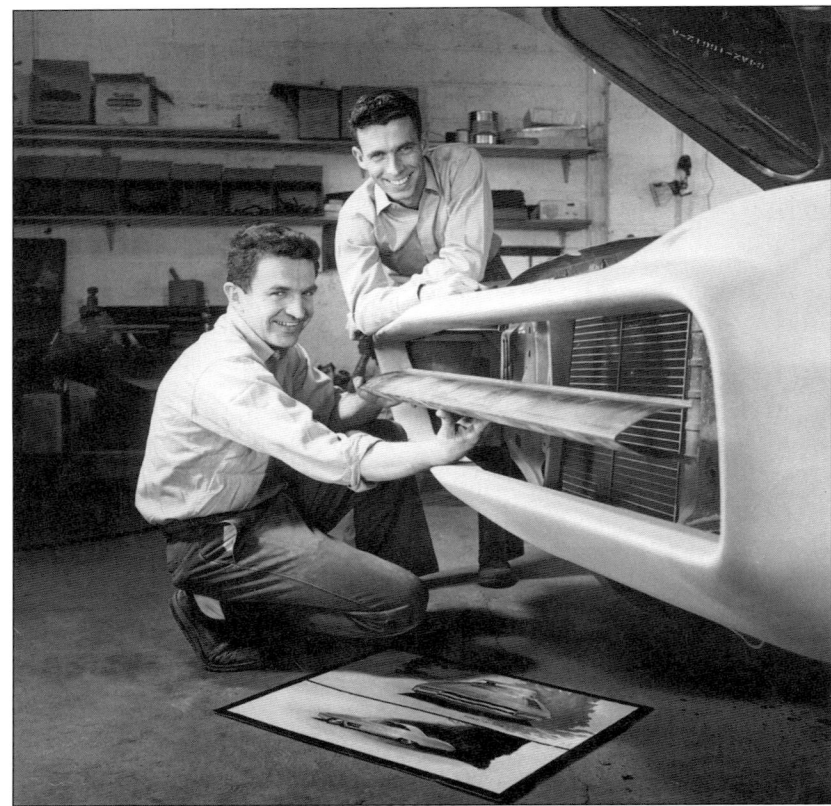

The Alexander brothers were famous for show-winning customs. Here with Larry (right) watching, Mike (left) holds a floating-type grille in position on the custom car under construction.

The Sexy Sixties

Opportunities came more slowly for Larry.

"I took a while to find my niche in life," said Larry, "but I eventually got back into the custom car business with Ken Yanez. After the '74 recession, I was a metal model maker for the Ford Motor Company, often secretly developing bodies for cars of the future, until retiring in 1995. I have been truly blessed with a wonderful life!"

Although the Alexander brothers worked as a team for only a little more than a decade, they left their mark as a pair of greats in customizing history.

Built for Don Vargo, this candy apple red 1934 roadster with a padded Carson-type top was the 1962–1963 ICAS Grand Champion. The body and doors are sectioned, the trunk corners rounded, and the contoured grille is hand-formed.

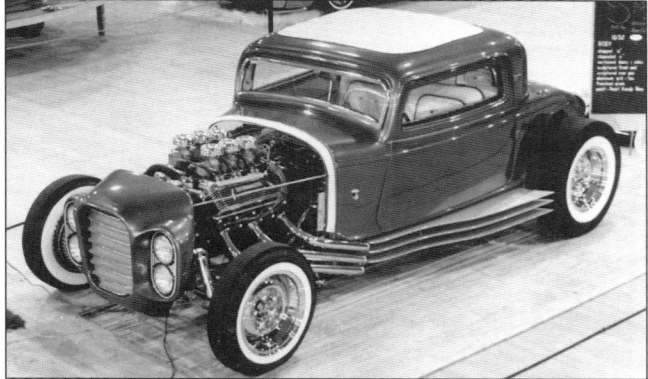

The "Silver Sapphire," a 1932 Ford five-window coupe, was another popular show stopper. It was chopped, sectioned, and channeled with unique aluminum fins where the rocker panels used to be.

The "Triden," a one-of-a-kind show car based on the 1971 Thunderbird, built by Kar Kraft, has a chopped top, skirted rear, and pointed snout that adds an illusion of length.

A candy lime paint enhanced this meticulous 1960 Pontiac hardtop with matching grilles, both front and rear. The headlamp brows were frenched, and the doors were shaved with handles removed. Sculpted tail lights had custom lenses.

Hot Rod Pioneers

Rear-end identification of this 1956 Chevy custom is nicely hidden by the extended rear quarter panels with deeply recessed tail light lenses. The side trim borders the accent stripe.

The "landaulet" roof and trunk of this restyled 1970 Maverick recalls another Golden Age of motor cars. The experimental custom was built by Kar Kraft under the direction of Mike Alexander.

Detroit "Muscles In"

The "muscle car," manufactured in the 1960s and early 1970s, was Detroit's quintessential answer to hot rodders' backyard daring. In those wild days of car lust—unhampered by federal emissions and crash standards—U.S. auto makers saw a huge moneymaking potential in offering monster-engined performance cars, which were akin to the hybrids that rodders had been building for themselves for years.

Each major car maker touted its own version of the muscle car, usually by shoe-horning in big competition-tuned power plants in lightweight coupes. They also added special goodies, such as four-speed floor shifters, hood scoops, rear spoilers, rally wheels, beefed-up suspensions, and even exotic paint colors and bold racing stripes. All of this added racing mystique and targeted the youth market.

The "top brass" of GM said it wouldn't sell, but the 1964 Pontiac, the first of the muscle cars, kicked off a decade-long romance with performance aficionados. Developed mostly under cover, John DeLorean and cohorts conspired to reach the youth-oriented market by plunking a big-block 389 c.i. engine into a Tempest. With a hood-mounted tachometer and four-on-the-floor, it was priced at $3,200.

PETE VAN IDERSTINE
Wheeler-Dealer

"Only in America can a peon start with an idea and make really big bucks." These words were spoken by Pete Van Iderstine who, similar to a few other old-time hot rodders, took the fast and rocky road.

In high school, Pete liked fast rods even more than loose girls. He willingly spent his 1952 wages on his first love, a 1932 Ford roadster; however, his obsession for Deuces grew as he acquired seven of them. He drag raced some of them in northern New Jersey, his beloved turf.

After securing a loan from the bank, Pete, who had acquired a marketing degree, rented a vacant store in East Hanover, New Jersey, with engine builder Tony Feil. By 1966, Pete had purchased a larger building, expanded the wholesale activity, and opened branch locations. Watching popular trends and offering dealers fast-selling products, Pete competed successfully with his own customers—normally a taboo in business. Actually, his retail stores competed with the same speed shops to which he wholesaled. By 1990, Pete had sold his thriving enterprise. However, three eastern stores continued to bear the Van Iderstine name.

On the West Coast, Pete began racing on the Santa Maria Speedway, but the entrepreneur bug bit him again in 1992 when he began catering the growing nostalgia street rod market—specializing in chromed classic disk wheels and more.

"The great hot rod industry will continue to grow in spite of Detroit forgetting about the 'car guy,'" said Pete.

Pete did not fare well in 1975 racing in Flemington, New Jersey, in what appears to be a real killer "demolition derby."

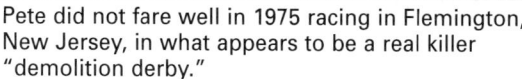

To advertise his speed shop, Pete regularly sponsored or ran fast stock cars or dragsters such as this fueler, which did 202 mph/7.47 e.t. at an eastern dragway in 1966.

Pete, in his 1932 Ford coupe, is ready for another run at the 1959 NHRA Nationals in Detroit. Unfortunately, he didn't win.

"New Look" Accessories

Awakened Sleepy Stock Designs and Put the "Oomph" Back into Aging Vehicles"

Remember foxtails, neckers knobs, and fuzzy dice? Similar to the cheap chrome trim gadgetry of the 1940s, they were replaced with less baroque aftermarket "dress-up" accessories of the 1950s that promised to give even the plainest of automobiles a sparkling personality of its own. Popular add-on goodies back then were fender skirts, fiberglass fins, custom grilles, louvers, dummy spotlights, glitzy wheel covers, and lakes pipes to go with the roaring dual exhaust systems. Continental tire kits and pinstriping, which had hit the East, soon fizzled, as did preformed paste-on fiberglass fins and bubble skirts.

The muscle car decade of the 1960s ended tail fins but intensified the "mag" wheel craze. Other sporty items included racing stripes, hood scoops, wings, wooden or leather-rimmed steering wheels, and enhanced interior decor that complemented floor shifter conversions. On the flip side, showy consoles and dozens of fancy gauges, walnut trim, and shifter knobs and handles also were "in" items.

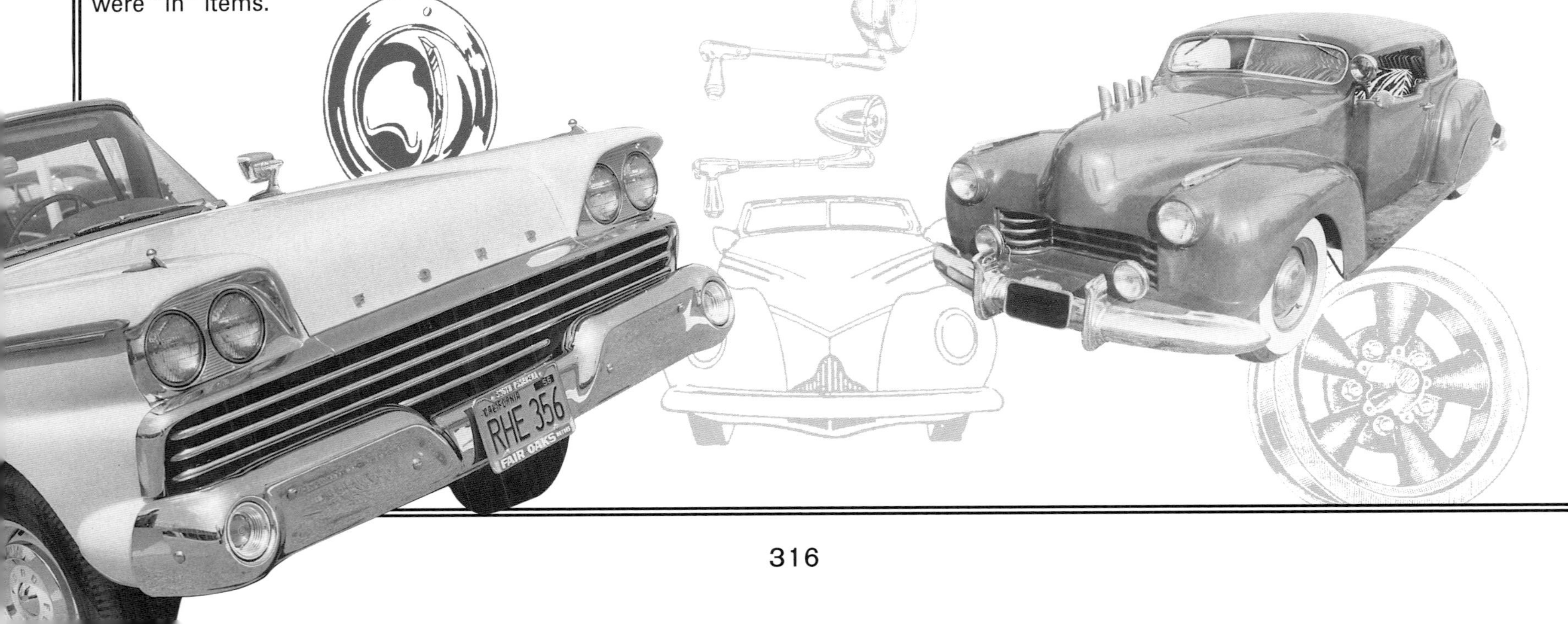

The Sexy Sixties

Despite the recession of the 1970s, "mag" wheels were available in dozens of styles and continued to be the hottest of all replacement parts. Aerodynamic spoilers, tinted glass, sunroofs, and more interior goodies became common add-ons, too. Even decorative hardware such as "superfly" headlamps, Rolls-Royce-styled radiator grilles, and Landau Irons on vinyl tops were visible in the developing luxury auto market.

The environmentally inspired wind-tunnel studies of the 1980s forced auto makers into a mode of more functional body streamlining, which brought about stark, look-alike "jelly-bean" styling that often cried out for help. The aftermarket responded with appearance-enhancing products that included even more "mags" and ground effects systems with spoilers, air dams, and more billets. The popularity of formerly "hot" van conversion treatments began being replaced with much-needed cosmetics for boxy sport utility vehicles and pickups.

What will be the next styling trend? Look at the next crop of concept vehicles, customizers' show cars, and aftermarket accessories because they often forecast the future.

HANK WEIDENHAMMER
Diehard Corvair Defender

Because speed shops have always projected an aura of automotive excitement and glamour, many hot rod hobbyists with an entrepreneurial bent have been attracted to the notion of starting their own speed businesses. Similar to most aspiring or new speed shop operators beginning in the 1950s, Hank Weidenhammer was heavy on enthusiasm and light on cash when he parlayed a tiny investment into a large wholesale distributor (WD) in Pennsylvania.

As his first wholesale supplier of speed equipment in 1953, I remember Hank's anxiety about taking the leap into the then-precarious marketing side of hot rodding, wherein he invested in a few small engine and chrome parts. Within a couple of years, Hank exceeded $100,000 in annual gross business, a true bonanza for a beginner in those days.

"For fun, we became the first big go-cart distributor on the East Coast around 1956 and remained so until the fad was over," said Hank, who built a 10,000-square-foot retail and wholesale operation.

By the 1960s, the sport of karting was all the rage, with karts such as this snarling little import driven by Eric Rickman (pictured) costing up to $3,000. Many famous drivers sharpened their talents racing similar single- or twin-engined go-carts.

After he put a fleet of trucks on the road to distribute products to retail outlets, the business became one of the largest in the East, and envious competitors dubbed him "Hustling Hank." Later, SEMA elected Hank as the first WD of its Board of Directors.

In the early 1960s, as the sporty Corvair gained popularity with young drivers, Hank established his firm as a Corvair performance specialist.

"Business was going great until Ralph Nader came out with his self-serving book, damning the Corvair in *Dangerous at Any Speed*," recalled Hank. "Despite the fact that our special equipment could make that car as safe as or safer than any on the road, the bad publicity killed both the car and our sales. Like countless others in the automotive industry, I was badly hurt financially by self-proclaimed consumer advocates such as 'Nader the Raider.'"

Today, both Corvair and the diamond-fingered Hank are gems in the memories of yesteryear.

After a struggling $60 start in 1953, Hank's tiny speed shop in Berwyn, Pennsylvania, grew into one of the largest WD on the East Coast.

CARROLL SHELBY
His Snake Slayed Giants

After almost going broke raising chickens in Texas, Carroll Shelby returned to racing in the 1950s—first as a driver and then as a builder of world-class competition cars.

With his big black hat and farmboy outfit, Shelby appeared everywhere from LeMans to Riverside. On torturous road racing circuits around the globe, Shelby learned what was needed to win before he proceeded to build America's first true sports car—the 427SC Cobra.

Shelby's legendary roadster was created in the early 1960s by stuffing a Ford V8 into a lightweight aluminum-bodied British AC two-seater. The original Cobras, which sold well as "component cars," are copied today as fiberglass "kit cars" and peddled by others.

In the mid-1960s, Ford made a deal with Shelby to modify a limited run of Mustangs for image-enhancing racing. The resulting Cobra GT Daytona Coupe program made history by beating Ferrari to steal the 1965 FIA World Manufacturers Championship, the first such title by an American team.

When I last spoke to Carroll Shelby in 1996, he was slowly returning to health after a kidney transplant. Even after a heart transplant in 1990, Shelby (then 73) still had the enthusiasm and drive of a much younger entrepreneur chasing another dream. And why not? Wasn't there a song in 1963, "Hey, Little Cobra," in his name?

After racing legend Carroll Shelby found "hidden horses" in the Mustang, he helped the sporty car "kick tail" on road courses everywhere.

Introduced in 1964, the Mustang was the best runaway sales success for Ford since the Model T. Shelby offered a complete line of performance products that really put a "jump" in the "Pony Car."

Hot Rod Pioneers

Hayden Proffitt

During the 1950s and 1960s, veteran driver and tuner Hayden Proffitt ran every make of drag machine, including dragsters, jets, super stock, and funny cars. His knack for reading the starter's mind made him a kingpin in beating competition out of the hole.

After driving both Pontiac and Chevy stockers to big 1961 gas class victories, Hayden took Stock Eliminator honors at the 1962 NHRA Nationals in a Chevrolet. Although Hayden prepared the chassis, the engine was completely reworked to factory specifications by Chevy racing specialist Bill Thomas.

Later, Hayden match-raced a compact American Motors funny car, unexpectedly defeating Big Three auto makers.

Gerry Grant

In the "Roaring Twenties," when most car engines had to be overhauled every 40,000 or 50,000 miles, opportunist Gerry Grant manufactured engine rebuild kits that eventually evolved into the famous line of Grant Piston Rings. To kindle sales, Grant helped back Indy and drag strip cars, including this compact Rambler Rebel SST funny car fueler, which Hayden Proffitt drove to six track records and a national 1/8-mile record of 136 mph in 1967. Grant wisely sponsored the prestigious 200 Miles Per Hour Club, which further aided product acceptance.

Dimitri "Dema" Elgin

After 25 years of drag racing and sports car competition, Dimitri "Dema" Elgin, shown here in this vintage Formula Jr., opted on designing camshafts mainly for SCCA road racing—a sorely neglected field. With the aid of legendary Ed Winfield and later a computer, Dema developed "hot" camshafts for Ford, GM, Indy, GT-1, Formula-V, and "Winston Cup" engine builders. Most applications now are so secretive that not even Dema is privy to performance results. Of his genius friend, Dema said, "Ed Winfield is gone now, but his timeless legacy of camshaft theory remains a benchmark."

Doug Anderson

Doug Anderson's fast 1931 Ford coupe is a show winner in its flawless white and candy apple red color coat. The top is reshaped with a rounded visor, and the souped 327 Chevy V8 engine delivers almost 1 hp per c.i.

After serious drag racing in the 1960s and then a stint with Bonneville racing teams, Doug, a New York native and a member of the Tri-City Flywheelers that was organized in 1955, says, "Plain and simple, a street rod costs less and is more fun."

Emery Cook and Don Garlits

Former West Coast dragster king Emery Cook (pictured) and Don Garlits, who had dethroned Cook more than a decade earlier, teamed up to run this altered "Ultra Shock Challenger" in the newest competition class for funny cars in 1968. The 18-foot topless fiberglass Dodge Dart body was mounted on an old 120-inch wheelbase Swamp Rat dragster powered by a 425 Dodge Hemi. After Cook turned 200 mph, the project was abandoned when other racers complained.

Shirley Muldowney

In the battle of the sexes, Shirley Muldowney, shown here in her winning pink Top Fuel dragster, was known for her competitive and daredevil spirit that helped her become the first female driver to win major NHRA Top Fuel championships against men (1977, 1980, 1982, and 1983). Shirley started racing in the mid-1960s and became a hot match-race attraction by 1970. She was nicknamed "Cha Cha," and her much publicized involvement with Connie Kalitta, a fiery funny car crash, and her remarkable comeback were legendary. Shirley's $40,000 car was called a technological hybrid with its 2000-hp supercharged aluminum engine developed from a basic Plymouth Hemi.

Ed Schartman, Jr.

One of the first drivers to break the eight-second barrier was "Fast Eddie" Schartman, who drove this speedy supercharged 1000-hp 427 OHVC Merc Cyclone exhibition car in 1968. The front wheels had no brakes, and the rear wheels had two individual LeMans-type disc brakes. Rear tires were 11.75 x 16 slicks. The car weighed less than 2000 pounds; its fiberglass flip-top body added only 205 pounds.

Ed was born in Cleveland and began racing in the early 1950s. He earned many super stock wins before taking the S/XS title at the 1966 NHRA Nationals. Ed continued racing after opening an auto research center in Olmsted Falls, Ohio.

ED CHOLAKIAN
The WD, "T-Shirt King," and Fanwear Pioneer

"One day, top racing drivers will earn more money endorsing products than driving cars," predicted Ed Cholakian, a happy-go-lucky car sponsor whose Chrysler-powered Corvair had once set a Bonneville record. "That day has come, but it's ironic that, in the past, former racing personalities who lent their star-studded image to a product, had to settle for a few cases of motor oil."

"Another prediction I made decades ago has come true," continued Ed, who at one time had taken a side job with Alex and Phil Krause, Eastern Auto and Cal-Custom founders and forerunners of sundry restyling parts. "Fanwear is now big business. Even carmakers today sell haberdashery like T-shirts, caps, and jackets. It's almost a flashback to old times when you could take a garment like a T-shirt, put a catchy logo on it, bag it, hang it in a store, and count the money."

As racing increased in popularity, so did fans' worship of their idols. To identify with their favorite race car driver and race car team, fans began purchasing fanwear clothing and memorabilia. Frank Smith, a post-war top official track photographer in the East, was one of the first to develop his own line of racing paraphernalia, including decals, jackets, emblems, and special license plates. One bumper sticker joked, "Crime doesn't pay...Racing doesn't either!"

Ed Cholakian is the pioneer WD who made T-shirt history.

Because Ed (similar to Midwesterner Chuck Bobins) foresaw an opportunity to capitalize on and rectify the young industry's detrimental and often cutthroat marketing tactics, he started EC Enterprises. The company was one of the first U.S. warehouse distributors (WDs) that wholesaled specialty products to more than 1,000 speed shops and auto stores nationwide. At one time, Ed owned one of the premiere silk-screen operations in the United States. Therefore, he was labeled the "T-Shirt King" of racing.

As Ed discovered, fanwear is a cheap form of advertising. As people circulate daily on the tracks, in the malls, and on the streets, they serve as colorful moving billboards with bold designs. What a smart marketing trick that has invaded even the political scene today!

JERE STAHL
The Racer's Header Innovator

In the early 1960s, when most American exhaust headers were of the familiar "Tri-Y" design (but rarely the most efficient), former drag racer Jerry Stahl opted for a new four-tube "Independent" design for 409 Chevys. After exhaustive dyno testing, Dave Strickler successfully ran the handmade prototype at a super stock meet in York, Pennsylvania, and became the first of the AFX'ers in the East and Midwest to go into the 11s and exceed 120 mph.

Since 1963, Stahl Engineering, one of the first to produce the vaunted four-tube equal-length headers, is said to have developed more header innovations than any other header company. Over the past decade, Stahl's optimistic expansion into computer-generated racing camshafts resulted in so many winners that he no longer needs to advertise. Jere said, "It's a marketing dream come true."

In 1966, young Jere Stahl stands beside a Plymouth at Indianapolis after winning the A/S NHRA record in his 1966 Chevy II. The car was never beaten in NHRA Eliminator or NASCAR Ultra Stock competition that year.

Jere Stahl's 1956 Chevy was one of the first wagons to attract national attention by lowering the G/stock record from 14.25 to 13.57 in 1964.

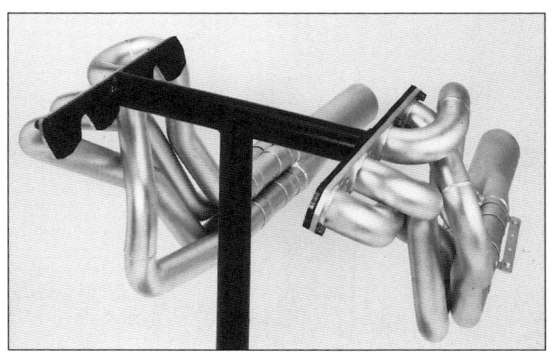

Stahl's spaghetti-like exhaust headers for a Mustang (1979-1993) featured free-flowing port angles with equal-length primaries.

THE SAD SEVENTIES

How times change! The horsepower that was king in the 1960s had to give up the throne to emission regulations in the 1970s. Then, the OPEC nations held America hostage at the gas pumps in 1973, with gasoline prices soaring as shortages and long gasoline lines created havoc.

Emission controls strangled performance and worsened gas mileage, with the average Detroit car getting only 11 mpg. Despite the risk of a penalty for anti-smog tampering, hip car owners desperately were "retuning" their engines to regain lost performance. The good news was that the Vietnam conflict ended.

President Richard M. Nixon ordered emergency allocation as gasoline prices jumped to more than 50 cents per gallon. Many motorists spent hours in line for only a few gallons of gasoline. Unleaded gas, which became the norm, outraged motorists who believed it harmed old engines.

"Made in Japan" was now taken seriously by America, as the foreign small-car invasion began grabbing a big chunk of Detroit's market share. Nobody wanted big cars, and by the mid-1970s, two of every five auto workers were laid off.

Tilt steering wheels, or the "fat man's wheel," first offered on a 1906 Peerless, returned as an option. Catalytic converters debuted in 1975 as motorists complained about poor quality and rusting in domestic cars—motorists' memories lasted longer than the paint on the cars.

324

The national speed limit dropped to 55 mph in January 1974, and motorists began using CB radios to combat "Smokey's" radar traps. High insurance rates and "gas hog" appetites further quashed the fervor of the muscle car era. The proud monster engine now seemed doomed.

"Red ink" flowed for the "Big Three" auto makers in Detroit in the 1970s. These were dark times for the performance aftermarket as well, and the "muscle car" period ended abruptly. Politicians seemed intent on burying the auto industry with tough smog regulations and anti-car propaganda. For them, speed became a dirty word. The U.S. Environmental Protection Agency (EPA) axed small companies that could not afford expensive and often unrealistic laboratory testing. However, major speed merchants fought back with a new crop of environmentally friendly products.

McDonald's first drive-through window opened in 1974 in Oklahoma City, and impromptu weekend rodding meets began during that year also. With the demise of the muscle car came the rebirth of the street rod with nostalgic T-buckets, retro rods, and street cruisers.

The flamboyant design direction of hot rodding continued to strongly influence the U.S. auto industry. The aftermarket's custom wheels, low-profile tires, and many sundry performance and appearance-enhancing products would become "must-have" options.

As a second energy crisis loomed in 1979, gasoline prices climbed to more than one dollar per gallon. Fuel economy also assumed the trappings of a cult, and small-car sales zoomed.

The decade of the 1970s marked the demise of big iron and the truly individual American automobile, the likes of which probably will not be seen again.

Forgettable cars, the loss of young car enthusiasts, and the end of the 1970s left people wondering, "What's next?" Will the glory days of the automobile ever return?

GRATIOT AUTO SUPPLY
The World's Biggest Hot Rod Shop

As a pre-teen, Angelo rode this Whizzer, which had a Weber high lift cam.

In 1945, the emerging new motorsport of happy highway hot rodders did not have much going for it yet. Therefore, it took a lot of guts and vision for young Bill Toia to open Detroit's first "speed shop." Soon Bill's customer list grew into four categories: serious rodders who built cars for competition; stock auto owners who bolted on speed equipment; chrome addicts who dressed up cars; and customizers who torched, leaded, and restored their chariots. The renegade little shop satisfied them all.

As the sport exploded in the 1950s, Bill's Gratiot Auto Supply expanded with additional area stores plus a mail-order business, while Bill's brother opened Leo's Hot Rod Shop in Tucson, Arizona.

"This was a challenging time to be in the speed parts business because the entire industry was beginning to boom, and we too were a growing biggie in the marketplace," remembers Angelo Giampetroni, who as a youngster worked weekends at the main store.

After high-school graduation, Angelo Giampetroni became a full-time store employee before being drafted into the U.S. Army in 1957. Upon his discharge three years later, Angelo became a manager of the main store, which now consumed an entire city block.

This is the way Gratiot Auto Supply appeared in the 1950s. Bill Toia (left) with Angelo Giampetroni (right) are standing behind the 1934 pickup built by Angelo.

The Sad Seventies

This chopped 1934 Sedan delivery built by Brizio has a Ford 351-SVO engine with specially made four-valve heads. It is Angelo's current street rod.

Linda Vaughn helped with the promotion of this snappy Gratiot track roadster, which was available as a kit car in the late 1970s.

In 1970, Angelo acquired half partnership, and in several years he owned the entire Gratiot enterprise. Soon the business mushroomed into nine stores in Michigan, Indiana, and Ohio, with an annual gross of more than thirty million dollars. However, by the 1980s, a floundering domestic economy, a weakened auto industry, and brutally high interest rates wreaked havoc with Gratiot Auto Supply. The end came when foreign investors, who were buying up Midwest banks, suddenly called in Gratiot's inventory loan. That was the death knell for Gratiot Auto Supply, a 53-year-old company that sadly disappeared into history.

However, Angelo again bucked the odds and became Ford Motor Company's national sales manager for Special Vehicle Operations. Today, Angelo remains true to his calling. He is "Ford's new hot rod answer man."

A.J. Foyt

Here, superstar driver and mechanic A.J. Foyt is making a trial run in a Ford 427 Torino before a Daytona 500 race. (He won in 1972 in a Mercury.) A.J.'s career spanned five decades as he raced everything with wheels—winning more than 90 features in midgets, dirt, spring, sports, and stock cars before winning four Indy 500s and six IndyCar championships. A.J. drove a rear-engined flathead-powered wing tank dragster to the finals but lost in the 1955 San Antonio Regional Drag Races. The likeable but reportedly cantankerous Texan battled with officials in defense of racing's "little guys," endearing him to fans. (Circa 1958)

PAULA MURPHY
The First Woman to Drive Around the World

Pretty Paula Murphy was the world's fastest woman racer when she circled the globe in 105 days in 1976 for the U.S. Bicentennial Global Record Run. She returned for her last leg of the world trip to the Daytona International Speedway, while Johnny Parsons ended his trip at the Minnesota State Fair Grounds.

Attractive and fun-loving Paula Murphy began sports car racing in the late 1950s after a devilish dare. After a stint as a successful stock car driver, she became a professional drag racer. In 1966, Paula was issued a license by NHRA and AHRA to drive a blown fuel funny car, but the license was later rescinded. After cries of discrimination and pressure from STP, Paula's license was reinstated and she became the first woman licensed by NHRA to drive rocket-powered dragsters.

On July 4, 1976, the "world's longest automobile trip" began, and race car drivers Paula Murphy and Johnny Parsons, winner of the 1950 Indy classic, revved their engines. The U.S. Bicentennial event was sponsored by Pontiac and the National Car Rental System in an attempt to have a world-endurance driving record for circumventing the world. (Actually, they covered more miles than the circumference of the earth along the equator.) Johnny departed from Minneapolis in a 1977 Pontiac Grand Prix V8 rental car that averaged 16.6 mpg and 47 mph. Paula left from Daytona in a four-cylinder Pontiac Sunbird that averaged 21.5 mpg and 47.5 mph.

"It was like a 30,000-mile-long Baja run," recalled Paula about the 105-day goodwill tour. "Gas was a headache—17 cents in Iran, and $2 in Turkey. The Russian gasoline was only 56 octane. The roads, at times, were unreal. Bikes, camels, water buffalo, elephants, and monkeys crossed the road."

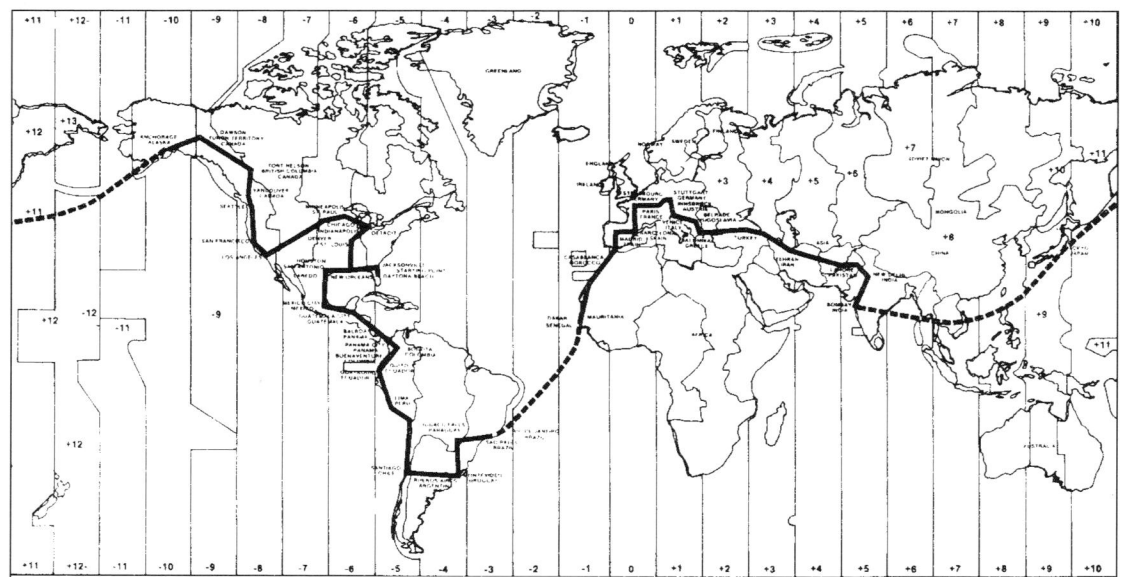

For the U.S. Bicentennial Global Record Run, the U.S. Auto Club, which sanctioned the 38,786.45-mile route, required that all local speed limits and regulations of the 29 visited countries be followed. There was no driving after dark.

The Sad Seventies

In 1973, Paula Murphy was honored by the Associated Press managing editors for being one of the most outstanding women athletes of the century.

Paula set a closed-course land speed record of 172 mph in a Studebaker Avanti in 1976 in Talladega, Alabama. She also captured the land speed records on the Bonneville Salt Flats, including a 250-mph world record from a standing start.

After drag racing stockers, Paula made a name for herself driving rocket-powered cars. She quit racing rockets and jets after her neck was broken due to a parachute failing to open at the end of a high-speed strip run.

As a favorite with drag strip fans, Paula does her "burn out" in a 1972 Duster funny car, which is the customary tire-heating process that precedes a timed run.

JOSEPH HRUDKA
Mr. Gasket

Before he became a multimillionaire, Joe Hrudka's teachers told him he would never amount to anything.

The CBS television show "60 Minutes" did a feature on Joe Hrudka and described him as one of "the most impressive success stories of the second half of the twentieth century." As a skinny teenager, Joe had no perception that his life would be a magic carpet ride, which started with only one gasket.

In 1965, with five dollars in his pocket, Joe started a business with his own adaptation of heat-resistant gaskets for use in high-performance cars. Six years later, Hrudka had netted more than $12 million from the sale of his company, an enterprise that had sky-rocketed to more than 1,000 different auto parts and accessories.

Joe had been a D-student from John Marshall High School in Cleveland, Ohio, and attended one semester at Bowling Green State University before quitting to pursue his drag racing hobby with his brother, Tom. In 1961 and 1962, with Joe as driver and Tom as mechanic, they were terrors on area drag strips with their Class D coupe/sedan gassers.

When racers were having frequent problems with header gaskets blowing out on highly revved Chevy engines, Joe found an asbestos composition made for diesels that withstood extreme pressures and temperatures. Joe began marketing his 283 Chevy exhaust gasket, and the first few hand-cut samples sold rapidly. The "Mr. Gasket" business expanded with special gaskets for valve covers, intake and oil pans, and even skin-packaged fasteners.

In 1965, the first year of operation, the tiny company barely survived, grossing only $70,000. By the fourth year, the business had taken in more than $1 million, and Hrudka's salary had swelled to $100,000.

Joe's drag racing background and his uncanny instinct for knowing what hot rodders wanted proved invaluable to the business. Because most of Joe's products retailed for less than $10, Joe bypassed the middle-man and sold directly to retailers. Without realizing it, Joe had broadened his product line and pioneered a high-volume, direct-distribution method with attractive prices.

In 1969, when the general economy was flat, Mr. Gasket Company went public at $12 per

From this 1967 Chevrolet, Joe first peddled gaskets to other racers.

The Sad Seventies

This mansion in Cleveland is only one of several mansions owned by Joe. At his estate in Palm Springs, he gave a farewell party for President Gerald Ford.

Towering Wall Street finally pays its respect to the hot rod industry in this "Wall Street" ad. Mr. Gasket ad appeared in 1970 in "Hot Rod Industry News."

share. By the time he was 30 years old, Joe's interest was valued at $5 million.

In 1971, after netting $1 million on $8 million of volume the previous year, the conglomerate W.R. Grace & Company acquired Mr. Gasket for $17.7 million in a well-oiled deal, of which Joe's share was a cool $12 million. Cushioned with a 10-year, noncompete contract, Joe continued as the salaried president of Mr. Gasket Company. In 1981, Joe repurchased his company for $4.5 million and within four years did $150 million in sales. With its climb back to prominence in 1983 with a second stock offering, the new Mr. Gasket/Performance People Companies purchased the assets and product lines of Hurst, Hays, Lakewood, Cyclone, and Cragar.

Ten years later, the renewed Mr. Gasket Company continued as a major force in the specialty aftermarket. In 1993, a new Mr. Gasket Performance Group formed with Echlin, its parent company.

In a time when drag racers and even high-performance cars were brutal on gaskets, Joe Hrudka capitalized on his idea and filled a void in the marketplace. This success story came from a young man who once read water meters in Cleveland's rat-infested ghetto basements.

John Wolf

Forever Ford-lover John Wolf (standing, right) built the nitro-burning blown Ford flathead engine in this modified roadster constructed by Phil Freudiger (standing, left), which did 188 mph in 1955. Also pictured are Bud Fox (standing, center) and driver Bob Berquette. Wolf built seven world record-holding engines, including one that helped him break the 1972 world's five-liter class hydroplane record at 152 mph.

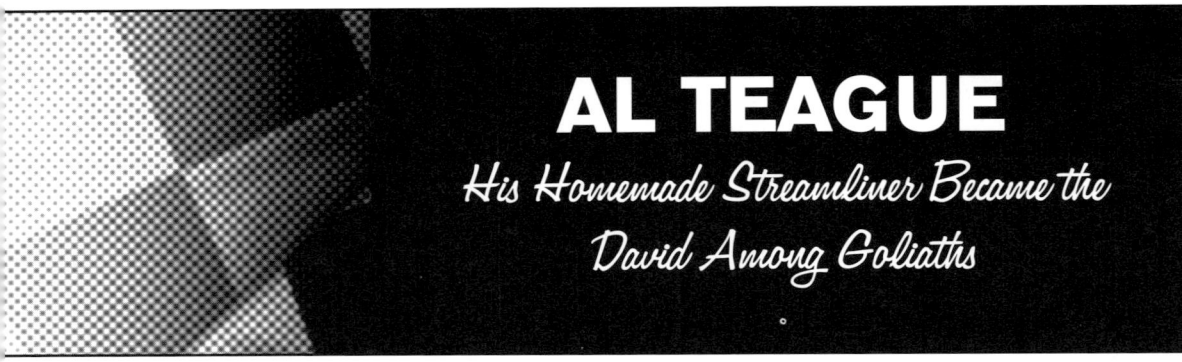

AL TEAGUE
His Homemade Streamliner Became the David Among Goliaths

As an upstart novice drag racer in the late 1950s, Elwin (Al) Teague never believed in doing things in a small way. Despite minuscule funds and little formal technical training, Al never gave up his dream of capturing a world's land speed record. However, it took Al almost three decades to accomplish his goal because, unlike competitors, he had no sophisticated test facility or big-time moneyed sponsor to support him. All Al had was rugged determination and what little was left each week from his paycheck as a mechanic to help him build the record car in his mother's ancient twelve-car garage.

Soon after serving in the U.S. Army in Vietnam, Al began serious racing at El Mirage Dry Lakes where he turned a record 205 mph with his red roadster pushed by a Chrysler Hemi. Later, at Bonneville, Al set a 1972 B/Blown Fuel Roadster record of 250.805 mph.

Al's streamliner, Spirit of '76 (which started as a Lakester), was stretched into a super-streamliner and first run at Bonneville in 1976. Over time, the Keith Black Hemi-powered car has undergone extensive remodeling, including moving the front wheel inside the body and covering the rear wheels. Each successive year, Al's trap speed over the measured mile climbed until August 1991. Then, he and Nolan White did 425.230 mph and a two-way average world record of 409.986 mph, then the fastest that any piston-driven vehicle had gone since the Summers brothers in 1965.

In August 1998, Al and his famous blue streamliner again trekked to Bonneville's Speed Weeks, where Al reclaimed the SCTA C/Blown title that was set in 1996 by Hoffman-Markley. Al's winning speed was 366.043 mph.

Al Teague (standing) built and raced this phenomenal A/Fuel streamliner without big-bucks sponsorship. The piston engine car carried him to a new FIA, SCTA, and BNI world record of 409.986 mph in 1991. Top speed was 432 mph one way.

The Sad Seventies

Ragtop Rage and Its Renaissance

Strip away the average person's rationale for buying a sensible family sedan, and you will find a heart that secretly longs for a sportier "personal" car—such as a flashy roadster designed for fresh-air fun on the open road.

Happy hot rodders have always been keen on ragtops and have kept the "topless" tradition alive. However, long before cantilevered or Carson-style tops were popularized by post-World War II customizers, removable glass-sided "Rex" tops and "California Tops" were available for open touring cars as far back as 1910.

By the 1930s, the call for comfort was answered when folding soft-top cars finally came with roll-up windows. In 1939, Plymouth offered the first production power-top option. Then, in 1949, General Motors introduced the harbinger of them all—a pillarless "hardtop convertible" that did not convert. This design concept remains today.

Even as convertibles faded away and the popular optional T-roofs and Targa tops threatened to pound the last nail into the coffin of the soft tops, ragtop lovers continued with alternatives such as simulated convertible tops, articulated hard tops, or convertible conversions.

In the late 1970s, American Sunroof Company (ASC) with the 1980 Buick Riviera convertible began its role in reviving the ragtop. This helped usher in a new era of open-air motoring.

A lower reshaped window frame was required for the permanent Carson-style top installation on Joe Wilhelm's car project.

The fabric-covered padded top on this Coupe de Ville was "split" so either the front section or the entire top could be lifted off and stored. A similar long-hooded Lincoln with fade-away front fenders and Carson top was built for Henry Ford II in 1947 by Coachcraft, Ltd. of Hollywood.

Hot Rod Pioneers

Street Rodding Makes a Comeback
How the NSRA Helped the Rebirth

Often called America's original sports car, the traditional type of hot rod was essentially a double-duty "daily driver," often used for both street and weekend racing. As drag racing became more specialized, competition became tougher and so costly that rod and custom car owners began parking their beloved machines while the essence of hot rodding started to wither. However, many gearheads continued to hang on, including a new crop of baby boomers.

In 1970, a group of passionate rodders—among them Tex Smith and Tom Medley—decided it was time for a national street rodding event, exclusively for street-driven hot rods. The first "rod run," promoted through *Rod & Custom* magazine, was produced in Peoria, Illinois. Six hundred cars from nearby car clubs participated, and winners received their awards on a flatbed hay trailer. After similar meets run by others elsewhere in the country, Dick Wells became involved with Gilbert Bugg and Vernon Walker, who had already formed the foundation for the National Street Rod Association (NSRA). It was to be the start of something big in the reemerging street rod movement.

In 1972, Dick Wells, who became the first president of the underfunded NSRA, was assisted by volunteers. "A spare bedroom in my house in Studio City served as the 'offices' of NSRA, where I prepared the first newsletters, typed out membership cards, did mailings on my own," said Wells, "and began *StreetScene* magazine." Wells even put a second mortgage on his home to pay for promotion of early nonracing events.

During his tenure, Wells said, "Hundreds wrote to say they had taken their once-raced hot rods out of storage, since there was now a meaningful, drive-'em-daily use for them. Street rodding then began to be promoted as a wholesome hobby. The Street Rod Nationals became a family's annual vacation, and safety was emphasized. Street racing was not allowed, but everyone was welcome as NSRA participants—from the owner of an in-progress home-built hot rod or custom or professionally built, to the one not yet built."

Because legislators seemed determined to pass laws against street rodders, *Rod & Custom* magazine and others deserve credit for jumping on the lobbying bandwagon to bring together early hot rodders, who would not allow the big lawmakers to chip away at the cornerstone of the sport but who would later provide credibility and educational information on hot rodding.

Annual regional NSRA meets have swelled now to unimaginable levels—with more than 50,000 NSRA members cruising to the slogan, "We put the FUN in FUN with CARS."

As the popularity of street rodding spreads throughout America, so will the diversity of street machines, with no two alike—from T-buckets, deuces, and sedans, to shoebox-type customs and factory muscle cars, to heaven only knows what!

Dick Wells (standing) is now communication vice president for SEMA. Here he is visited by old timers Bill Burke (left), Tom Medley (rear), and Don Francisco (front). These former staff members of *Hot Rod Magazine* certainly enjoy "bench-racing."

National Street Rod Association logos.

The Sad Seventies

'Jungle Jim' Liberman

"Jungle Jim" Liberman of West Chester, Pennsylvania, began match-racing this blown hot Chevy II funny car with fiberglass body and lengthened Logghe chassis in 1969. Later, he won the 1976 "Big Go" at Bakersfield with 6.22/226.70.

Shirley Muldowney and Bobby Riggs

"I'm going to blow your doors off, Bobby Riggs," shouted spritely professional drag racer star Shirley Muldowney before their 1976 showdown. The racing star called his bluff when Riggs said he could "beat any of those women drivers...because they are so bad." Both Riggs and Muldowney drove new Mercury Bobcat "S" cars in a slalom competition. Guess who won? (From a Ford press release.)

Ed McCulloch

In a funny car final, durable Ed McCulloch (right) triumphs with his 6.85/217 victory over the smoking Candres & Hughes entry. Without a doubt, 1972 was a great year for Ed, who won the NHRA Winternationals, Bakersfield, and the Nationals. Ed had started racing in 1957 as a teenager. In 1969, he switched from dragsters to funny cars. Popular as a match racer, Ed reportedly did 100 or so events a year for $1,500 guarantees. Major funny car wins were Bakersfield in 1974; IHRA Winternationals in 1985; NHRA Gator in 1986, 1989, and 1990; NHRA Finals in 1990; and the 1992 Nationals (dragster).

RON COVELL
A Creative Metalsmith

As a young "eager beaver" with a talent for sculpting metal, Ron Covell was lucky when he met legendary chassis builder Kent Fuller. At the time, Fuller was looking for someone to fabricate dragster bodies, and he believed Ron was that man. Fuller provided the tools and a nearby shed, and he "bought all the bodies."

Of that golden dragster era, Ron said, "Kent Fuller was one of the most creative persons I have ever known. He was my mentor and my teacher during my early twenties."

Midway in the 1970s, Ron opened his own shop. With the occasional help of gifted designers such as Don Varner, Ron began to custom-build works of automotive beauty. Since then, many of Ron's spectacular street rods have won worldwide praise and top honors at major shows.

Recently, Ron has been giving metalworking workshops around the country. He also has been plugging videotapes and a line of metalworking tools by mail order.

"I truly love teaching others the technique it took me years to master," says Ron. "The more you work with metal, the more it will teach you."

Ron Covell completely hand-formed the California Star, with its sleek aluminum body. The car has 10 machined aluminum grilles for air intake and exhaust through the body. The space-frame chassis was made from chrome-moly tubing with fully independent suspension. The engine is a Chevy V6 with a turbocharger.

Ron Covell, shown with the "English Wheel" for shaping metal panels, continues to use a hammer as his primary tool. Ron says, "It's not how hard you hit it; it's how you hit it hard."

This trackster sports the frontal styling and suspension system popular in the 1960s and 1970s.

The Sad Seventies

Is this the hot rod or street rod of the future? The magnificent California Star concept rod, designed earlier by Don Varner, was a ground-up fabrication that was started in 1979 by Ron Covell. Later, the car won the coveted "America's Most Beautiful Roadster" award at Oakland and was featured in all hot rod magazines.

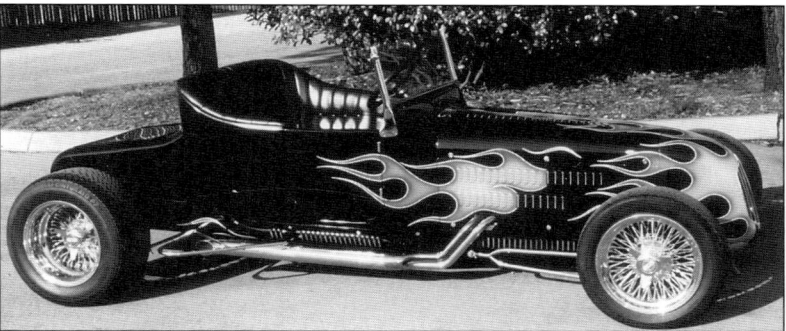

This traditional Track T was scratch-built by Ron Covell in 1974. It has an aluminum hood, nose, and belly pan with a custom-built frame, nerf bars, and outside exhaust system. The engine is a Ford Capri V6.

Tom McEwen

Tom McEwen was nicknamed the "Mongoose" for his ability to hustle and to competitively hassle his friend, the Snake Prudhomme. Both characters attracted big-money sponsorships. (Note the signs on the car.) Tom was one of the first to exceed 200 mph in the quarter, and his major wins included NHRA Bakersfield Top Fuel in 1972, NHRA Nationals funny car in 1978, IHRA in 1977 and 1984, and AHRA in 1979, 1982, and 1984. The inset (right) shows big Tom sandwiched in a speedster's tight cockpit. (Alan Earman photos.)

Hot Rod Pioneers

T-Buckets for Show and Go

Long after Henry Ford's marvelous Model T track racing days, Kooky's T-bucket from the old " '77 Sunset Strip" TV show helped rejuvenate a roadster style that remains a favorite of rodders today.

Similar to their racing predecessors, the little 1923 through 1927 Model T bodies (stripped of fenders and non-essentials to lighten them) continued to adorn hot rodders' roadsters for both street or competition. Power plants that evolved from four-bangers to flathead V8s and OHVs now usually have mouse motors with snakelike outside headers and are cloaked in sparkling chrome and billet.

These photographs compare original Model T roadster racers with later street rods and their contemporary clones.

The poster on the wall advertises the first hot rod show in Los Angeles in 1951, and it touts $10,000 in prize money. Meanwhile, Don Blair and a buddy sit in a T-bodied rod, waiting patiently for a tow to the track.

The Sad Seventies

Ed Roth and Liz Kitzul talk about her T-bucket that logged more than 50,000 miles in five years, as she rode coast to coast for rodding events. A rodder since the 1970s, Liz tracks criminals via computer for the police in Winnipeg, Canada.

DOUG THOMPSON
A Master Craftsman

Similar to many other gifted customizers, Doug Thompson was born with a creative talent that hungered for artistic expression. Doug's talent manifested itself in some of the most beautifully sculpted custom automobiles in the country, eight of which received international acclaim.

Doug, a congenial artisan in the custom car field since the mid-1950s in the Kansas City area, is somewhat reclusive. He seldom advertises. Nonetheless, he gets hefty commission projects that take months, or even years, to complete.

"When I do a complete ground-up custom," explains Doug, "I try to create a car that individualizes or makes a statement about some aspect of the owner's personality."

Judging from the response and awards Doug's cars have won, they obviously touch the public's appreciation for style and aesthetics.

This blue 1950 Ford features a Carson top, louvered hood, custom grille, and headlight scoops. Here, Doug Thompson holds a Bradley Design Award.

The fastback in primer paint shows off the handmade fender skirts, filled seams, split window, and clean-shaven rear that adds classic beauty.

Doug Thompson named his subtly restyled 1946 Packard "Evening Shadows" to emphasize its deep glossy finish. The top was lowered, the headlights frenched, and the body sides reshaped and smoothed.

The Sad Seventies

Doug Thompson built this Hirohata Mercury clone for Jack Walker. An exact copy of the original built by the Barris brothers, the car stands out at any auto show.

This 1958 Chevy Impala was inspired by the movie "American Graffiti."

The front fender of this 1950 Chevy flares gracefully into the rear quarter panels. The hood of the car is louvered and the top is chopped, but the Cadillac-like fins date the car.

The String of Pearls is a 1941 Ford, beautifully customized by Doug Thompson. The Carson top and exotic blue pearl-like finish enhance its gracefully sculpted lines.

This completed body in base metal (no plastic) shows the laboriously hand-formed front end with grille openings, flared fenders, and sweeping side panels.

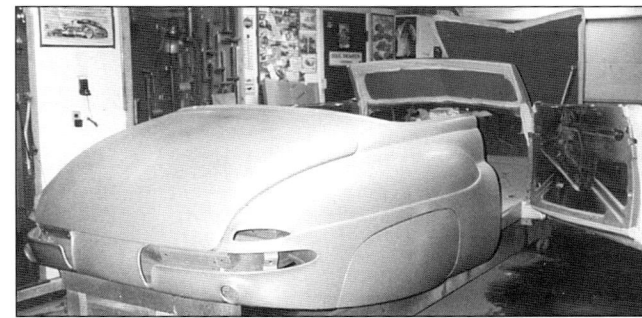

This convertible in base coat will receive dozens more coats of paint before being Carson topped.

Durable Deuces
The "Hot Wheels"

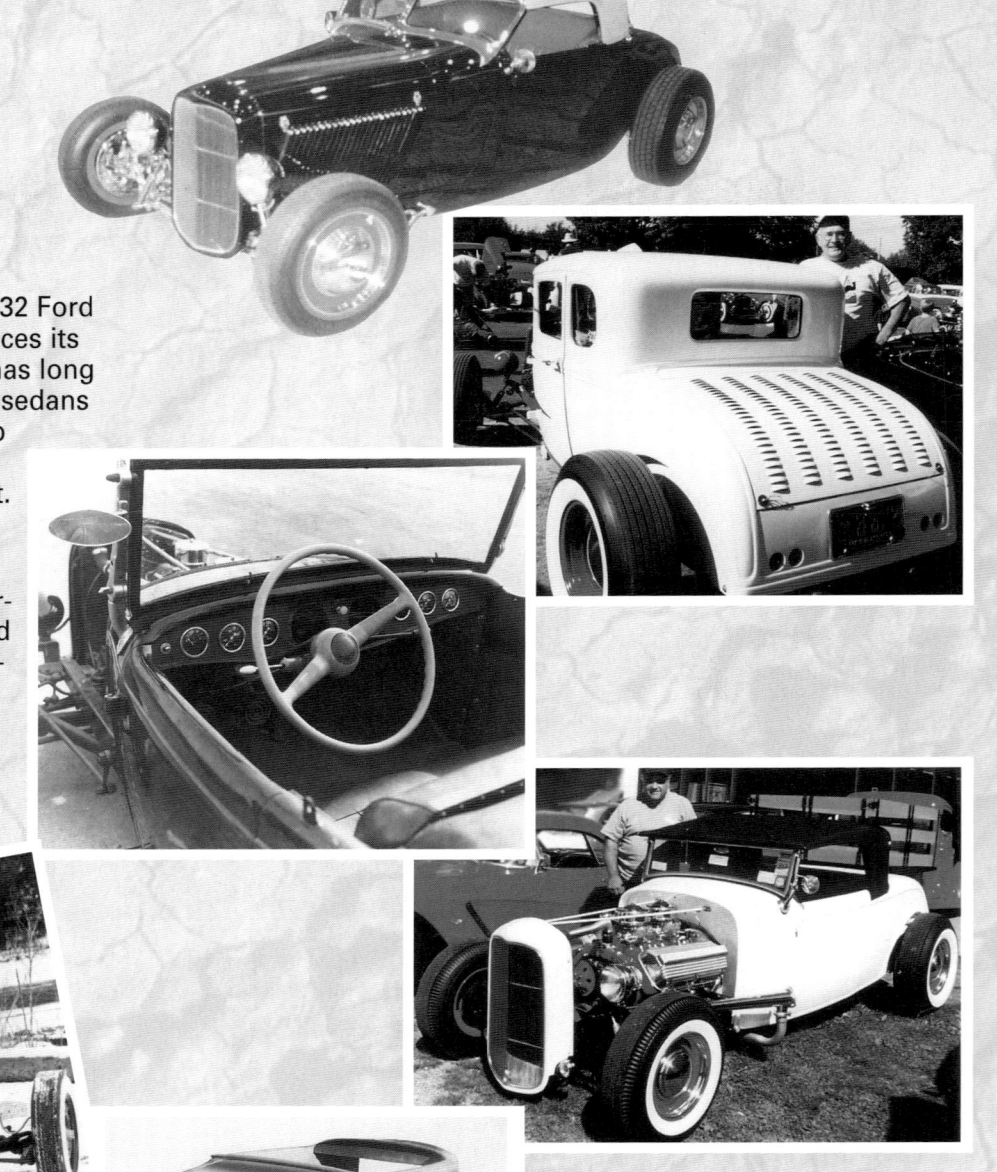

Often called the "Quintessential Hot Rod," the 1932 Ford Deuce, especially the High Boy roadster that traces its heritage back to the post-war SoCal dry lakes, has long been the rodder's traditional favorite. The coupes and sedans that were preferred in cold climates frequently had top chops of up to five inches, and those that were fendered sometimes had "floating-cycle" types in the front.

After the Deuces met their sad fate from rust or the junkyard crusher in the same way as their older Model T kin, several companies began producing fiberglass replicas for another generation of rodders. Priced at $179, the Deuce body shown below—made in color-impregnated fiberglass by Almquist—had smooth sides with dual pods for lights or exhaust. Its low 45-pound weight made it ideal for competition.

The Sad Seventies

Hot Rod Pioneers

Tony Feil

In the 1960s, Tony Feil successfully raced this neat A/Gas 1948 Anglia, which was powered by a small-block Chevy engine. It helped promote Tony's engine shop, which is now a high-tech multiplex facility in Raritan, New Jersey. The shop has become a major supplier of championship engines for street, strip, and dirt track in the Northeast.

Tony Feil set the C/MP class record (e.t. 9.62 sec./141.5mph) in 1979 at Maple Grove Dragway in Reading, Pennsylvania, with this 1960 Corvette. The 600-horsepower engine reached 9,000 rpm.

Car-lover Tony Feil's first garage was under a backyard tree, and there he built many of his own parts. Tony ran this fast-charging big-block 1968 Camaro to many Pro-Stock wins, including a class win at the NHRA Divisional meet in 1970 at Raceway Park in Englishtown, New Jersey (9.92 e.t./149.23 mph).

KEITH BLACK
The Horsepower Man of Drag Racing

When the aftermarket aluminum Chrysler Hemi blocks were initially produced in 1974, a virtual three-way tie existed between Keith Black and two other builders for racer acceptance. A decade later, the KB engine had almost total dominion of the market. Then, Keith established somewhat of an aftermarket cue in 1985 when he introduced the aluminum KB 500 big block Chevrolet engines and KB 600 big block Olds engines. Six years later, Keith died, but not before he had become the most famous engine builder in drag racing.

Keith was born in Los Angeles in 1926. By the age of 14, he had overhauled the family's car engine, much to his dad's surprise. After World War II, Keith had become a wagon peddler, plugging his wholesale auto parts business on wheels. With his mind set on making mighty horsepower, Keith, who had been building boat engines, decided to build Ford flathead engines. By 1959, he was big time into larger racing engines.

Drag racing history was in the offing when Keith collaborated with Tommy Greer and Don Prudhomme, then an unknown driver.

Robert "Keith" Black was famous for his aluminum block Hemi, Olds, and Chevy racing engines. The Chevys developed more than 1000 hp and were fully water jacketed with dry cylinder walls. By the 1980s, a KB engine cost $20,000 to $25,000.

The Greer-Black-Prudhomme fueler with a Keith Black Chrysler engine and a Fuller chassis brought back West Coast drag racing supremacy with hundreds of victorious runs.

The Sad Seventies

With Don on the pedals, the Greer-Black-Prudhomme dragster acquired 236 wins in 243 record attempts. That record is unsurpassed even today in the NHRA recordbook.

"In the early 1960s in San Gabriel, we raced Garlits but had to win the Eliminator first," said Keith. "We beat everybody and blew Garlits' socks off the first time that we had raced him."

In the late 1960s and early 1970s, the later model Hemi "Elephant Motor" was becoming the chosen power plant for top fuelers. Moving his working environs to South Gate, Keith worked on further developing his "Elephants" that were the dominant force for top fuelers and funny cars until the aluminum block onslaught.

For Keith, the early 1980s became a time to draw on the previous experiences of boat racing, drag cars, Indy cars, and off-road racing. Keith transitioned his Chevy block, crankshaft, and gear drives into a 1980s hot rod application that included the Black/Camaro joint project and Shirley Muldowney's comeback car.

For more than 25 years, Keith Black, who was the greatest hands-on builder of blown, injected, nitro-burning engines, had wrenched on legendary top fuelers and funny cars for the likes of Garlits, Force, Amato, Ivo, and others. For this SEMA Hall of Fame inductee, the name of the game was rip-roaring, gut-wrenching horsepower that catapulted him to well-deserved fame.

Keith (right), with a technician, examines a magneto connection on this Hemi rear engine dragster. A log intake manifold with six Ford carburetors was popular in the 1950s. Exhaust headers that "zoomed" upward also enjoyed brief notoriety.

Gasoline Alley

Called the "world's smallest speed shop," this 10-foot-wide storefront known as Gasoline Alley was located near New York City and was once packed with vintage speed equipment and dress-up accessories from the 1940s and 1950s.

"When I bought the remains of a 1971 auction," said George Monte, "I threw away tons of obsolete Almquist cams, flywheels, pistons, and rare engine parts, not realizing their future worth in today's nostalgia market. I could kick myself now."

The photo here shows Gasoline Alley in 1972.

Trick Dream Trucks

Should the truck mania of today be blamed on baby boomers or long-ago customizers' sporty pickups that paradoxically may have helped change the small-truck image from a redneck's workaday tool to a personal-use vehicle for yuppies?

In the late 1950s, the hauler-lovers resurrected a clever idea from the Great Depression days—converting a passenger car into a sleek compact pickup by cutting out a "flower-car" style bed or by grafting a box to its chopped-off rear. Both helped promote interest in new truck classes for show circuit competition.

A big hit on the show circuit, the Dodge Deora was a futuristic "sports pickup truck" built from a wrecked Dodge compact pickup before its caterpillar-to-butterfly transformation in 1967 by the Alexander brothers. Features included a retractable windshield, center-hinged front door, and swing-away steering wheel. The hidden tail lights, until lit, made it difficult to tell whether the car was coming or going. The Deora won eight awards at the Detroit Autorama before being leased by Dodge for a nationwide tour.

This modernistic long-nosed 1974 Ford van was built by Darryl Starbird.

Darryl Starbird's Trik Truck has six wheels, and they all function. The blown Chevy engine located in the center of the truck is covered with a clear plastic bubble to show it off.

The Sad Seventies

Dream Trucks such as the snappy little custom shown here received the full George Barris treatment, including canted fenders and a finned pickup bed complemented by decorative scalloping and pinstriping that is used today by hobbyists to enhance the street rod look.

The idea for a modern convertible sports pickup was in Bob Kaiser's head long before he had built 14 of them, similar to the modern extended cab Ford illustrated. Note how the ground effects and spoiler help accent the rear wing and sports bars.

"Hauler-Heads" say Studebaker's first 1937 pickup truck was one of the most stunning pickups ever created. Almquist supplied OEM floor shifters for the right-hand-drive export 1961 models.

Hot Rod Pioneers

Dick Moroso

After 12 years in the drag racing circuit, Dick Moroso decided to try his winning ways with a speed shop business. However, it was a tough road in the 1960s because wholesale distributors were unwilling to supply Dick's tiny business that he ran from his parents' basement. One morning before dawn, Dick and his partner sneaked to the local Chevy dealership and hung an enormous "Speed Associates" sign on the building. The resultant photograph appeared in a directory that allowed direct buying from manufacturers. Since then, Moroso has become one of the largest makers of hardcore racing parts in the country. The speed equipment aftermarket lost one of its foremost innovators when Dick Moroso, 59, succumbed to cancer in 1998. Dick's family continues to operate all the Moroso enterprises, including Moroso Motorsports Park in Florida.

Dick Moroso's speed shop business offers deep sump oil pans, valve covers, and lightweight front tires, to name only a few items. Pictured here is Don Garlits'/U.S. Navy co-sponsored dragster.

Joe Fontana

Racing modified stock cars in Connecticut, before setting a Bonneville record in the early 1970s, convinced engineer Joe Fontana to follow his hunch. Realizing the hidden potential of a "new generation" 8.3L Hemi, Joe purchased Nick Arias' engine division. This led to improvements, plus a line of cylinder heads, manifolds, and other products, including the four-banger Ford-based aluminum block racing engine shown here. Similar to his predecessor Ed Donovan, who was the first to produce aluminum block Hemis, Joe is a high-tech force in the racing world.

Jerry Pennington

This snout-nosed custom Corvette was made by Jerry Pennington for a TV show. Pennington was a Michigan body-shop operator who built other show-car award winners in the 1970s.

Mike Estlack

This sleek, bobtailed pickup truck was made from a 1972 Corvette by Mike Estlack of Oklahoma City, Oklahoma.

The Sad Seventies

Lou Baney

Lou Baney drove his hot four-carbureted Baron-equipped Ford flathead coupe regularly at the dry lakes, often exceeding 135 mph. Popular Lou Baney was one of the most influential men in drag racing. After moving from Jamestown, New York, Lou started running flatheads at SCTA meets in the 1940s and became president of the Russetta Timing Association. In the early 1950s, Lou co-owned (with Louis Senter) the Saugus Drag Strip, one of the first drag strips in the United States. Later, Lou became famous for Baney-built racing engines. He sponsored many top competition cars, including those driven by Tom McEwen and Don Rackemann. From 1972 to 1976, Lou was Technical Director for SEMA. In 1992, he was inducted into the SEMA Hall of Fame. Lou Baney died in 1993.

Ready for a run at the Saugus Drag Strip in the early 1950s are (standing, from left to right): starter unknown, Lou Baney, Lou Senter, Howard Hudson, and Bob Corbett.

Bob Glidden

Until he made his strong Pro Stock debut in 1972, Bob Glidden was a rather obscure Super Stocker. However, he shocked the drag racing world by beating the elite of the tough Pro Stock class at the 1973 NHRA Nationals before continuing to win more big NHRA events than any other racer. (The inset shows Bob with the SS/JA trophy won at the 1969 Springnationals with a 1969 Mustang.) A born hustler, Bob teamed with his wife Etta to make racing a true family affair. Note the jumbo-size slicks on Bob's little Pro Stocker taking off at the 1979 Gatornationals.

THE SWINGING EIGHTIES

By the 1980s, the American auto industry had just emerged from the darkest decade in its history, and had come to grips with tougher emissions, fuel economy, and safety legislation as well as the disenchanted public's demand for better quality. All these factors forced technology, such as computerized engine controls, to leapfrog at a pace never seen in the past.

Look-alike "jelly bean" styling and gutless new-vehicle performance carved the way for a new generation of bolt-on dress-up accessories and high-tech power products to enhance both appearance and performance—all from aftermarket entrepreneurs who were quick to act on car makers' mistakes. Unfortunately, the small entrepreneurs, from where mainstream innovations had come, were pushed aside by large manufacturing company CAD/CAM facilities.

Lee Iacocca, the godfather of the Mustang, joined teetering Chrysler and secured a government "bail-out" that saved the company and enabled the revolutionary minivans to be introduced in 1983. Customizing the vans, which were dubbed "sin bins" that contained TVs, wet bars, beds, and other comforts of home, soon became a "big niche" business in the aftermarket.

Even the new generation of auto junkies was returning to the roots of hot rodding, and muscle cars of the 1960s were "in" again. Flashy billet aluminum hardware, crate motors, and nostalgic repro rods helped street rodding remain in high gear. Bizarre continued to be "cool," with more monster trucks, road-scraping low-riders, and radical one-off street machines featured at big-time shows.

The Ford F-series pickup truck, born in 1963, became the best-selling U.S. vehicle in the 1980s, forecasting the future might of "four by fours" and sport utility vehicles (SUVs). Ironically, most old cars being junked today are more fuel efficient than modern "petrol porkers" classified by EPA as "light trucks."

Big-buck-sponsored superteams began to dominate the Top Fuel, Pro Stock, and Funny Car classes. Don Garlits grabbed the most big dragster wins, followed by future Top Fuel star Kenny Bernstein. Ageless Eddie Hill made history in 1988 with Top Fuel's quickest quarter-mile run of 4.99 seconds. By the end of the decade, drag racing (similar to its stock car cousin) was the fastest-growing motorsport in the world.

KEN "POSIE" FENICAL
Reviving the Golden Oldies

Ken Fenical stands at the podium, telling about his next project.

Ken Fenical, a talented man with the unlikely nickname of "Posie," is living proof that automobile customizing exists today as a vividly classic American art form. From his humble beginnings as a pinstriper in the early 1960s, Ken has risen to become one of the premier custom car builders in America.

Ken comes from the little town of Hummelstown, Pennsylvania, where his creative shop specializes in timeless restyling and special-interest vehicles. There, Ken also produces the Posies line of suspension parts and special springs for both early and present-day vehicles.

Said to be a primary force behind the custom car revival of the 1980s, Ken regularly awards trophies for outstanding cars at shows. One of the trophies is "The Posies Statement," which honors creativity in auto reconstruction. The inscription on the trophy best describes the philosophy of the customizer: Building a special-interest vehicle can be immensely satisfying if you are willing to forego a few luxuries—such as money and companionship.

After 35 years, Ken continues to cruise in high gear, doing dazzling design tricks on award-winning street rods, customs, and sport trucks. His one-of-a-kind work continues to grace the covers of auto magazines.

This Pro-Street 1934 Chevy Coupe is smartly restyled for Tony Giangrande.

This striking 1932 Ford Club Sport Coupe became one of Posie's "contemporary" statements. It was dramatically restyled by chopping the windshield, lowering and reshaping the top of the suicide doors, and relocating the custom top rearward over the re-lidded turtleback. The zig-zag striping over a yellow smoothed body gave the car a one-of-a-kind character. (Courtesy of Posies, 1989)

This "big-butt" 1932 Deuce street rod was built for Dean DuCray.

The Swinging Eighties

Fat-fendered Fords, such as this 1937 convertible, can be made beautiful again with a "Posies Statement"—by lowering and reshaping the top and by removing door handles and excess ornamentation. The red-and-black paint job further lowered the silhouette of the car. (Courtesy of Joe Mayall)

The clean and flowing lines of this blue 1932 Ford hardtop add "new edge" freshness to some old customizing tricks. The reshaped doors, the slanted windshield, and the low sculpted metal top that extends well beyond the rear rump produced an elegant fastback appearance. (Courtesy of A. Odeholm, 1980)

Bill Guentzler

Dr. William Guentzler didn't forget his midwestern hot rodding past when he became a professor and then a recruiter for San Diego State University. Bill now builds dream cars such as this Corvette Grand Sport, an exact duplicate of the rare factory Grand Sport models made for racing in the 1960s. During the last gasoline crisis, "Dr. G" collaborated with Ed Almquist in developing several fuel-saving inventions.

Hot Rod Pioneers

This swept-back 1937 Ford has a resculpted top, sides, and fenders, *sans* chrome.

This lemon-yellow 1936 Ford convertible has teardrop skirts, dummy spots, and DeSoto bumpers.

The black Phunkie 1932 Ford roadster won the Posie's Statement trophy for 1997. The rear fender (inset) shows clever use of fabric stretched over a frame similar to the convertible top. A neat idea from the days of horses and buggies.

This red and black 1938 Ford four-door convertible has a cut-down top.

The Flattop Flyer 1933 Ford Coupe has cycle fenders and a chopped top.

The Museum of Drag Racing
Keeping History Alive

"The novel idea for a museum came over 30 years ago when we realized that the history of drag racing should be preserved in its own museum," says Don Garlits. "It wasn't easy. It took a lot of prayer and all the money we could muster to purchase the property and start a nonprofit corporation."

Don and his wife Pat labored more than seven years without pay to establish what is now the only drag racing museum in the world. The doors opened in 1984, and the museum traces the evolution of the drag racing motorsport from its very beginning to the present. Shown are many of the fastest and most unusual cars in drag racing, such as Mickey Thompson's record-breaking Assault, the fastest Chevy engine in racing, and Bob Sullivan's Pandamonium, which was the 1960 AHRA Top Eliminator.

The Don Garlits Museum of Drag Racing is open daily and is located in Ocala, Florida.

Hot Rod Pioneers

NHRA Motorsports Museum

Called "One of the most knowledgeable hot rod historians on the planet," Greg Sharp, (left), pictured with Tommy Ivo (right), has written hundreds of automotive articles and is now the curator for the NHRA Motorsports Museum.

The roots of hot rodding come alive with historical memoribilia and vintage cars such as Chrisman's open-wheeler now showcased at the NHRA Museum in Pomona, California. The first director, Steve Gibbs, has been a major player in the growth of drag racing for almost 30 years.

The Swinging Eighties

Joe Sigretto

"I'm gonna race at Bonneville once before I die," vowed Joe Sigretto, a veteran New Jersey car afficionado. Joe has produced dozens of street rods in New Jersey, such as this 1934 roadster.

Pinky Randall

Pinky Randall is at the wheel of his 1926 Chevy "Mercury Body" speedster. His rare 46-year collection of Chevrolet memorabilia includes almost every "bow tie" emblem, car ornament, sign, badge, pin, literature, and item ever made by Chevrolet.

Eddie Hill

Texan Eddie Hill blasts off in his blown twin-Pontiac-engined dragster at the 1961 Nationals. Eddie was the first racer to run the quarter-mile in the 4's (4.99 seconds), and he won the NHRA Gator in 1988. A former speedboat record holder, veteran Eddie Hill is still going strong after more than four decades as arguably the most colorful innovator, builder, and driver in drag racing. (Photo courtesy of Don Garlits Drag Racing Museum)

THE HIGH-TECH NINETIES

As the winds of change continued in the 1990s, the effects were profound. Hot rodding's technological leaps in the past 25 years encompassed almost everything in the high-performance world. Unlike old-time speed equipment makers, entrepreneurs today deal mainly with computer software rather than hardware. Even top tuners wishing to optimize the performance of late-model "pavement pounders" must use their newfound computer savvy.

Choosing the right speed parts for a hot street car or a winning race machine has always been bewildering. Therefore, smart suppliers of the 1990s pushed coordinated systems that matched heads, induction systems, camshafts, and other components with each other for optimum performance. The aftermarket even offered turnkey restyling packages to make cars and trucks cool again.

A magic milestone in the quarter-mile record book was achieved in 1992 when Top Fueler Kenny Bernstein became the first driver to break the 300-mph barrier with a blistering run of 301.70 mph. Pro Stock and Funny Cars continued strong with their awesome e.t.'s.

Everything old was new again, with the baby-boomer generation now able to afford nostalgic street rods. They bolted cloned old-style replica bodies on high-tech chassis to make their "dream cars." Even car makers' smart concept cars combined the sporty retro appearance with new mind-boggling technology.

What lies ahead for hot rodding beyond the year 2000? Will future hot rod expression be akin to today's small-truck mania, which many admit is a rebellion against underpowered econocars and tinny, ho-hum family sedans?

Will our unique car culture endure, and will hot rodders continue to spawn innovations that ultimately help make family cars and trucks more efficient, safer, better performing, and more stylish? Even with governmental "green" mandates, will a practical pollution-free car ever exist? What will be the next vehicle rage: another hot SUV, pickup, or Prowler? As drag racing competition becomes tougher, speeds become faster and elapsed times beome quicker. Will another form of propulsion be found, such as electric vehicles? Or will our present internal combustion engine continue its starring role? No one can accurately foresee the end of one era and the beginning of another.

However, whatever the future brings, the hot rod motorsport is sure to be front and center.

Hot Rod Pioneers

International Drag Racing Hall of Fame

Each year, the International Drag Racing Hall of Fame honors and recognizes those inductees who have made significant contributions to the sport. Founded as a not-for-profit organization by Don Garlits and friends, the Hall of Fame was endorsed wholeheartedly by racing associations after receiving its official seal of approval by the Secretary of the State of Florida in 1984. The following honorees for the twentieth century were impartially selected by an independent board:

1998 International Drag Racing Hall of Fame Honorees are: (top row, left to right) Tom McEwen, Doug Thorley, Dode Martin, Gary Beck; (bottom row, left to right) Tom Lemons for Connie Swingle, Teresa Long for Les Lovett, C.J. Hart for Peggy Hart, Dick Landy. Not pictured: Bob Cahill and Roger Coburn.

Hall of Fame Members
(Through 1999)

Sydney Allard
Art Arfons
Zora Arkus-Duntov
Joaquin Arnett and the Bean Bandits
John Bandimere
Lou Baney
Raymond Beadle
Gary Beck
Keith Black
Willie Borsch
John Bradley
Bob Cahill
Candies & Hughes
Don Carlton
Larry Carrier
Chi-Town Hustler Farkonas, Coil and Minick
Art Chrisman
Jack Chrisman

Roger Coburn
Emery Cook
Bud Coons
Buster Couch
Bob Creitz
Bruce Crower
Ted Cyr
Ed Donovan
Vic Edelbrock, Sr.
Scotty Fenn
Kent Fuller
Don and Pat Garlits
Woody Gilmore
Ray Godman
Dale Ham
Barb Hamilton
Jack Hart
C.J. Hart
Peggy Hart
Ernie Hashim
Chet Herbert

Doris Herbert
George and Ruth Hoover
Joe Hrudka
Vance Hunt
George Hurst
Ed Iskenderian
Tommy Ivo
Bill Jenkins
Howard Johansen
Chris Karamesines
Dick Landy
Bobby Langley
Jim and Alison Lee
Tommy "T.C." Lemons
Jim Liberman
Leslie Lovett
Art Malone

Buddy Martin
Dode Martin
Tom McEwen
Bob Metzler
Ak Miller
George Montgomery
Lefty Mudersbach
Paula Murphy
Tony Nancy
"Jazzy" Nelson
Jim Nelson
Don Nicholson
Jimmy Nix
Danny Ongais
Gary Ormsby
Palamides & Smith
Herb Parks

Wally and Barbara Parks
Frank Pedregon
Peters & Frank
Robert E. Petersen
Ed Pink
Joe Pisano
Setto Postoian
Calvin Rice
Eric Rickman
Marvin Rifchin
Pete Robinson
Paul Schiefer
Marvin Schwartz
Lloyd Scott
Ralph Seagraves
Shirley Shahan
Lee Shepherd
Mike Snively
Gene Snow
Ronnie Sox

Bob and Don Spar
Speed Sport Special Bush, Fisher, Greth & Maynard
Stone, Woods and Cook
The Surfers
Skinner, Jobe and Sorokin
Connie Swingle
Richard Tharp
Mickey Thompson
Doug Thorley
Jim Tice
James Warren
Phil Weiand
John Wiebe
Jack Williams
Ed Winfield

The High-Tech Nineties

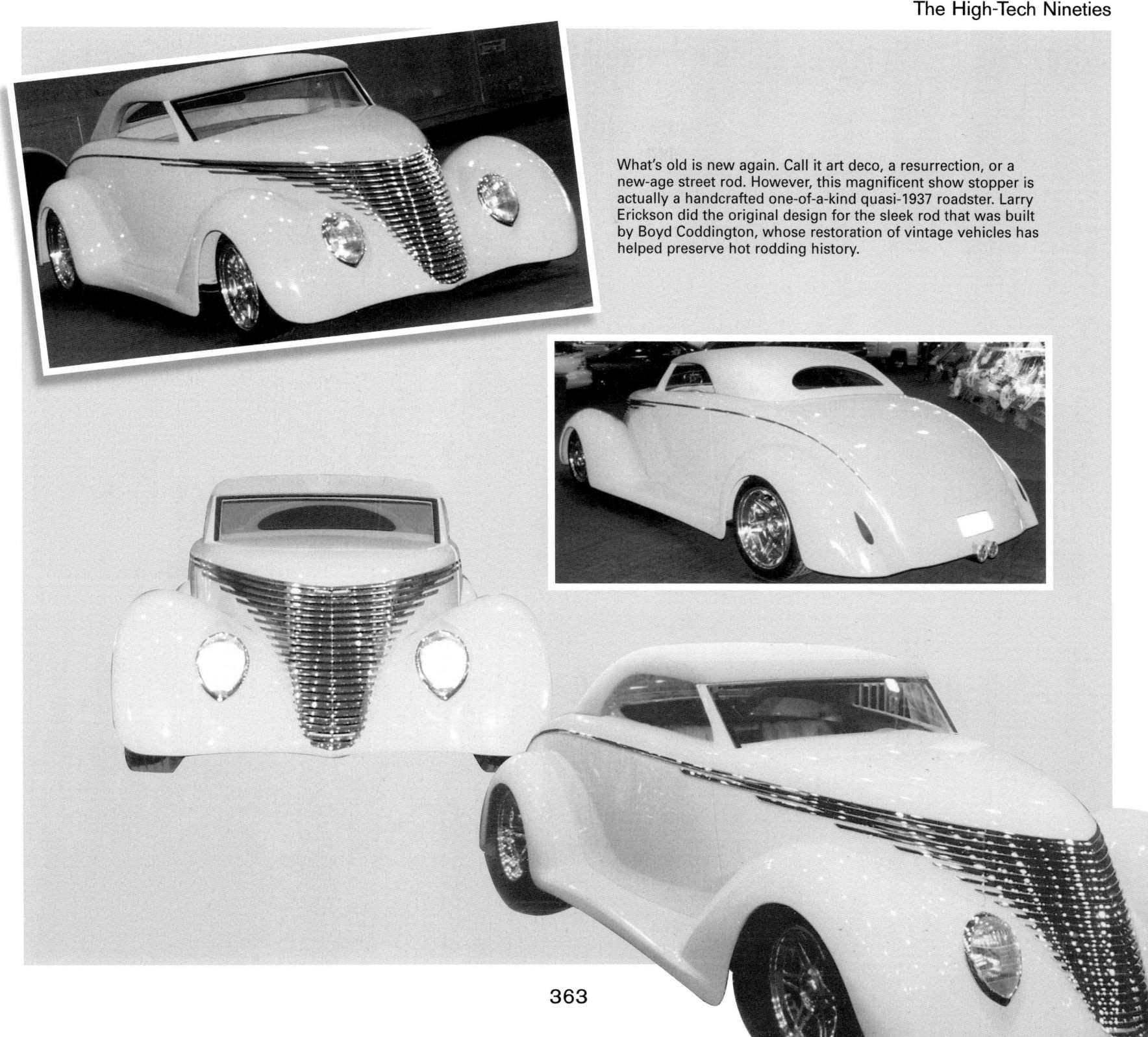

What's old is new again. Call it art deco, a resurrection, or a new-age street rod. However, this magnificent show stopper is actually a handcrafted one-of-a-kind quasi-1937 roadster. Larry Erickson did the original design for the sleek rod that was built by Boyd Coddington, whose restoration of vintage vehicles has helped preserve hot rodding history.

Hot Rod Pioneers

Old Timers' Reunion

Bench Racing Memories

Some of the hot rod heroes of yesterday were photographed at special historical gatherings in the late 1990s. Although many of them are now in the twilight of their adventurous and productive lives, they remain modestly active in motorsports, assured of their place in history.

 Don Garlits
 Tommy Ivo
 Bill Burke
 Ak Miller

Old-time champs receive belated trophies at a recent Bakersfield reunion: (left to right) Hayden Proffit, Don Garlits, and Gene Mooneyham.

At the Peterson Museum in 1997, veteran racers Bob Morton (left), Carroll Shelby (center), and Lou Senter (right) rehash the good ol' times.

The High-Tech Nineties

Leland Kolb

Bobby Meeks

Bill Summers

Bobby Spere

Holly Hedrich

Former competitors "Cam Father" Ed Iskenderian (left) and Ed Almquist (right) renew their friendship at a recent SEMA show in Las Vegas. Both now admit to secretly using unsuspecting customers as 'guinea pigs' when developing experimental cam grinds for early drag and stock car racing.

Lee Blaisdell, who was hot rodding's first official photographer, also participated in early California Roadster Association track racing and the first SCTA time trials. Lee's photographs helped give *Hot Rod Magazine* its start in 1948. Lee recalled, "We hawked 3,000 of the first issue ourselves for twenty-five cents each at the Gardena race track." Lee (right) here stands with successor Eric Rickman (left) at a 1999 "Old Timers" party. Many of their early photographs appear in this book.

Two generations of rodding greats include: (back row, left to right) Ray Brock, Wally Parks, Bruce Meyers, Alex Xydias, and Kong Jackson; (front row, left to right) Craig Breedlove and Al Teague.

INDEX

Abbreviations are used after the page number to indicate photo (p), figure (f), or color photo (c); color photo page numbers are 1 through 16 of the color insert.

AAA. *See* American Automobile Association
AAR. *See* All-American Racers
Accessories for cars, 316–317, 316p–317p
ACCUS. *See* American Automobile Competition Committee
Aerodynamics and racing
  Breedlove designs, 280
  Johnson designs, 257
  streamlining for speed, 268, 268p–270p
Aftermarket hot rod industry
  background, 98–99
  sales reps and, 245
  speed equipment for, 134–137
AHRA. *See* American Hot Rod Association
AiResearch, 273
Air-Ram, 137p
Alderson, Don, 306, 306p
Alexander, Mike and Larry, 312p, 312–313, 313p–314p
All-American Racers (AAR), 275
Almquist, Ed
  aftermarket hot rod industry and, 98–99
  background and career, 48p, 48–50, 49p, 50p, 148p, 379
  camshaft designs, 112
  encounter with Henry Ford, 1
  first hot rod book, 22
  partnership with George Hurst, 210–211
  reunion photo, 365p
  with Russ Dodd, 156p
Almquist Engineering
  fiberglass bodies, 202p–205p
  innovations from, 48–49
  Silencer muffler, 125
Amato, Joe, 248p
AMC. *See* American Motors Corporation
American Automobile Association (AAA)
  changes in imposed limits, 12
  Contest Board, 12, 111
  disassociation from racing, 13, 111
  stock car racing program, 156
  timing instruments used, 111p
American Automobile Competition Committee (ACCUS), 13
American Hot Rod Association (AHRA), 13

American Hot Rod (Batchelor), 86
American International Airways, 294
American Motors Corporation (AMC), 52, 102
American Seat Belt Council (ASBC), 77
American Sunroof Company (ASC), 312, 333
Ames, Danny, 229p
Anchel, Jonas, 49, 210
Anco Industries, 214
Anderson, Doug, 320p
Andrews, Jack, 53
Andrews, Keith, 128p
Ansen Automotive, 53, 54, 55
Ardun OHV conversion kit, 154, 155
Arfons, Art and Walt, 196, 242p, 242–243
Argetsinger, Cameron, 191
Arkus-Duntov, Zora, 154p, 154–155
Arnett, Joaquin, 180–181, 181p
Arnold, Billy, 12p
ASBC. *See* American Seat Belt Council
ASC. *See* American Sunroof Company
Ascot Speedway, 4
Assault cars, 221p
Attempt I streamliner, 165p
Auto shows, 252–253, 258
Avanti, 63p
Aztec Chevy, 57p

B&M Automotive, 256
Bagnall, Art, 25
Bailon, Joe, 185, 185p
Baker, Buck, 157
Baldwin, Alex and Keith, 84, 86p
Baney, Lou
  background and career, 351, 351p
  with Ed Pink, 289p
  pop-up piston and, 29
Banks, Henry, 24p
Bannister, Ralph and Fran, 170, 170p, 171p
Baron, Frank, 29, 30, 31p
Baronian, Norris, 108
Barringer, George, 15

"Barris Coachworks Treatment," 58
Barris, George, 57–58, 59p, 379p
"Barris Kustom," 57
Batchelor, Dean, 84, 86, 86p, 117p
"Batman" car, 59p
Bean Bandits Car Club, 180, 181, 299
"Beast III" streamliner, 151
Beatnik Bandit, 174, 175p
Bell Auto Parts, 24–25
Bell helmet, 24, 25, 25p
Bell-Toptex, Inc., 24
Belly tank racers, 30, 30p, 31p
Belond, Sandy, 124, 125p, 270p
"Beverly Hillbillies" car, 59p
Black, Keith, 183p, 346p, 346–347, 347p
Blair, Don, 94, 95p, 338p
Blair's High Performance, 94
Blair's Speed Shop, 94
Blaisdell, Lee, 365p
Blowers
  first use of, 129p
  Mooneyham use of, 240
  purpose and use, 230, 231p
"Blue Flame Special," 270p
Blum, Charles "Chuck," 304p
"Bob Jones Skyland Ford Special," 72
Bonneville Salt Flats
  history, 23, 117
  National Speed Trials, 67
  speed records. *See* Speed records
Bonney, Ted, 161p
Borgeson, Griff, 134
Borick, Louis, 77
Boucher, Charlie, 309p
Bradley, Harry, 312
Bragdon, Bud, 92
Braun, Adolph F. "Brownie," 127, 127p
Breedlove, Craig
  background and career, 280
  reunion photo, 365p
Bretches, Harold, 82p
Brock, Ray
  background and career, 176–177, 176p–177p
  with Bill Burke, 75p

Hot Rod Pioneers

Brock, Ray *(continued)*
   with Ak Miller, 66p
   reunion photo, 365p
   road racing and, 191
Brown, Ray, 76p, 76–77
"Bubble Top King," 260
Buckel, Mike, 236p, 237
Bugatti, 34p
Bugg, Gilbert, 334
Buick Bug racer, 17p
Burke, William
   background and career, 74p, 74–75
   belly tank racer, 30
   reunion photo, 334p, 364p
Burman, Bob, 17
Burnett, Rex, 121p
Businesses. *See* Hot rod businesses
Byron, Red, 115, 157

C & T Automotive, 168, 169
Cadillac 16-cylinder pace car, 13p
Campbell, Bill
   background and career, 210, 214p, 214–215
   partnership with Ed Almquist, 49
Campbell, Malcolm, 268p
Campbell, Wally, 157p
Camshafts
   Crane designs, 290, 291
   Crower designs, 141
   Duntov designs, 155
   Harman & Collins designs, 130
   Isky production and marketing, 108–109
   roller designs by Chet Herbert, 151
   speed innovations, 136
   typical flathead modifications, 148
   Weber designs, 112–113
   Winfield grinder, 5
Carburetors
   dual design, 29, 29p, 131
   four-barrel invention, 127
   fuel efficiency efforts, 103
   types, 135
   typical flathead modifications, 148
   Winfield modifications to, 4, 5p
Card, Charles "Honest Charley," 182, 182p
"Carpet Sweeper," 41p
Carrillo, Fred, 250p, 250–251, 251p
Carrillo Rod, 251

Carson-style roof tops, 333, 333p
Cartoons about hot rods, 121, 292, 293
Casper, Carl, 281p
"Cast Iron Wonder," 87
Challenger I streamliner, 165, 165p
Champlin, Doug, 251p
Chaparrals, 300p
Chappel, Lee, 117p
Charbonneau, Ray, 86p
Chassis kits for dragsters, 212–213, 213p, 218
Chassis Research Company, 212
Chayne, Charles A., 17p
Chevrolet
   265/283 models, 33, 92
   Aztec custom, 57p
   cast iron engine, 87
   "Legends of Performance" group, 110
   six T-roadster, 41p
   stock car designs by Henry "Smokey" Yunick, 119
   V8 engines, 155, 234–235, 235p
Chevrolet, Louis, 7, 17
Cholakian, Ed, 322, 322p
Chrisman & Cannon Hustler, 219p
Chrisman, Art and Jack
   backgrounds and careers, 152p, 152–153, 153p
   with Ed Pink, 289p
   races with Francisco "Fran" Hernandez, 73
   speed records, 123p, 150, 219p, 247p
"Christmas Tree" countdown starter, 110p, 111
Chrondek timers, 110p, 111
Chrysler
   Eight Roadster, 12p
   hemis, 92, 142, 183, 236, 265
   V8 engine design, 1951, 142
Clark, Don, 168p, 168–169
Clark, Rex, 133p
Clarkson, Howard, 147p
Clay Smith Engineering, 89
Claybaugh, E.E., 226p
Clifford, Jack, 300p
Clutches
   for automatics, 256
   blowup-proof design, 92, 113
   "Crowerglide" kit, 140–141, 141p
   safety designs, 232
Clymer Motorbook Special, 71
Cobb, John, 117

Coddington, Tom, 238p
Coffin car, 59p
Cole, Ed, 119, 235
Collett, Gordon "Collecting," 300p
Collins, Cliff, 130
Coltren, Pete, 176p
Comet funny car, 73
Compression ratios
   hop-up tricks, 135
   typical flathead modifications, 148
Convertibles, 333, 333p
Cook, Emery, 321p
Cook, Terry, 253p
Cook-Bedwell dragsters, 181p
Cool air induction speed innovations, 137p
Cooper, Gary, 107p
Cooper, Jackie, 97p, 191
Corbett, Bob, 351p
Cortopassi, Ed, 227p
Corvettes, 154, 262
Costs of racing
   in 1960s, 265
   Allard J-Type roadster, 155p
   camshaft designs, 130
   camshaft regroundings, 112
   chassis kits for dragsters, 212
   conversion kits, 10, 11p
   current, 188
   customization, 14c, 105
   Deuce body price, 342
   fiberglass bodies, 202, 203
   Fordallac, 60–61
   fuel, 23, 143, 188, 265, 325
   Grancor-Fords, 60
   hop-up prices, 166
   ignition systems, 44
   KB engine, 346
   Kurtis Kraft kits, 36
   manifolds, 131, 135
   Model T, 7p
   modifications in 1950s, 149
   OHV, 1930, 18p
   racing cars, 1920s, 14
   restoration investments, 104
   show-quality coupes, 247
   stock cars, 158
   tires, 188
   trends in 1940s, 22–23
   trends in 1950s, 142–143
   trends in 1960s, 264–265

Costs of racing *(continued)*
    trends in 1970s, 324–325
    turbocharger, 1965, 273
Coughlin, Jeg "Captain Quick," 295*p*
Coupes, 160, 247, 247*p*
Covell, Ron, 336*p*, 336–337
Cragar Industries, 25
Crane Engineering, 290
Crane, Harvey J., 290*p*, 290–291, 291*p*
Crankshafts
    Crower designs, 141
    Senter modifications to, 53
Crash barriers, 254, 255*p*
Crocker, J. Otto, 111
Cross Ram manifold, 33
Crower, Bruce, 140*p*, 140–141
Crower Cams & Equipment Company, 141
"Crowerglide" clutch kit, 140–141, 141*p*
Cunningham, Briggs, 191
Cushenberry, Bill, 310*p*, 310–311
Customization
    add-on accessories, 316–317
    by Alexander brothers, 312–313, 313*p*–314*p*
    Barris modifications, 57–58, 57*p*–59*p*, 349*p*
    costs of, 14*c*, 105
    by Bill Cushenberry, 310–311, 311*p*
    fenders, 2*c*–3*c*, 78*p*–81*p*
    by Ken "Posie" Fenical, 354*p*–356*p*
    front ends, 10*c*–11*c*, 64, 64*p*–65*p*
    by Bob Kaiser, 349*p*
    by Frank Kurtis, 36, 37*p*
    of Mercurys, 8*c*–9*c*
    metal works by Covell, 336*p*–337*p*
    paint schemes, 178–179, 185
    philosophy of, 354
    popularity in 1950s, 143
    rear ends, 12*c*–13*c*
    Starbird cars, 260, 261*p*
    by Doug Thompson, 340*p*–341*p*
    trends and techniques, 1*c*
    of trucks, 16*c*, 348*p*–349*p*
    by Joe Wilhelm, 276, 277*p*
    by Gene Winfield, 105, 105*p*
Cyclops, 243*p*, 244*p*
Cylinder heads
    Edmunds designs, 131
    Kenz modifications to, 70
Cyr, Ted, 255*p*

Dailey, Bill, 86*p*
Daniels, Bob, 233
Darren, James, 139*p*
Darrin, Bobby, 238*p*
Davison, John, 147*p*
Daytona International Speedway, 115
De Bisschop, Robert, 273, 273*p*
De Saknoffsky, Alex, 262
Debolt, Minor T. "Fat," 129
Deist, Jim, 225, 225*p*
Del Roy, Frankie, 245*p*
DePaolo, Pete, 27*p*, 119
Derham, 107
DeRoy, Frank, 166*p*
Deuces (Model A Ford)
    1932 Brown roadster, 76*p*
    1932 Orr roadster, 83*p*
    popularity of, 4*c*–5*c*, 9*p*, 18*p*, 23, 342, 342*p*–344*p*
    racing attributes, 18
DeWaters, A., 17*p*
Distributors, 40
Dodd, Russ, 156*p*
Dodge Deora, 348*p*
Dominianni, Frank, 120, 120*p*
Dondero, Ken, 302–303
Don's Speed Shop, 194*p*
Douglass, Howard and Ruth, 82*p*, 82–83, 83*p*
Douglass Muffler Shop, 82
Drag chutes, 225, 282, 283*p*
Drag racing
    Chrisman entries, 152
    East vs. West, 196–197
    evolution to a sport, 186–188, 196, 223
    first commercial strip, 162
    first NASCAR race, 220
    first official race, 114
    Granatelli brothers and, 61–62
    international interest in, 308
    museums, 357, 358
    NASCAR association with, 13, 163, 220
    official meet beginnings, 142
    rear-engined dragsters, 171*p*
    safety. *See* Safety in racing
    speed records, 181*p*
    strip layout, 172
    tire designs, 206*p*, 208–209
    use of Weber equipment, 113*p*
    see also Dragsters

Dragmaster Company, 138–139
Dragsters
    aerodynamics and, 270*p*
    Bustle Bomb, 190*p*
    chassis kits, 212–213, 213*p*, 218
    designs by Kent Fuller, 218
    nitromethane use, 188
    slingshot, 159, 164, 305
    Swamp Rat, 192*p*, 193–194, 259*p*
    Willys coupes, 160*p*
    see also Drag racing
Drake, Albert, 38–39, 48, 169*p*
Drake, Dale, 19
Dry lake racing
    history of, 38–39, 38*p*–39*p*
    safety rules creation, 66
    speed records, 28, 94*p*, 100
Dry Lakes Hall of Fame, 35
Duntov designs, 262

Eagle racers, 275
East African Safari, 73
Eckstrand, Al, 212*p*
Economaki, Chris, 279
Eddie Meyer Engineering, 19, 20, 160
Edelbrock Company, 33–34, 34*p*
Edelbrock, Vic, Jr., 32–34, 33*p*
Edelbrock, Vic, Sr.
    background and career, 32*p*, 32–34, 33*p*, 34*p*, 35*p*
    Thickstun dealer, 29
Edmunds, Eddie, 131
Eelco Company, 286–287
El Mirage, 39
"Elephant Motor," 183
Elgin, Dimitri "Dema," 320*p*
Elliot, Frank, 19
Ellis, George "Hot Rod," 245
Emory, Neal, 84
Engines
    aircraft engine use, 242
    by Keith Black, 346
    compression ratio hop-up tricks, 135
    conversion kits by Robert Roof, 10, 11*p*
    flatheads. *See* Flathead V8 engines
    hemis, 183
    Kenz modifications to, 70
    multi-engined cars, 299

Hot Rod Pioneers

Engines (continued)
    rear-engined sports car, 255p
    speed innovations, 137
    supercharging, 230, 231p
    turbochargers, 273
    Wayne Chevy, 87
Equipment for hot rods, 134–135, 134p–135p
Estes, Robert, 100, 100p, 101p
Estes Special, 101p, 169p
Estlack, Mike, 350
Evans, Earl, 56
Evans Racing Equipment, 56
Exhaust systems
    backpressure problems, 124–125
    clover-leaf, 285
    gaskets for, 330
    speed innovations, 136
    typical flathead modifications, 148
Exhibition cars, 272p

Fanwear, 322
Faust, "Jungle Larry," 240
Federation Internationale de l'Automobile (FIA), 12, 13, 196
Feil, Tony, 345p
Fender customizations, 2c–3c, 78p–81p
Fenical, Ken "Posie," 354p, 354–356
Fenn, Gordon "Scotty," 212–213
Fenton, Aaron, 131, 190, 190p
FIA. See Federation Internationale de l'Automobile
Fiberglass car bodies, 202, 202p–205p
Fire protection gear, 282
Firestone tires, 208
Fitch, John "Fearless," 254, 254p, 255p
"Flame Thrower" ignition, 41f
Flat Out: California Dry Lake Time Trials (Drake), 39
Flathead V8 engines
    Baron Racing Equipment, 31p
    Don Blair modifications to, 95p
    Chevrolet, 155, 234–235, 235p
    Chevy "Mouse Motor," 234–235
    Chrysler 1951 design, 142
    cylinder and manifold modifications, 10, 11p
    decline of market, 92
    designs in mid-1950s, 142–143
    dual-carburetor design, 29, 29p
    in first Ford cars, 2
    Grancor modifications, 60, 62p
    Carl "Pop" Green modifications to, 8–9
    hemis, 92, 142, 183, 236, 265
    horsepower enhancements by Harry Miller, 14–15
    Barney Navarro modifications to, 51
    in Novi racing cars, 6
    in police cars, 17p
    popularity of, 27
    replacement by Chrysler, 142
    on So-Cal Special, 71
    on speed boats, 19
    in stock car racing, 157p, 158
    typical modifications to, 29, 45, 148
Flock, Bob, Fonty, and Tim, 27p, 115, 157, 166
"Flying Milkman," 156p
Flynn, Jack, 297p
Flywheels
    typical flathead modifications, 148
    wear-resistant design by Schiefer, 92
Fontana, Joe, 350
Ford, Bensen, 100p, 103p
Ford, Edsel, 15
Ford, Henry, 1, 1p
Ford Motor Company
    999 racer, 1p
    flathead V8 engines. See Flathead V8 engines
    F-series pickup, 353
    manifold modifications, 19
    Miller-Fords, 14p, 15p
    stock cars, 16p
    Bill Stroppe race car designs, 122
    Henry "Smokey" Yunick's work for, 119
Fordallac, 60–61
Foyt, A.J., 275, 327p
France, Bill Sr., 114p, 115, 156p
Francisco, Don
    with Howard Douglass, 82p
    first rear-engined tanker, 74p
    magazine work, 89p
    reunion photo, 334p
Frank, Nye, 213, 293p
Freeland, Don, 101p
Frenching, 1c
Front end customizations, 10c–11c, 64, 64p–65p
Frontenac equipment, 7, 7p
Fuel injection systems. See Injectors
Fuels
    in Class C speed record, 1952, 85p
    costs of, 23, 143, 188, 265, 325
    efficiency efforts, 102, 103, 122
    experimental mixture by Rufi, 28
    hydrazine, 229
    injectors. See Injectors
    NHRA ban on nongassers, 222
    nitromethane reinstatement, 264
    nitromethane use, 35, 73, 137, 188
    nitrous oxide use, 118
    use by Vic Edelbrock and Bobby Meeks, 32, 35
Fuller, Kent
    association with Ron Covell, 336
    background and career, 213, 218p, 218–219, 219p
Funny cars
    Bruce Larson wins, 248
    name origin, 73, 298
    popularity of, 264, 265
    Ramchargers and, 236–237

Gabelich, Gary, 270p
Gardena, 27
Garlits, Don "Big Daddy"
    background and career, 93p, 192p, 192–194, 193p, 194p
    chassis kits and, 213
    drag racing museum, 357
    partnership with Emery Cook, 321p
    partnership with Ed "Isky" Iskenderian, 109
    partnership with Art Malone, 198
    reunion photo, 364p
    speed records, 143, 193
Garner, Willie, 184
Gasoline Alley, 8, 347p
Gassers, 206
Gear boxes, quick change, 106–107
Giampetroni, Angelo, 326p, 326–327
Gibbons, Ned, 186p
Gibbs, Steve, 358
Gilmore, Woody, 213
Glidden, Bob, 350, 351p
GMC G-71 superchargers, 46
Golden, Bill "Maverick," 272p
Goosen, Leo, 6
Gorlick, John, 124
Granatelli, Andy, Joseph, and Vince, 6, 60–63, 61p, 62p, 63p
Grancor, 60–61

Index

Grant, Gerry, 320p
Gratiot Auto Supply, 326–327
Great Bend Motorplex, 146
Green, Carl "Pop," 8p, 8–9
Green Monster, 242, 243p
Greenwald, Lawrence, 211
Greer-Black-Prudhomme dragster, 346p, 347
Grille restyling, 64p–65p
Grove, Tom, 309p
Guentzler, William, 355p
Gurney, Dan, 191, 274p, 274–275
Gwynn, Fred, 59p

Halibrand Engineering Company, 106
"Halibrand Shrike," 107p
Halibrand, Ted, 106p, 106–107
Hall, Jim, 300p
Harman & Collins, 130
Harman, Kenny, 130
Harris, Leonard, 285
Harroun, Ray, 12
Harry Miller Racing Facility, 4
Hart, C.J. "Pappy," 162p, 162–163
Hart, Peggy, 163
Hartman, John, 87
Hashim, Ernie, 129, 129p
Heacox, Frank, 25
Headers
    back pressure problems, 124
    speed innovations, 136
Hedman, Bob, 124
Hedrich, Holly, 365p
Helmets for racing, 24, 25, 25p, 282
Hemi Chryslers, 92, 142, 183, 236, 265
Herbert, Chet, 151, 151p
Herda, Bob, 306
Hernandez, Francisco "Fran"
    background and career, 73, 73p
    with Bobby Meeks, 33p
    partnership with Fred Offenhauser, 73, 96
"High Weiand" manifold, 47p
Hilborn, Stuart, 68p, 68–69, 69p, 273
Hill, Eddie, 359
Hill, George, 165
Hinnershitz, Thomas, 42, 42p, 43p
Hirohata Mercury, 57p

Hi-Speed Power Equipment, 120
History of hot rods
    aftermarket parts industry, 98–99
    bad publicity effects, 318
    beginnings of hot rodding, 2, 9
    cartoons in, 121, 292, 293
    dry lake racing, 38–39, 38p–39p
    East vs. West, 196–197
    fiberglass bodies, 202, 202p–205p
    first automobile race, 12
    first use of "hot rod" term, 66
    local race popularity, 222–224
    misconceptions about hot rodding, 239
    museums. See Museums
    positive-image efforts by NHRA, 67, 138, 144
    racing organizations and clubs, 12–13
    road racing, 191
    safety issues. See Safety in racing
    speed equipment, 134–137
    sports rods, 262, 262p–263p
    stock car racing, 156–158, 215
    street rodding comeback, 1970s, 334
    trends in 1940s, 22–23
    trends in 1950s, 142–143
    trends in 1960s, 264–265
    trends in 1970s, 324–325
    trends in 1980s, 352–353
    trends in 1990s, 360–361
Hobbins, Jim, 214
Hollywood Speed Shop, 167p
Holman, John, 271, 271p
Honest Charley Speed Shops, 182
Hooper, Mal, 269p
Hopper, Bill, 255p
Hop-up tricks
    costs of, 166
    typical flathead modifications, 148
Horn, Ted, 14p
Horning, Wayne, 87
Hot rod businesses, 146
    Almquist Engineering, 48–49, 125, 202p–205p
    Anco Industries, 214
    Ansen Automotive, 53, 54, 55
    B&M Automotive, 256
    Bell Auto Parts, 24–25
    Bell-Toptex, Inc., 24
    Blair's High Performance, 94
    Blair's Speed Shop, 94
    C & T Automotive, 168, 169
    Clay Smith Engineering, 89

Cragar Industries, 25
Crane Engineering, 290
Crower Cams & Equipment Company, 141
Don's Speed Shop, 194p
Douglass Muffler Shop, 82
Dragmaster Company, 138–139
Eddie Meyer Engineering, 19, 20, 160
Edelbrock Company, 32–34, 34p
Eelco Company, 286–287
Evans Racing Equipment, 56
Gasoline Alley, 8, 347p
Grancor, 60–61
Gratiot Auto Supply, 326–327
Halibrand Engineering Company, 106
Harman & Collins, 130
Hi-Speed Power Equipment, 120
Hollywood Speed Shop, 167p
Honest Charley Speed Shops, 182
Hot Rod Shop, 285
Hurst-Campbell, 210–212, 214, 215
Isky production and marketing, 108–109
J-B Car Care Products, 67p
Jenkins Competition, 303
Kansas Custom Shop, 310
Kar Kraft Design Center, 312, 314p
Kenz and Leslie Ltd., 72
Lakewood Industries, 232
Leo's Hot Rod Shop, 326
M&H tire, 208, 209
Moroso's, 350
Morton & Brett, 18
Mr. Gasket Company, 330–331
M/T Tire Company, 165
R & R Manufacturing Company, 10, 18
Ray Brown Automotive, 76–77
Schiefer/Harman & Collins, 92
So-Cal Speed Shop, 84, 84p
Sparkomatic Corporation, 49, 98
Speedway Motors, 189
Stahl Engineering, 323
Star Kustom Shop, 260
Superior Industries, 76p, 77
Thickstun Manufacturing Company, 29
Thunderbird Products, 228
Trans-Dapt Corporation, 184
Weber Camshaft Company, 112, 113
wholesale distributors, 322
Windy's Custom Shop, 105
Winfield Carburetor Company, 4
Hot Rod Industry News, 85
Hot Rod Magazine
    cartoons in, 121
    cost in 1948, 22

371

Hot Rod Pioneers

Hot Rod Magazine (continued)
   first technical editor, 89p
   founding of, 67, 90
   marketing ads in, 99
   Eric "Rick" Rickman's photography
      work for, 173
Hot Rod Shop, 285
Hot Rodder! From Lakes to Street (Drake), 39
Hough, Roscoe "Pappy," 16, 16p
Howard, Kenneth "Von Dutch," 178
Hrudka, Joseph, 330p, 330–331
Hudson, Howard, 351p
Hunt, Joe "Magneto-Man," 128, 128p
Hunter, Creighton, 162
Hurricane Hot Rod Association, 61
Hurst, George, 49, 210p, 210–212, 211p
Hurst-Campbell, 210–212, 214, 215
Hurst/Olds 4-4-2, 216, 216p
Hurtubise, Jim, 54
Hustler I, 153p
Hydrazine as fuel, 229
Hydrogen as fuel, 103
Hydro-Stick, 256

Iacocca, Lee, 352
ICAS. *See* International Championship Auto Show
Ignitions
   innovations, 136
   Kong systems, 44, 45
   magnetos, 128
   modifications by Spalding, 40, 41p
   typical flathead modifications, 148
IHRA. *See* International Hot Rod Association
Immerso, Ermie, 228, 228p, 229p
Indianapolis Speedway
   1947 race, 30
   first 500 mile race, 12
   first stock engine to race, 62
   Granatelli brothers' entries, 62
   history, 116
   Novi runs, 6
Injectors
   Hilborn designs, 68–69, 69p
   speed innovations, 136
   for turbochargers, 273
   vapor, by Almquist, 49
International Championship Auto Show (ICAS), 253, 258

International Drag Racing Hall of Fame
  inductees
    Art Arfons, 243
    Joaquin Arnett, 181
    Kent Fuller, 219
    Ernie Hashim, 129
    "TV" Tommy Ivo, 201
    listing of, 362
    Art Malone, 198
    "Ohio George" Montgomery, 207
    Ed Pink, 289
    Eric "Rick" Rickman, 273
    Marvin "Giant Killer" Rifchin, 209
International Hot Rod Association (IHRA), 13
Ireland, Bill, 309p
Iskenderian, Ed "Isky"
   background and career, 108p, 108–110, 110p
   reunion photo, 365p
Ivo, "TV" Tommy
   background and career, 200p, 200–201
   with Greg Sharp, 358p
   multi-engined cars, 299p
   reunion photo, 364p

Jackson, Charles "Kong"
   background and career, 44p, 44–45, 45p
   at Muroc Dry Lake, 39p
   reunion photo, 365p
Jackson, Johnny, 305
Jade Idol, 105
James, Joe, 100p
Jaws of Life, 212, 214
J-B Car Care Products, 67p
Jeffries, Dean, 179, 195, 195p
Jenkins, Ab, 117
Jenkins, Bill "Grumpy," 302p, 302–303
Jenkins Competition, 303
Johansen, Howard, 150
Johnson, Bob, 54p
Johnson, Jim, 133p
Johnson, Junior, 115
Johnson, Robert "Jocko," 257
Jones, Parnelli, 63
Justice brothers, 67p

Kagan, Leo, 211
Kalitta, Conrad "Connie," 294, 294p, 300p

Kaltz, Milt, 312p
Kansas Custom Shop, 310
Kar Kraft Design Center, 312, 314p
Karamesines, Chris "The Golden Greek," 229
"Katzenjammer Kids." *See* Granatelli, Andy, Joseph, and Vince
Keaton, Dee, 153p
Kelly, Jack, 101p
Kenz and Leslie Ltd., 72
Kenz, William, 70–71, 70p
Kenz-Leslie cars, 71, 117p
Kerr, "Smiling Jim," 211, 245p
Kitzul, Liz, 339p
Kladis, Danny, 62
Knapp, Dan, 238p
Knudsen, "Bunkie," 165
Kolb, Leland, 365p
"Kopper Kart," 58p
Kraft, Dick, 186p
Kurtis, Frank, 36–37, 262
Kurtis Kraft 500s, 36p, 36–37

Lakewood Industries, 232
Landy, Dick, 298, 298p
Larivee, Bob, 253, 258, 258p
Larson, Bruce, 248, 248p
Lavely, Ray, 25
Lawton, Bill, 301p
"Legends of Performance" group, 110
Leo's Hot Rod Shop, 326
Leslie, Roy, 70, 71, 72
Liberman, "Jungle Jim," 335
Lindsay, Bob, 90, 90p
Lockhart, Frank, 114p, 268
Lohn, Els, 55, 286, 286p
Lombardo, Larry, 302–303
Lund, Tiny, 189

M&H tire, 208, 209
Mag wheels, 25, 106, 316, 317
Magazines, 22
Magnetos, 128
Mail-order marketing, 24–25, 98–99
Malone, Art, 198, 198p, 199p

Mancini, Dan, 237, 238p
Manifolds
　Almquist version, 49
　Cross Ram, 33
　dual intake design, 29
　by Edelbrock Company, 33–34
　Edmunds designs, 131
　four-carburetor, 31p
　isothermal production models, 103
　Kong systems, 44–45
　Meyer system, 19
　Navarro modifications to, 51–52, 52p
　Offenhauser design, 96, 97p
　Slingshot, 33p
　by Thickstun, 29
　types, 134, 135
　typical flathead modifications, 148
　"U-Fab" kit, 140
　Weiand systems, 46, 47p
Mantaray, 195p
Mantz, Johnny, 100
Marion, Milt, 115
Marmon Wasp, 12
Marquez, Dave, 146, 146p, 147p
Martin, Buddy, 301p
Martin, Dode, 138p, 213
Martin, Richard, 126p, 126–127, 127p
Martinez, Ben, 147p
Maserati two-seater race car, 13p
McCahill, Tom, 60
McClure, Doug, 139p
McCormack, Jack, 251p
McCulloch, Ed, 335
McEwen, Tom, 337p
McGrath, Bob, 133p
McGurk, Frank, 87p
McNicholl, Pete, 284p, 284–285
Meadors, Gary, 253
Mechanix Illustrated, 202
Medley, Tom
　background and career, 121, 121p
　street rodding races and, 334, 334p
Meeks, Bobby
　background and career, 35, 35p
　with Dean Batchelor, 86p
　reunion photo, 365p
　work with Edelbrock, 32, 33p, 35
　work with Alex Xydias, 84
Mendenhall, Jack, 226, 226p, 227p

Mercury
　1946 V8, 30
　customization of, 8c–9c
　drag racing promotion, 73
　flathead modifications, 29
　Hirohata custom, 57p
　manifold modifications, 19
　Stroppe race car designs, 122–123
"The Mermaid," 123p
Mexican Road Races, 73, 100, 144
Meyer, Eddie "Pappy" and Louis
　backgrounds and careers, 19p, 19–20, 20p
　Ray Brown's work with, 76
Meyer, Louis "Sonny," Jr., 20
Meyers, Bruce, 365p
Mid-Atlantic Timing Association, 222
Midget racing
　by Edelbrock, 32
　Roscoe "Pappy" Hough and, 16
　Justice brothers' cars, 67p
　by Kurtis, 36, 37p
　by Johnny Parsons, 21p
　racing popularity, 27
　by Roy Richter, 24
　by Tom Sparks, 160, 161p
"Mileamore" camshafts, 109
Millar, Pete, 292p, 292–293
Miller, Akton "Ak"
　background and career, 144p, 144–145
　hot rod ad, 291p
　Pike's Peak win, 66p
　reunion photo, 364p
　road races, 191
　work with Ray Brock, 176p
Miller, Eddie, 69p
Miller, Harry, 14–15, 18
Miller-Fords, 15p
Milliken, Bill, 34p, 191
Miss Elegance car, 185p
Mobile Economy Run, 102p, 103p
Model A Ford
　1932 Brown roadster, 76p
　1932 Orr roadster, 83p
　popularity of, 4c–5c, 9p, 18p, 23, 342, 342p–344p
　racing attributes, 18
Model B Ford
　popularity of, 9p, 10, 10p, 11p
　racing attributes, 18
　speed record in, 4

Model SR carburetor, 4–5, 5p
Model T Ford
　Frontenac modifications to, 7, 7p
　as hot rod parts, 338, 338p–339p
　number produced, 7p
　photos of, 4c–5c, 4p
　raced by Meyer, 19
　speed equipment for, 10
Modern Moroso Motorsports Park, 172p
Monopoli, Tony, 14c
Monte, George, 347
Montgomery, Don, 287p
Montgomery, "Ohio George," 206p, 206–207, 207p
Moody, Ralph, 271, 271p
Moon, Dean, 266p, 266–267
"Moonbeam" car, 267p
"MoonEyes USA," 267
Mooneyham, Gene
　background and career, 240, 240p, 241p
　with Jack Chrisman, 153p
　reunion photo, 364p
Moore, Lou, 106p
Moroso, Dick, 350
Morris, Ollie, 96, 96p
Morton & Brett, 18
Morton, Robert
　background and career, 132, 132p
　reunion photo, 364p
Moss, Jay, 157p
Motor Life, 86
Motor Monarchs of Venturi, 146
Motor Trend, 177
Motorsports Hall of Fame
　Art Arfons, 243
　Art Chrisman, 152
　Conrad "Connie" Kalitta, 294
Mr. Gasket Company, 330–331
"Mr. Horsepower," 88p
M/T Tire Company, 165
Mudersback, Lefty, 213
Mufflers
　Douglass manufacture of, 82
　noise reduction for, 125
Muldowney, Shirley, 294, 321p, 335p
"Munsters Coach," 59p
Muntz Road Jet, 36, 37p
Muroc Dry Lake, 6, 38, 39
Murphy, Jim, 26p

Hot Rod Pioneers

Murphy, Paula, 328p, 328–329, 329p
Muscle cars, 265, 314
Museums
    drag racing, 357, 358
    Motorsports, 358
    National Rod and Custom Hall of Fame, 260
    Petersen's Automotive, 91, 221p
    Petroleum, 226, 227p
    Mustang, 265, 319p

Nadar, Ralph, 318
Nalon, Duke, 166p
Nancy, Tony, 278–279
NASCAR. See National Association of Stock Car Automobile Racing
National Association of Stock Car Automobile Racing (NASCAR)
    drag racing entrance, 13, 220
    drag racing sponsoring, 163
    founding of, 12, 115, 157
National Hot Rod Association (NHRA)
    championship results, 1959, 197
    first drag racing competition winner, 180
    first license for a woman, 328
    first technical director, 138
    founding of, 13, 67, 142
    fuel ban, 222
    Motorsports Museum, 358
    nitromethane reinstatement, 264
    positive-image promotion efforts, 67, 138, 144
    tire rules, 206p, 209
National Rod and Custom Hall of Fame museum, 260
National Street Rod Association (NSRA), 334
Navarro, Barney, 51–52, 52p
Nelson, "Jazzy" Jim, 171p
Nelson, Jim
    background and career, 138p, 138–139, 139p, 213
    with Danny Ongais, 300p
NHRA. See National Hot Rod Association
Nicholson, "Dyno Don," 73, 307, 307p
Nicoli, Nick, 187p
Nitromethane fuels, 35, 73, 137, 188
Nitrous oxide use, 118
Noble, Richard, 270p
Norris, Ken, 251p

Nosing, 64
Novi racing cars, 6, 6p, 63
NSRA. See National Street Rod Association

Offenhauser engine
    fuel injection improvements to, 68
    Louie Miller and, 19–20
    turbochargers for, 273
Offenhauser, Fred C., 96p, 96–97
OHV manufacturers, 18
Oil additives
    Power-X, 50
    STP, 63
Oldfield, Barney, 1p
Olds, Ransom E., 114
Olinger, Robert, 147p
Ongais, Danny, 300p
Orr, Karl and Veda, 83
Ostick, Nathan, 177p
Owen, Cotton, 157

Packard, Sam, 158p
Painting for customization, 178–179, 185
Parks, Wally
    background and career, 66p, 66–67
    comments on Linda Vaughn, 217
    reunion photo, 365p
Parsons, Johnny, 21p, 128p, 328
Paxton, 62, 63
Pennington, Jerry, 350
Petersen Automotive Museum, 91, 221p
Petersen, Robert "Pete"
    background and career, 90p, 90–91, 91p
    with Ray Brock, 176p
    hiring of Eric "Rick" Rickman, 173
Petroleum Museum, 226, 227p
Petty, Lee, 115, 157
Phillips, Judd, 100p
Pickup trucks, 16c, 353
Pierson, Bob and Dick, 104, 104p
Piggins, Vince, 234p, 235
Pike's Peak Hill Climb, 145
Pink, Edward, 288p, 288–289, 289p, 291p

Pistons
    pop-up concept, 29, 53
    speed innovations, 136
    typical flathead modifications, 148
Platt, Hubert, 309p
Police cars
    use of Thickstun equipment, 29
    V8 engines in, 17p
Popular Hot Rodding, 134
Porter, Herb, 273
Posey, Doug, 251p
Power steering, 119
Power-X oil additive, 50
Presley, Elvis, 58p
Proffitt, Hayden, 320p, 364p
Prudhomme, Don "The Snake," 219p, 296p, 296–297

R & R Manufacturing Company, 10, 18
Race Car Vehicle Dynamics (Milliken), 34p
Racetracks
    Bonneville Salt Flats, 23, 67, 117
    Daytona International Speedway, 115
    Great Bend Motorplex, 146
    Harry Miller Racing Facility, 4
    Indianapolis Speedway, 116
    Modern Moroso Motorsports Park, 172p
    Muroc Dry Lake, 6, 38, 39
    Redding Drag Strip, 133p
    Santa Ana Drag Strip, 162–163
    Saugus Drag Strip, 54, 351
Racing organizations, 12–13
Racing records. See Speed records
Ragtops, 333, 333p
Ramchargers club, 236–238
Ramirez, Carlos, 180, 181p
RamJet "gassipper," 49
Randall, Pinky, 359
"Rapid Transit" streamliner, 151p
Rathmann, Jim, 119
Ray Brown Automotive, 76–77
RCO Flash motor, 10
Rear-end styling, 12c–13c
Reavie, Ed, 253
The Red Baron, 259p
Redding Drag Strip, 133p
"Redhead" streamliner, 133p

Index

Reider, Paul, 167p
Replogle, Bill, 129
Revell, 174
Rice, Calvin, 182p
Richter, Roy, 24p, 24–25
Rickenbacker, Eddie, 13p
Rickman, Eric "Rick"
  awards, 173, 258p
  background and career, 173, 173p
  Kart photo, 318p
  reunion photo, 365p
Rifchin, Marvin "Giant Killer," 208p
Riggs, Bobby, 335p
Riley, George, 5, 18
Riley, O.V. "Ollie," 111
Ring-Free Miller, 20p
Rishell, W.D., 117
Road and Track, 22, 86
Road racing, 191
"Road Rebels" club, 74
Roadster racing
  Model T bodies for, 338, 338p–339p
  popularity of, 26–27, 26p–27p
Roaring Roadsters, 26
Roberts, "Fireball," 157
Roberts, Floyd, 27
Robinson, Bob, 132p
Robinson, L.R. "Pete," 197, 305
Rod Action, 177
Rod and Custom, 334
Roll cages, 138, 157p
Romney, George, 102p
Ronda, Gaspar "Gas," 281p
Roof, Robert M., 10p, 10–11, 18
Rose, Mauri, 27, 68
Roth, Ed "Beatnik"
  background and career, 174p, 174–175, 175p, 179
  with Liz Kitzul, 339p
"Rotofaze" ignitions, 44
Rubio, George, 132, 132p
Rubirosa, Porforio, 176p
Rufi, Robert, 28, 28p
Rule-making organizations, 12–13
Russetta Timing Association, 351

Ruttman, Troy
  background and career, 190p, 191
  roadster racer, 27p
Ryssman, Otto, 159, 159p

Sachs, Eddie, 107p
Safety in racing
  clutch designs, 232
  crash barriers, 254, 255p
  dangers to spectators, 159
  drag chutes, 225, 282, 283p
  fire protection gear, 282
  helmets, 24, 25p, 282
  promotion of, 67, 138
  racing rules creation, 66–67
  roll cages, 138, 157p
  Safety Wall invention, 119
  seat belt use, 76–77, 225
"Safety Safari," 67, 138
Santa Ana Drag Strip, 162–163
Saugus Drag Strip, 54, 351
SCCA. *See* Sports Car Club of America
Schartman, "Fast Eddie," 73, 321p
Schiefer, Paul and Carl, 92p, 92–93, 93p
Schiefer/Harman & Collins, 92
Schubeck, Joe, 232, 232p, 233p
Schuman, Mort, 256p
SCORE shows, 85, 164
Scott, Lloyd, 190p
SCTA. *See* Southern California Timing Association
SCTA Racing News, 83
Scully, Shane, 55
Seat belt use, 76–77, 225
SEMA. *See* Specialty Equipment Market Association
SEMA Hall of Fame inductees
  Lou Baney, 351
  Keith Black, 347
  Ray Brown, 77
  Harvey Crane, 291
  Jim Deist, 225
  C.J. "Pappy" Hart, 162
  Stuart Hilborn, 69
  Ed "Isky" Iskenderian, 110
  "Smiling Jim" Kerr, 245p
  Els Lohn, 286
  Dick Martin, 126
  Dean Moon, 267

  Paul Schiefer, 93
  Lou Senter, 55
  Bill Simpson, 282
  Bill Smith, 189
  Linda Vaughn, 217
  Phil Weiand, 47
  Alex Xydias, 85
Senter, Louis
  background and career, 53p, 53–55, 54p
  partnership with Lou Baney, 29, 351, 351p
  reunion photo, 364p
  in roadster race, 26p
  work with George Barris, 59p
Senter, Marsha, 55
Sharp, Al, 241p
Sharp, Greg, 358p
Shedden, Tom, 25
Shelby, Carroll
  background and career, 319, 319p
  partnership with Dan Gurney, 275
  reunion photo, 364p
Shifters
  Almquist innovations, 49, 50
  floor conversion kits by Ansen, 54
  Hurst design, 210
  speed innovations, 137
Shoebox Fords, 6c–7c
Sigretto, Joe, 359
Simpson, Bill, 282, 282p, 283p
Singer, Rodney, 221p
Slingshot dragsters, 159, 164, 305
Slingshot manifold, 33p
Smiley, Pete, 24p
Smith, Clay, 88p, 88–89, 122
Smith, D. William "Speedy Bill," 189, 189p
Smith, Frank, 322p
Smith, Tex, 334
"Smithy" mufflers, 125
Smothers, Dick, 25p
"The Snake" Prudhomme, 219p, 296p, 296–297
So-Cal Special, 71, 84p, 117p
So-Cal Speed Shop, 84, 84p
Southard, Andy, 178p
Southern California Timing Association (SCTA)
  dry lake races, 38–39
  founding of, 12
  presidents of, 144
  racing records, 19
  reactivation in 1937, 66
  safety promotion, 67, 138

375

Hot Rod Pioneers

Sox, Ronnie, 301p
Spalding, Bill and Tom, 40p, 40–41
Spar, Bob and Don, 256, 256p
Sparkomatic Corporation, 49, 98
Sparks, Tom
   background and career, 149p, 160, 160p, 161p
   work with Tony Nancy, 278
Specialty Equipment Market Association (SEMA)
   founding of, 25, 267
   hall of fame. See SEMA Hall of Fame inductees
   mission of, 304
Speed Age, 22, 26
Speed and Mileage Manual (Almquist), 48
Speed Equipment Manufacturers Association. See Specialty Equipment Market Association
Speed records
   100-mph average, 27p
   100-mph speedboats, 19
   180-mph mark for dragsters, 143
   200-mph barrier, 159, 264
   250-mph barrier, 242
   300-mph barrier, 360
   600-mph run, 280
   999 racer, 1903, 1
   AHRA Winter Championship, 1962, 55
   beach drag racing, 114
   Bonneville, 1950, 86p
   Bonneville, 1951, 144
   Bonneville, 1952, 77p, 85p, 151, 151p
   Bonneville, 1955, 129
   Bonneville, 1956, 228
   Bonneville, 1957, 71
   Bonneville, 1960, 221p
   Bonneville, 1972, 75p
   Bonneville, 1973, 220p
   Bonneville, 1976, 329p
   Bonneville, Class D, 1940, 15p
   Bonneville, E/SR, 1963, 145p
   Bonneville speed trials, 117
   Class B roadster, 1953, 177p
   Class B roadster, 1972, 332
   Class B streamliner, 1948, 68, 69p
   Class B streamliner, 1949, 87
   Class B streamliner, 1952, 169p
   Class B streamliner, 1966, 133p
   Class C/Altered, 1959, 236
   Class C coupes, 1953, 85p
   Class C lakester, 1951, 56
   Class C lakester, 1952, 77p, 85p
   Class C record, 1949, 30p
   Class C record, 1952, 77p, 85p, 169p
   Class C roadster, 1950, 132p, 177p
   Class C roadster, 1951, 168, 250
   Class C streamliner, 1952, 100, 269p
   Class C streamliner, 1953, 269p
   Class C streamliner, 1996, 332
   Class D coupe, 1949, 104p
   Class D lakester, 1956, 228
   Class D streamliner, 1960, 75p
   Class F streamliner, 1952, 75p
   coupes, 1948, 73, 129
   coupes, 1952, 126
   coupes, 1953, 75
   D/Gas roadster, 1991, 226
   dragsters, 182p
   dry lake races, 1930, 100
   dry lake races, 1940, 28
   dry lake races, 1945, 94p
   engines by Meeks, 35
   FIA B Class, 1953, 269p
   four-lap average at Indy, 1946, 62
   Granatelli brothers' Studebakers, 62
   land speed, 269p–270f
   Mexican Road Races, 1952, 100
   Model B, 1927 T, 74
   Model B, 1933, 4
   at Muroc Dry Lake, 38
   NASCAR, 1961, 62
   single-engine, land, 114p
   by Spalding brothers, 40
   streamliner class, 1948, 132p
   streamliner class, 1959, 257p
   streamliner class, 1965, 270p, 286
   streamliner class, 1997, 268
   Swamp Rat dragster, 193
   XF/GMR class, 101p
   Xydias-Batchelor Streamliner, 23
Speedway Motors, 189
Spere, Bobby, 365p
"Spirit of America," 280p
Sports Car Club of America (SCCA), 12, 191
Sports cars, 262, 262p–263p
Sprint cars, 42
Spurgin, Charlie, 28
Stack, Robert, 18p
Stahl Engineering, 323
Stahl, Jere, 323, 323p
Star Kustom Shop, 260
Starbird cars, 260, 261p
Starbird, Darryl, 260
Starting line system lights, 110–111, 172
Stillwell, Ray, 246p, 246–247
Stilwell, Frank, 162
Stock car racing
   boom after World War II, 156–158
   early opinions of, 215
   Fords, 16p
   increasing popularity of, 142, 158
   NASCAR founding, 115
   restrictions on modifications, 158
"Stove Bolt Six," 87
STP, 63
Streamliners
   aerodynamics and, 268, 268p–270p
   Attempt I, 165p
   Challenger I, 165, 165p
   Kenz-Leslie cars, 71, 71p
   modifications to, 28
   by Bob Rufi, 28p
   by So-Cal, 71, 84, 84p
   by Spalding brothers, 41p
   speed records. See Speed records
Streetmaster manifold, 34
StreetScene, 334
Striegel, George "Honker," 89p
Strip Star, 105p
"Stroker McGurk," 121
Strokers Club, 223p
Stroppe, Bill
   background and career, 122p, 122–123
   in road competition, 191
   work with Bob Estes, 100
Studebaker, 62
Sturdy, Clyde, 85p
Stutz Special, 268p
Sullivan, Ed, 103p
Summers, Bill and Bob, 270p, 365p
Sunbeam Corporation, 212
"Super Cyclone," 153p
"Super Duty Group," 216
Supercharging, 230–231
Superior Industries, 76p, 77
Swamp Rat, 192p, 193–194, 259p
Sweikert, Bob, 273p
Swingle, Connie, 213, 259

Tanker cars, 74
Tattersfield, Robert, 29, 30p, 30–31

Index

Tattersfield Special, 30
T-buckets, 4c–5c, 338, 338p–339p
  see also Model T
Teague, Al
  background and career, 332, 332p
  reunion photo, 365p
Teague, Marshall, 157
Tebow, Clem, 168p, 168–169, 176p
They Call Me Mister 500 (Granatelli), 63
Thickstun Manufacturing Company, 29
Thickstun, Tommy, 29, 29p, 30
Thomas, Herb, 157
Thompson, Doug, 340p, 340–341
Thompson, Mickey
  background and career, 164p, 164–165
  Ernie Hashim and, 129
  speed records, 264
  tire needs, 208p
Thornton, Jim, 236p, 237, 238p
Thunderbird Products, 228
Timing devices and practices, 110–111, 172
"Tin Lizzie" speedsters, 1, 7, 7p
Tires, 188, 206p, 208–209
Toia, Bill, 326, 326p
Trans-Dapt Corporation, 184
Transmission adapters, 184
Transmissions, automatic, 256
Travers, Jim, 68
Travis, Jim, 220, 220p, 221p
Trend, Inc., 67
Truck customizations, 348p–349p
Tucker, Preston, 14
Tuning for speed, 137
Tunnel Ram manifold, 34
Turbine engines restrictions, 63
Turbochargers, 273
Turner, Curtis, 157
Turner, "Roscoe," 24
Tuthill, Bill, 115
"Twin Bear" gas dragster, 150p
TY-RODS, 223p

"U-Fab" manifold kit, 140
United States Auto Club (USAC), 12, 13
Unser, Jerry, 190p
U.S. Bicentennial Global Record Run, 328

USA-1 funny car, 248
USAC. See United States Auto Club

V6 engines, 207
V8 engines. See Flathead V8 engines
Van Iderstine, Pete, 315, 315p
Vans, 16c, 352
Vaughn, James, 211
Vaughn, Linda, 217, 217p, 327p
Vertex Scintilla magnetos, 128
Viland, Les, 100, 102p, 102–103, 103p
Vogt, "Red," 115
Voight, Fritz, 165
Volumetric efficiency modifications, 148
Vukovich, Bill, 27, 37, 68p

Waite, Don, 177p
Walker, Vernon, 334
Wallace, Mike, 113p
Walling, Roger, 222p
Ward, Glen, 150
Watkins Glen, 191
Watson, Jack "Doc," 215, 216, 216p
Wayne Chevy Track engine, 87
Weber Camshaft Company, 112, 113
Weber, Harry, 112p, 112–113, 113p
Weiand, Phil and Joan, 46p, 46–47
Weidenhammer, Hank, 318
Wells, Dick, 334p
Westerdale, Don, 237p
Western Timing Association, 74
Wheel turning tracer machine, 54p
Wheels, 106
Whipp, Roger, 133
Wholesale distributors, 322
Wight, George, 24
Wilcox, Larry, 123p
"Wild Dream" custom car, 277p
Wilhelm, Joe, 276, 276p, 277p
Willys coupes, 160p
Wilson, Scott, 309p
Windy's Custom Shop, 105
Winfield Carburetor Company, 4

Winfield, Ed
  background and career, 4p, 4–5
  Novi engine design, 6
  in roadster race, 26p
  work with Charles "Kong" Jackson, 44
Winfield, Gene, 105, 105p
Winfield, W.C. "Bud," 4, 6p, 6–7
Winterbottom, Bill, 249p
Wolf, Joe, 166p, 166–167
Wolf, John, 331p
"World's Wildest Willys," 206p

Xydias, Alex
  background and career, 84p, 84–85, 85p
reunion photo, 365p
  So-Cal Special, 86p, 117p
  speed record, 1949, 23
Xydias-Batchelor Streamliner, 23

Young, Willy, 71, 117p
Yunick, Henry "Smokey," 118p, 118–119, 119p

Z28 Camaro, 234p
Zeushel, Dave, 218p, 219p

ABOUT THE AUTHOR

Ed Almquist, who has played a major role in the growth of post-World War II motorsports, realized the American dream when his racing hobby evolved into an exciting innovative career filled with what he calls "imagineering." Soon after Ed copyrighted the nation's first hot rod manual, he discovered a bonanza in supplying Almquist speed equipment to the rapidly emerging motorsports of stock car and drag racing spurred on by the new car culture called hot rodding.

Ed's backyard enterprise quickly grew to become the largest manufacturer of racing equipment in the East—often the choice of champions. Today, many of Ed's inventions—such as Sparkomatic plugs, floor shifters, and wear-proofing lubricants—are still in use on both road and track.

An active member of the Society of Automotive Engineers, Ed now plans to reproduce a few of the goodies that made him the mail order icon of hot rodding from the 1940s through the 1960s.

Ed currently lives in Milford, Pennsylvania, and drives a souped up 1931 Ford pickup.

When famous pioneer custom-car builder George Barris (left) spoke to Ed Almquist (right) at an auto show years ago, he correctly predicted the nostalgic return of retro styling concepts that would make the best of the old look new again. Barris also said, "Good customizing will never die...and like most hot rodding, will continue to be a lap or two ahead of the automakers."